LAYOUT DESIGN AND VERIFICATION

Advances in CAD for VLSI

Volume 4

Series Editor:

T. OHTSUKI
*Department of Electronics
and Communication Engineering
School of Science and Engineering
Waseda University
Tokyo, Japan*

NORTH-HOLLAND
AMSTERDAM · NEW YORK · OXFORD · TOKYO

Layout Design
and Verification

Edited by

T. OHTSUKI
*Department of Electronics
and Communication Engineering
School of Science and Engineering
Waseda University
Tokyo, Japan*

1986

NORTH-HOLLAND
AMSTERDAM · NEW YORK · OXFORD · TOKYO

ISBN: 0 444 87894 7
ISBN set: 0 444 87890 4

Published by:
ELSEVIER SCIENCE PUBLISHERS B.V.
P.O. Box 1991
1000 BZ Amsterdam
The Netherlands

Sole distributors for the U.S.A. and Canada:
ELSEVIER SCIENCE PUBLISHING COMPANY, INC.
52 Vanderbilt Avenue
New York, N.Y. 10017
U.S.A.

Library of Congress Cataloging-in-Publication Data
Main entry under title:

Layout design and verification.

   (Advances in CAD for VLSI ; v. 4)
   Includes indexes.
   1. Integrated circuits--Very large scale integration--
Design and construction. I. Ohtsuki, T. (Tatsuo),
1940-   . II. Series.
TK7874.L318 1986   621.395   85-32544
ISBN 0-444-87894-7 (Elsevier Science Pub. Co.)

INTRODUCTION TO THE SERIES

VLSI technology has matured to the extent that hundreds of thousands or even millions of transistors can be integrated in a single silicon chip, and VLSIs are now the key to the design of efficient electronic systems. It therefore follows that the problems of designing integrated circuits are also becoming increasingly complex. A wide variety of topics on computer aided design (CAD) have emerged. This is a period when no-one can be a specialist in all of the topics in CAD for VLSI, and the whole area is beyond the scope of a single volume. The requirement for information and communication is increasing.

In 1982, Dr. R. Morel, at that time Acquisition Editor at North-Holland, conceived a range of projects to relieve this problem. Initially this resulted in "INTEGRATION, the VLSI journal", a quarterly aimed at speeding up communication among VLSI designers. It was further decided to launch a book series in the field, an idea enthusiastically supported by Dr. L. Spaanenburg, Editor-in-Chief of Integration. Dr. Morel approached me to edit this series.

It was agreed that the book series should be aimed towards a comprehensive reference for those already active in areas of VLSI CAD. Each volume editor was asked to compile a present state of the art, scattered in many journals. The book series therefore should help CAD specialists to get a better understanding of the problems in neighbouring areas by reading particular volumes. At the same time, it should give novices a foothold for doing research in areas of VLSI CAD, although a basic knowledge of VLSI technologies and design methods will aid understanding.

The book series, entitled "Advances in CAD for VLSI" consists of the following seven volumes:

Vol. 1	Process and Device Modeling	W. Engl (RTWH Aachen, FRG)
Vol. 2	Logic Design and Simulation	E. Hörbst (Siemens, Munich, FRG)
Vol. 3	Circuit Analysis, Simulation and Design (Part I and II)	A. Ruehli (IBM, Yorktown Heights, USA)
Vol. 4	Layout Design and Verification	T. Ohtsuki (Waseda University, Tokyo, Japan)
Vol. 5	VLSI Testing	T. Williams (IBM, Boulder, USA)
Vol. 6	Design Methodologies	S. Goto (NEC, Kawasaki, Japan)
Vol. 7	Hardware Description Languages	R. Hartenstein (Univ. of Kaiserslautern, FRG)

The first five volumes deal with major phases of VLSI design separately. The following two volumes are devoted to recent approaches which span the whole design phase. The integrated approach combining system design and VLSI design is also treated in these volumes, and it will suggest a major trend in the future. Each volume is reasonably self-contained so that it can be read independently.

All of the volumes were intended to include up-to-date results, and the latest developments with a good balance between theory and practice. Moreover, emphasis was placed on basic techniques, methods and algorithms rather than on descriptions of existing design tools. This, I hope, will prevent obsolescence at the time of publication. Only the readers however, can judge to what extent our intentions were successful.

Selecting the volume editors was not an easy task. In order to produce a quality book series, it was necessary to utilise authorities well known in their respective fields who in turn would attract outstanding contributors. Those active in VLSI design had other commitments. I was relying on their volunteering spirit, and they in turn faced the same dilemma with their authors. It was a great pleasure to me that ultimately we were able to attract such an excellent team of editors and authors.

The Series Editor wishes to thank all of the volume editors and authors for the time and effort they put into the book series, and especially Prof. W. Engl for his assistance in arranging other volume editors. I also thank Dr. E. Fredriksson, Dr. J. Julianus and Dr. R. Morel of North-Holland for their continuing effort to bring the book series from the initial planning stage to final publication.

Waseda University
Tokyo, Japan
1985   Tatsuo Ohtsuki

PREFACE

With the rapid evolution of VLSI technologies, design of integrated circuits is becoming increasingly complex and difficult. Among the various design phases of VLSIs, "layout" takes a major portion of design turn-around-time. This volume of the book series on Advances in CAD for VLSI is solely devoted to "layout design and verification", thus distinguishing it from other CAD books. In view of the present status, results obtained from fully automated layout tools are still discouraging compared with manual design. And there is still an increasing number of problems to be overcome in producing usable chips with a much shorter design turn-around-time. This background was the motivation for devoting an entire volume to discussing major topics in layout design.

This book was edited with the intention of covering all important topics relevant to layout design and verification ranging from practical aspects on VLSI chip floor plan to theoretical analysis of basic sub-problems. Its character leans towards a reference for those already active in this field by compiling the present state of the art including up-to-date results, which is scattered in various journals. Similarly it should help others working in neighbouring VLSI CAD areas to get a better understanding of layout problems. It should also provide novices with a stepping stone for doing research in this area, but some knowledge in VLSI technology and combinatorial algorithms will aid understanding. The main emphasis was put on basic approaches, methods and algorithms rather than a description of existing design tools, in order to prevent obsolescence at the date of publication.

The contents of this volume have been structured with nine reasonably self-contained chapters. Enough space was assigned for the first chapter to give the readers a practical background for understanding various layout styles such as full-custom, building-block, standard-cell, gate array, etc., which should be chosen by means of a trade-off between two conflicting cost factors: reduced design cost through layout standardization and increased yield through high integration density. The chapter also describes how the layout sub-problems being discussed in the following chapters are derived from the practical background. The second chapter deals with the subproblems (partitioning, assignment and placement) encounterd in the first major step of layout design, in which the abstract (logical) circuit description is mapped onto the physical one including size, shape and internal structure of component blocks.

The following three chapters (Chapters 3-5) are devoted to routing. In Chapter 3, rather classical maze-running and line-search algorithms are reviewed with their recent improvement and related controversial topics, as they are still used either extensively or partially in the current VLSI design. This is followed by Chapters 4 and 5, in which the two-stage routing approach consisting of "global routing" and "channel routing" for large-scale circuits is treated, and various routing algorithms are reviewed.

Chapter 6 deals with layout compaction, which is needed as post-processing for the results of automated building block LSI design or symbolic layout. This problem will also be relevant to the silicon compilation technique in order to get usable chips. Then the layout verification algorithms are described in Chapter 7. This design phase is still indispensable, particularly for full-custom LSI design in which error-prone human intervention is involved. Particular emphasis is put on those algorithms suitable for extremely large-scale geometrical problems considering the recent and near-future VLSI mask patterns.

Finally there are two chapters focusing on relevant mathematical topics. In Chapter 8, typical subproblems derived from the original layout problems by some abstraction and simplification are shown to be NP-complete, which encourages us to use heuristics for layout design. In Chapter 9, computational geometry algorithms proposed by computer scientists in other fields are introduced. The authors have selected those algorithms which seem to be nicely applicable to VLSI design. As design rules are becoming more complicated, such a geometrical approach will be a major trend in the future for routing and placement as layout verification.

Recent topics, such as hierarchical layout design, symbolic layout, and hardware engine, are not explicitly discussed in this volume. They will be included in Vol. 6 (Design Methodology) of the book series.

The contributors to this volume are all leading authorities in their fields. Many were extremely busy with other commitments, and I relied on their generosity in setting aside extra time to write their contributions. I would like to extend my grateful thanks to each author for having accepted my offer, and for all the time and effort which has resulted in this first class volume.

Waseda University
Tokyo, Japan
1985   Tatsuo Ohtsuki

TABLE OF CONTENTS

LAYOUT DESIGN AND VERIFICATION
T. Ohtsuki (Editor)
© Elsevier Science Publishers B.V. (North-Holland), 1986

Chapter 1

LAYOUT STRATEGY, STANDARDIZATION, AND CAD TOOLS

Kazuhiro UEDA, Ryota KASAI, Tsuneta SUDO

NTT Atsugi Electrical Communication Laboratories
3-1 Morinosato wakamiya, Atsugi-shi, Kanagawa, 243-01 Japan

This chapter describes LSI layout design problems and their solutions from various points of view. There are trade-offs between many factors in deciding layout design styles; hand-crafted, symbolic, PLA, gate array, standard-cell, hierarchical etc. Many layout CAD algorithms and methodologies have been exploited, and based on them, a number of layout design systems have been developed. Typical CAD tools and systems are explained in relation to layout algorithms. Further, standardization of layout interfaces is discussed and the future outlooks in this field are also described.

1. INTRODUCTION

As VLSI gets more complex, the design effort and turnaround time increase at a higher rate particularly in the layout design phase. Traditionally, the VLSI design is mainly performed manually. The Z8000 microprocessor was a typical example of such VLSIs. In its design, very little CAD was used. As a result, about 6,600 man-hours, or 50 % of the whole design effort, was required in its layout design phase. Altogether, approximately 13,000 man-hours was needed to produce functional/logic/circuit designs, mask designs, and testing and characterizing of the resulting design.[1]

As soon as the large scale integration concept emerged at the beginning of the 1960's, CAD was recognized to be indispensable in support of LSI design for the purpose of cutting design effort and turnaround time, eliminating human errors, and as a result, reducing design cost. Since then, various layout design styles and strategies have been proposed: hand-crafted, symbolic, PLA, gate array, standard cell, hierarchical, and so on. There are many factors to be considered in deciding which layout style should be employed. They include design effort and time, packing density, performance (speed, power, noise margin), yield, reliability, etc. There are trade-offs between these factors. In a hand-crafted layout design, maximum flexibility can be used; there is no need for standardization. In order to effectively introduce CAD tools into the design environment, however, standardizations must be introduced in many aspects.

Different layout styles require standardization in different aspects, and as a matter of course, need different CAD approaches and tools.

Interface problems involved with layout design have become significant as the usage of CAE/CAD tools spreads into a wider area. Consequently, effort for interface format standardization is currently being expended. The emphasis in this paper is on a review of the VLSI layout design problems and their solutions from various points of view.

2. LAYOUT STYLE AND STRATEGY

2.1 Layout Design Factor and Tradeoff

The use of LSI chips has increased and is now required in a wide variety of digital systems. Future development in the field will call for many innovations in design. There are several directions involved, each with its own features. The task of design selection should be based upon such considerations as tradeoffs among design factors for individual styles of design, the time and effort required for completion of such designs, chip size and circuit performance.

The following discussion takes up broad, general trends and tradeoffs among design factors which, due to their complexity and interrelation are, in practice, subject to quantitative analysis.

To begin with, let us consider tradeoff between design efforts and chip size. LSI cost per chip C_T can be expressed as:

$$C_T = C_D/N + C_P/(y \cdot n), \hspace{3cm} (2.1)$$

where N denotes production volume, C_D, primary investment mainly consisting of design costs of logic, circuit and layout, C_P, processing cost per wafer including testing and packaging costs, y, average yield and n, the number of chips on a wafer. Equation (2.1) indicates that the primary investment should be reduced for small production volume LSIs, and that the yield and the number of chips should be increased for large production volume LSIs. The number of chips is proportional to the reciprocal of the chip size. The yield is reduced exponentially as the chip size increases, as shown in Fig. 2-1 [2]. Thus to reduce the chip size, technologies for solving the following three problems are needed: minimizing the number of circuits per function, minimizing the number of devices and the size of devices in a circuit, and the reduction of dead space in the circuit placement and routing. Generally, these problems do not have clear solutions. They must

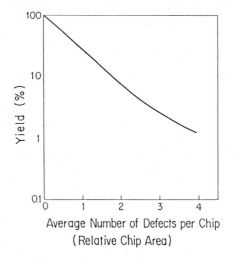

Fig. 2-1   Yield of chip vs. average number of defects per chip.

be treated heuristically or through the experiences and intelligence of the skilled designers. When these problems are treated on a computer with numerical analysis, the processing time drastically increases as the number of circuits increases.

Next, let us consider the tradeoff between circuit performance and chip size. Circuit performance is primarily affected by device structure and circuit configuration.   Nevertheless, layout influence on circuit performance cannot be neglected for LSIs which specifically require high speed or low power. Circuit delay time Tpd is expressed as:

$$Tpd = Tpd_0 + V_L(C_W + C_G)/Ip, \qquad\qquad (2.2)$$

while dissipation power Pd is given by

$$Pd \propto \begin{cases} Ip\ V_{DD} & ;\ \text{for ratio circuits,} \\ f(C_W + C_G + C_0)V_L^2 & ;\ \text{for ratioless circuits,} \end{cases} \qquad (2.3)$$

where V_L denotes logic swing amplitude, C_W, wiring capacitance, C_G, total gate capacitance of fanouts, Ip, active transistor peak current, V_{DD}, supply voltage, C_0, total self capacitance and f, operating frequency. C_G and Ip are proportional to the size of the active device (emitter area or channel width) and C_W is proportional to the wiring length, which depends on a square root of chip size.

It is clear that an increase in I_p and C_G leads to high speed and high power, while reduction of C_W brings on high speed and low power. Therefore, a compact layout design to achieve a shorter wiring length is needed for high performance LSIs. Clearly compact layout design requires a great deal of design effort and design time.

2.2 Process technology and design rule

Chip design must be discussed in the light of the process technology employed. Layout design strongly depends especially on process technology level, which is generally translated into design rules (or layout rules) that can be easily understood by designers or draftmen.

Basically, design rules are determined from the following factors 1) the minimum line width to be exposed and the minimum tolerance of registration between two layers, 2) the minimum device dimension determined by electrical or physical constraints such as punch-through, junction break-down and electromigration. As these values decrease, in other words, as process technology is improved, chip size appears to decrease proportionally. In practice, however, these values do not vary uniformly, because of the uneaveness and the congestion of patterns of the under-layer. Thus, complicated, case-by-case design rules must be provided if a high density, high yield chip design is required. In that case, a great deal of additional design cost and time must be expected. On the other hand, in the case of LSIs with a relatively small production volume, a simplified design rule can be used, because the circuit density is not a main factor. Therefore, the objectives of LSI design must be considered in deciding the extent to which design rules should be simplified.

As process technology advances, another important problem must be faced: how many layers should be employed for connections between devices? Multilayer metal technology makes it possible to reduce chip area significantly, because logic LSIs generally need wider areas for wiring than for devices when they are used with single or double layers.

Let us consider a chip with double metal wirings in which the wiring area usually occupies about 80% of the total chip area. If the same circuit is designed with triple metal wirings (where the third wiring pitch and direction are equal to the first), 40% decrease in the chip area is assumed possible, because the first metal wiring pitch is virtually reduced to one half. An actual study on effectiveness of triple metal layer wirings as compared with double metal layer wirings [3] results in a 30% area reduction, as shown in Fig. 2-2, where the third wiring pitch is the same as the first one.

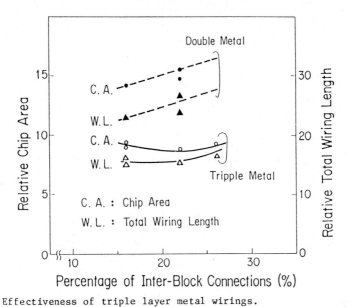

Fig. 2-2   Effectiveness of triple layer metal wirings.

A 5,000 gate logic network was laid-out with double and triple wirings
using an automatic hierarchical layout program based on the standard
cell design approach. Chip area was reduced by about 35% and total
wiring length minimized by about 25%, thanks to triple wirings.

The increase of wiring layers brings on an increase of process steps, and as a
result yield lowering. In some cases, the benefit of low processing cost due to
the area reduction can not be attained. Therefore, there must be an appropriate
choice in the number of wiring levels to a given process technology.

Design tools must be updated to meet the concurrent process technology so that
any solutions can be chosen from various process technologies for the tradeoff
between process technology and layout style.

2.3 Layout Style: Classification and Comparison

This section discusses layout styles classified from the point of view of standardization level differences. Standardization refers to the means of designing a
target LSI at a minimum cost; it leads to the saving of the design efforts,
though it sacrifices packing density and circuit speed. Table. 2-1 shows the
standardization in each level of six layout hierarchies, which are arranged in
ascending order of hierarchy level. The standardization listed in lower column
of the table requires much smaller design efforts than those in the upper column.
The higher level of standardization includes lower ones in most cases. The table

also lists layout styles corresponding to the standardization levels. For example, PLA and ROM have the highest standardization level and require the smallest design efforts, while the full-custom approach has the lowest standardization level but the highest packing density.

Table. 2-1 CLASSIFICATION OF STANDARDIZATION IN LAYOUT

Standardization Level	Standardization Item	Layout Style
Element	Design rule	Full custom
Device	Symbolization of transistors, via holes, connections, etc.	Symbolic design
Cell	Normalization of cell height, terminal location and wiring pitch	Standard cell
Block	Normalization of block terminal location and wiring pitch	Building block
Placement of cells and blocks	Standardization of cell and block placement	Gate Array
wiring	Standardization of wiring pitch and connection between devices	PLA, ROM

The details of each layout style will be described in the following, focusing on definition, merits/demerits, CAD tools supporting the layout style and their application.

2.3.1 Full-custom Approach

1) Definition: This approach manually packs the entire chip, if design rules allow, using an interactive graphic system, which generally offers various editing commands such as Insert, Move, Stretch, Erase, and Copy. Layout data can be seldom used in common; not even when the logic functions are equivalent to one another, because area minimization and performance optimization precede design efforts. Accordingly, either no standardization or only design rule standardization such as Mead & Conway's λ is employed [4].

2) Merits and Demerits: As for merits, the approach can obtain high packing density and high performance thanks to layout flexibility. On the contrary, design efforts and time increase drastically with the increase of the number of devices on an LSI, because of its low layout productivity (5-10 devices /man day), inevitable layout errors and circuit complexity. This means that the method is inappropriate for a large scale and complex network. In the case of random net-

works, the circuit size limitation would be about several thousand gates. For the repeatable structure logic networks such as array processor [5],[6] and RISC (Reduced Instruction Set Computer) [7], however, this method is worth using, even though large scale integrated circuits, because the number of actually hand-honed devices is significantly small thanks to the ease of reproduction of the structure.

3) CAD tools: Verification CAD programs are indispensable for full-custom design, as inevitable errors caused by designers must be corrected. Examples of available verifiers are: design rule check programs (DRC) [8], logic network connection check programs (CMAT, PAS-1) [9],[10] and timing simulators (MOTIS-C) [11]. There is no layout CAD program which can take the place of manual layout.

4) Applications and Examples: This approach has been applied to (up-to 16 bit) general-purpose microprocessors, general-purpose high-speed arithmetic execution LSIs, array processors and RISCs. Figure. 2-3 shows a photomicrograph of Z8000 [12]; the last chip with random network alone for a gate size of several thousands. Figure. 2-4 shows a photomicrograph of AAP [13] (Adaptive Array Processor) containing 64 one-bit processors in an 8 x 8 array.

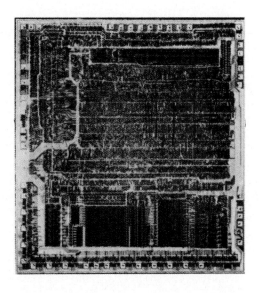

Fig. 2-3   Photomicrograph of Z8000 chip.(Reprinted with permission of Zilog Inc. Copyright © 1982, Zilog Inc.)

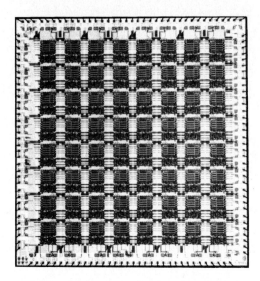

Fig. 2-4   Photomicrograph of adaptive array processor chip.
           (Copyright © 1982, IEEE.)

2.3.2 Symbolic layout approach

1) Definition: This method makes LSI layout images using symbolized components
such as transistors, via holes and connections. The layout image is designed
manually. The conversion of symbols into physical patterns and their compaction
are generally automated. There are two cases for symbolic layout: with simplified design rules and without. A symbolic layout image and the physical pattern
converted from the symbolic image are shown in Fig. 2-5 [14].

2) Merits and Demerits: Design efforts are significantly reduced, while maintaining high layout flexibility, as compared with the full custom design approach.
Productivity of generating a layout image using symbols still remains low, while
the frequency of layout error is significantly reduced. High packing density can
be obtained in the next place to the full-custom approach. If global layout
planning is not optimized, packing density may be extremely low, due to the
immaturity of compaction functions of current programs. The method depends on a
process technology to a great extent. Programs have to be updated whenever a new
process technology is developed.

3) CAD tools: Generation and edit of layout images of symbols, physical pattern
conversion and compaction are systematically supported by CAD systems (STICKS,
SLIM) [14], [15]. A verificatian program for design rule check is unnecessary.

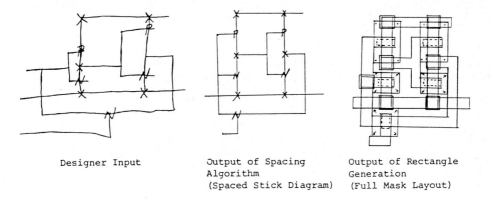

| Designer Input | Output of Spacing Algorithm (Spaced Stick Diagram) | Output of Rectangle Generation (Full Mask Layout) |

Fig. 2-5   Example of symbolic layout.

(Reprinted from Williams, 1978. Copyright © 1978, AFIPS press.)

4) Applications and Examples: This method covers the same application area as the full-custom approach. It can be also considered as a method for enlarging the extent of full-custom approach coverage. However, a VLSI designed on the basis of this method alone has not been reported yet.

2.3.3 Standard cell approach (polycell approach, cell library approach)

1) Definition: Logic, circuit and physical pattern libraries for 30 or 40 types of basic logic functions have been developed, and used in common for various chip designs. As for physical patterns of cells, cell height, terminal direction and location, and power line location are standardized, so that cells can be arranged in rows. An example of a cell physical pattern is shown in Fig. 2-6. Routing for intercell connections is performed after normalized wiring grid. Placement and routing are automated in most cases. Figure. 2-7 shows a pattern image of an automatically laid-out standard cell block.

2) Merits and Demerits: Design efforts and time are significantly reduced by employing CAD programs and libraries. A 20K-gate CMOS VLSI designed within 2 man-months has been reported [16]. Circuit performance is relatively high, though not to the extent of the full custom approach. As for demerits, (1) the method needs a time-consuming cell library update whenever process technology is improved, (2) some part of the cells inevitably remain unused, due to a limited number of cell types and (3) a redundant space can be easily generated in the wiring area, because of one-dimensional cell arrangement. These demerits are, however, not so serious as those for gate array.

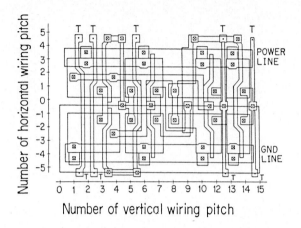

Fig. 2-6   A typical pattern of standard cell.

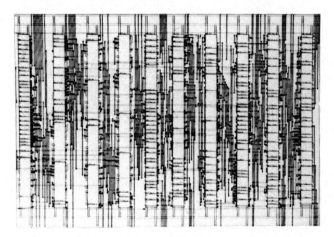

Fig. 2-7   A logic network block laid out automatically using standard cell
            approach.

3) Recent trends: (1) Development of CAD programs which automatically update the
physical cell pattern in the library [17]. (2) Hierarchical automatic layout
systems which make it possible to reduce a great amount of layout design effort
introducing top-down approach [16]. (3) Improvement of the method by introducing
macro blocks such as RAMs, ROMs, PLAs and data path with a limited condition[18].
(Without the limited condition, the method becomes identical to the building block
approach.) (4) Development of CAD programs for multi-layer metal wirings [19].

4) CAD tools: There are many layout programs for practical use, which are based
on the standard cell approach. It is remarkable that the channel routing algorithm [20] which certifies a complete inter-cell connection without unaccomplished

wiring can be used. As for verification programs, a timing simulator is the most important program because gate delays vary widely due to wiring length dispersion tendencies.

5) Applications and Examples: The method can be applied to LSIs with medium production volume and those with low production volume requiring a relatively high performance, like LSIs, for an in-house communication use. An example of a chip designed on the basis of this method will be shown in Chapter 3.2, which employs a hierarchical automatic design procedure.

2.3.4 Building block approach (General cell assembly approach)

1) Definition: This is an approach which places rectangular blocks of various sizes and aspect ratios in a two-dimensional array and routs wirings among the blocks as shown in Fig. 2-8 [21]. Block size and terminal position are normalized by horizontal and vertical wiring pitches. The building block approach is sometimes recognized as method including the standard cell approach. We treat them as different methods, since they have different standardization levels.

2) Merits and Demerits: If blocks are registered in their library, a chip layout can be finished within a relatively short design time, with little design effort.

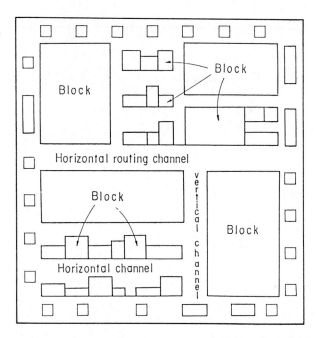

Fig. 2-8   Chip image of building block approach.

The blocks can be designed for higher performance than those designed using the standard cell method or gate array because they can be laid out independent of others, using a sophisticated layout scheme like the manual or symbolic methods. However, the most serious bottle-neck is that a great amount of design effort and time is required for an update of block physical patterns.

3) CAD tools: There is no layout program which can perform a complete connection for inter-block routing, since the direction of block terminals is not standardized. Currently available programs are based on a line-search [22] or maze algorithm [23] for the routing process. It is necessary that a small percentage of wiring is inserted or modified manually to accomplish a complete connection.

4) Applications and Examples: This method is a little more suitable for general -purpose LSIs thanks to its high performance. An example of a chip designed using this method is shown in Fig. 2-9.

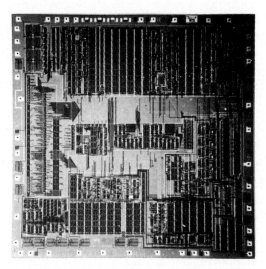

Fig. 2-9   Photomicrograph of a LSI designed using building block approach.
           (Courtesy of Nippon Electric Co.)

2.3.5 Gate array approach (Masterslice, Uncommitted logic array)

1) Definition: This method affords individualization of logic networks, by simply routing the connections among cells and devices inside the cell on uncommitted chips (called master chips). The devices on the master chip are placed in a simple arrangement, without wirings to the other devices, so as to be used in common over the various logic networks (See Fig. 2-10). Accordingly, only the

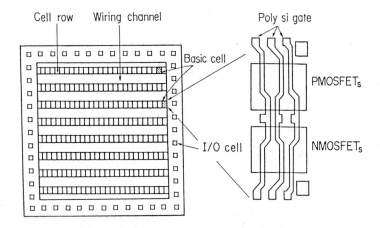

Fig. 2-10 Chip image of gate array approach.

mask patterns for via holes and metal wirings are customized. Logic function cells are provided as a cell library just as in the standard cell approach. It is common to build a family of master slices applicable to various sizes and performances of logic networks.

2) Merits and Demerits: The outstanding feature is the remarkably short time it takes from the beginning of the chip design to the end of chip fabrication or testing. Other merits are the ease of CAD support thanks to the standardization of the basic cell arrangement and the flexibility of design in physical cell patterns, given the small number of masks to be customized.

As for demerits, a logic wiring area must be prepared between adjoining cell rows to accommodate various numbers of inter-cell connections for various kinds of networks. Accordingly, it is not unusual that a considerable area remains unused. The regular arrangement of devices naturally makes for waste areas and redundant layout. These demerits result in, low packing density of about 1/2 of the average packing density of a standard cell approach as well as considerable circuit speed reduction.

3) Recent trends: Chip size and circuit speed of the masterslice show remarked improvement through positive use of fine process technology. In order to increase design flexibility various kinds of new gate array approaches have been developed; gate isolation array [24], channelless array [25] and gate array with fixed or variable macro [26].

4) CAD tools: This method offers many layout programs for practical use as does

the standard cell approach [27],[28]. Requirements for gate array layout programs are (1) technological independence, (2) minimization of the number of
unacomplished routings and of design time.

5) Application and Examples: This method is suitable for LSIs with small production volume and experimental LSIs. Examples of master chips are shown in
Figs. 2-11 (Bipolar 5K gate masterslice) [29], 2-12 (CMOS 6K gate array) [26] and
2-13 (CMOS channeless gate array - pair transistor array) [25].

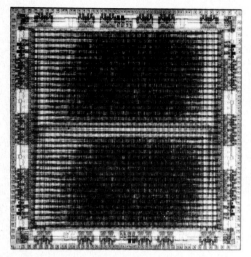

Fig. 2-11 Photomicrograph of 5K gate
bipolar gate array chip.

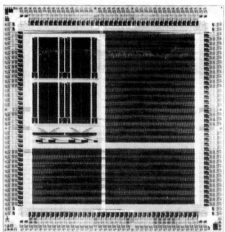

Fig. 2-12 Photomicrograph of 4.3K gate
CMOS gate array with 2K bit configurable
RAM.(Copyright © 1983, IEEE.)

Fig. 2-13 Photomicrograph of
CMOS pair-transistor arry chip.

2.3.6 PLA and ROM

1) Definition: The desired network can be realized by connecting or disconnecting horizontal/vertical lines terminating at input/output circuits and devices such as MOSFETs and diodes arranged in a two-dimensional array (See Fig. 2-14). ROM prepares $m \cdot 2^n$ devices for the network with n inputs and m outputs. PLA basically prepares AND plane with $2n \cdot t$ devices and OR plane with $m \cdot t$ devices for the network with n inputs, m outputs and t product terms. Only connections among devices and horizontal/vertical lines are customized.

2) Merits and Demerits: The layout of ROM and PLA except for connection customization is very simple, because of its highly regular device arrangement. Connecting points are easily customized using CAD programs which generate them from the original logic network description data. Those features greatly reduce the design time and effort. ROM and PLA LSIs are very inexpensive since they can be produced on a large scale in an unfinished state; remaining connection process. Field programmable structure can be also employed, leading to lower cost, because the field programmable logic array (FPLA) can be produced on a very large scale thanks to its common usability.

As to demerits, a large scale logic network cannot be accommodated because of the large number of unused devices originating from their redundant logic implementation. Operation speed is very low because of heavily loaded wirings.

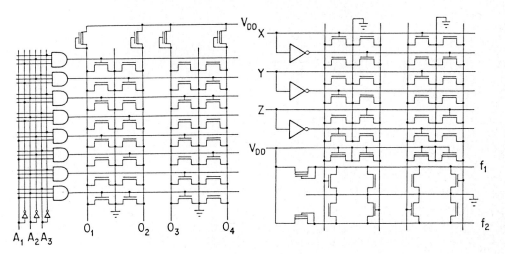

Fig. 2-14 MOS ROM and PLA configurations.

3) Recent trends: Utilizing the regular structure and the rewritability of ROM and PLA, many semiconductor factories try to reduce design time and efforts of complex logic VLSIs, especially the time required for debugging microprograms and sequencers. To widen the application area of those on-chip PLAs, folded structure is being positively employed.

4) CAD tools: Full top-down design systems [30], [31] have been developed. The automatic generation programs of physical patterns for folding structure PLAs are also available.

5) Application and examples: This method is suitable for inexpensive, relatively small scale, low performance ICs for toys and TV games. Recently, the method has been applied to replace some portions of logic blocks in large scale ICs. Figure 2-15 shows a photomicrograph of a bipolar FPLA chip [32].

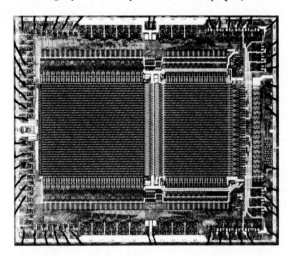

Fig. 2-15   Photomicrograph of bipolar field programmable logic array.

2.3.7 Other approaches

In this section, we shall describe some methods which do not belong among the previous design approaches.

(1) One dimensional array approach [33] (Weinberger layout style)
This is a method which arranges the input and output terminals of each basic gate in a vertical straight line, as shown in Fig. 2-16. These terminals can be swapped with each other. According to logic network data, logic gates are arranged in a one-dimensional array and inter-gate connections run straight horizontally.

The width of the block laid out is constant independently of the placement order of the gates. This method is well suited to automatic layout because wiring routes can be determined without considering the procedure of connections to the gate terminals.

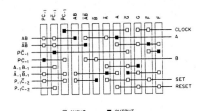

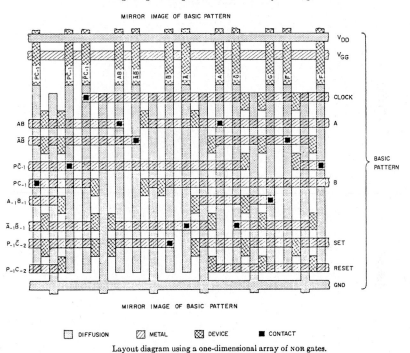

Logic diagram using a one-dimensional array of NOR gates.

Layout diagram using a one-dimensional array of NOR gates.

Fig. 2-16 One-dimensional transistor array approach. (Reprinted from Weinberger, 1967. Copyright © 1967, IEEE.)

(2) Gate matrix method [34]

This method was developed by Bell Labs. The basic concept of the gate matrix style is to superimpose all transistors of the standard cell method onto the wiring channel so that all of the transistors that have a common input are placed on a common polysilicon line. This line acts both as the gate of the transistor and as the connection among all commonly gated transistors. The polysilicon lines, which are equally spaced and parallel, become the columns of the gate matrix. The rows are formed by grouping transistor diffusions which associate with one another either in serial or parallel fashions. Gate matrix can be regarded as a generalized form of the Weinberger layout style. Figure 2–17 shows a CMOS logic network, its schematic layout image conforms to the gate matix method and its physical pattern is generated automatically. The layout of the gate–matrix ALU is shown in Fig. 2–18. This method has been successfully applied to various CMOS logic VLSIs such as WE32100.

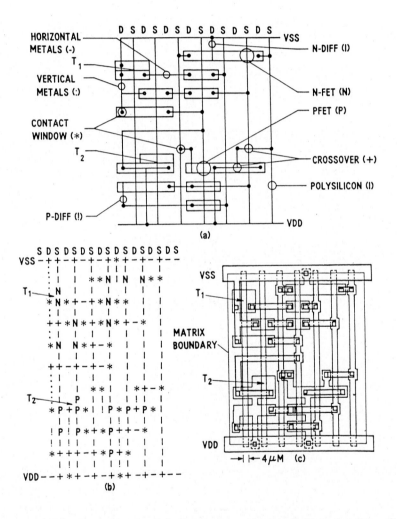

Fig. 2–17   Gate matrix approach. (a) The representational line drawing. (b) and (c) are the symbolic description and final artwork of (a). (Reprinted from Lopez et al., 1980. Copyright © 1980, IEEE.)

Fig. 2-18   32bit ALU designed using gate matrix method. (Reprinted from Lopez
et al., 1980. Copyright ⓒ 1980, IEEE.)

2.3.8 Summary

Tradeoffs among the aforementioned layout styles are summarized in Table 2-2.

Table 2-2   SUMMARY OF TRADEOFFS AMONG VARIOUS LAYOUT STYLES

Layout style	Design product- ivity	Process product- ivity	Area per function	Circuit speed	Occurence of design error	Test- ability	Redesign ability
Full-custom	✗	—	◎	◎	✗	△	✗
Symbolic	△	—	○	○	—	△	—
Stand. cell	—	—	—	○	○	—	○
Build. block	△	—	○	○	△	—	—
Gate array	○	○	△	—	○	—	○
PLA(ROM)	◎	○	✗	✗	◎	◎	◎

◎very high, ○high, — medium, △low, ✗ very low.

2.4 Basic Layout Flowchart

If layout is defined as conversion of circuit configurations into physical
patterns, the flow of layout procedure can be generalized as is shown in Fig.
2-19.

As circuit scale increases, area estimation and floor planning become the key
process of control in packing density of object LSIs. In the conventional layout, area estimation and floor plan were perfected after the physical pattern
design of most components. Recently, area estimation and floor plan before the

physical pattern design of components have been studied to obtain a higher packing density and automatize a full top-down layout. That requires an accurate
area estimation model. Practical systems which perform area estimation and floor
plans automatically have been developed for highly standardized layout styles,
such as PLA, ROM, gate array and standard cell. Full automation of the layout
procedure for ROM and PLA is of significance.

In a hierarchical layout design, of which the details will be presented in Section 3.1, the procedure from area estimation to verification is   performed   re-
peatedly in each hierarchy level.

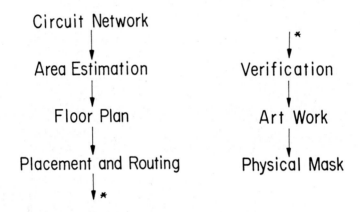

Fig. 2-19   Basic layout flowchart.

2.5 Complex Layout Strategy

As the size and complexity of logic LSIs increases, the use of a single design
approach in the entire chip design is considered uneconomical. In order to
resolve the problem, several authors have reported complex layout approaches
which integrate different design approaches in optimum combinations. In this
section, we will review two examples.

2.5.1 Integrated modular and standard cell design approach [35]

Logic VLSIs, especially processors, consist of bit-slice-structure data path
logic, random control logic and memories for register files and control storages.
The nature of this circuit structure in logic VLSIs necessitates the following
criteria for design method assignment: 1) hand-honed design should be applied to
bit slice or array structure (modular structure) networks such as ALUs, shifters,
registers. 2) automatic standard cell approach should be used for random-struc-

Table 2-3   DESIGN METHOD ASSIGNMENT FOR EACH FUNCTION BLOCK

Block Name		Network Structure	
		IMSA chip	SCA chip
DAT	SPM	b.s.m.	RAM
	Data Reg.	"	s.c.
	ALU	"	"
	Data I/O	"	"
	PSW CC	"	"
CNT	IRDEC	PLA	"
	CS	ROM	ROM
	uDEC	PLA+s.c.	s.c.
MEM	ATT	RAM	RAM
	ATL	b.s.m.+s.c.	s.c.
	Add. I/O	b.s.m.	"
BCK	CLK	s.c.	"
	BC	"	"
Others	MISC	PLA	"
	Counter	s.c.	"

b.s.m : bit slice module, s.c. : standard cell

Table 2-4   LAYOUT RESULT OF IMSA CHIP AS COMPARED WITH SCA CHIP

Unit Name	No. of Devices		Area (mm^2)	
	IMSA chip	SCA chip	IMSA chip	SCA chip
DAT	6,410	8,420	4.75	8.29
CNT	44,760	31,660	10.10	11.41
MEM	5,960	7,130	4.45	6.43
BCK	2,740	2,610	4.53	4.52
Others	6,910	5,800	16.88	23.31
Total	66,180	55,620	40.71	53.96

Area includes power buses and peripheral wirings.

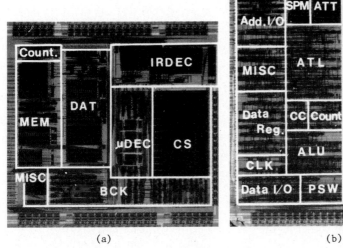

(a)   (b)

Fig. 2-20   (a) Photomicrograph of chip designed with the integrated modular and standard cell design method ; (b) photomicrograph of the chip designed with fully automatic standard cell approach : magnification is same as in (a). (Copyright ©1984, IEEE.)

ture logic network such as sequencers and controllers. 3) PLAs should be used for a random network if the network can be implemented on planes with high bit utilization.

A design automation (DA) system is indispensable for the integrated modular and standard cell approach (IMSA) to integrate the different design methods effectively and rapidly.

To verify the effectiveness of IMSA, two similar 16-bit microcomputers were designed. One used IMSA, the other, the fully automatic top-down standard cell approach (SCA). Chips were fabricated by a 2um Si gate CMOS process employing double Al wiring layers. They used the DA system, including a data base and various CAD programs linked to the data base through a data base manager. The fully automatic top-down SCA is supported by CHAMP/ALPHA system which will be presented in Section 4.1.4.

The IMSA chip designed according to the aforementioned criteria uses design methods shown in Table. 2-3 for each function block. Interconnections between the MCA, PLA and SCA blocks are carried out hierarchically and interactively. The SCA chip used 2528 predefined cells (41 types) and laid out in a similar manner as 32-bit processor chip which will be described in Section 3.2.
Photomicrographs of the two chips are shown in Fig. 2-18, layout results are summarized in Table. 2-4. The size of the IMSA chip is about 25 percent smaller than that of the SCA chip. Both chips are confirmed for operating at a machine cycle of 5 MHz, even in the worst cases. The circuit and layout design efforts were about 20 man-months for the IMSA chip (10 man-months for hand-honed modular blocks, 4 for PLA and memories). On the other hand, 4 man-months, including 3 man-months for ROM and RAM design, were required for the SCA chip. Design work for standard cell library preparation is excluded.

IMSA offers a high grade layout quality in comparison with SCA. The drawback of large design effort requirement can be reduced when sufficient functions are registered in the modular cell library and their physical pattern data are automatically renewed.

2.5.2 Mixed gate array and custom design approach [36]

The next example shows the improvement in power-performances density achieved through optimal combination of power supply voltage and physical design selections. The chip designed is a 32-bit hard-wired microprocessor. A 2um NMOS technology is employed.

Table 2-5   MASTER IMAGE VARIOUS HAND-HONED
            DESIGN

            (Reprinted from Erdelyi et al.,
            1984. Copyright © 1984, IEEE.)

Control Logic	Data Flow	Chip Size (mm)	Performance (ns)	Power (W)	Physical Design Time (Months)
Master Image	Master Image	9.5 x 9.5	390	3.5	5
Master Image	Hand Honed	7.6 x 7.6	150	2.2	10
Hand Honed	Hand Honed	6 x 6	110	2.0	20

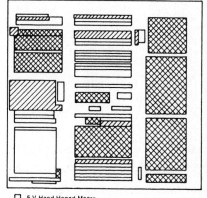

☐ 5-V Hand-Honed Macro
▨ 3.4-V Hand-Honed Macro
▩ 3.4-V Master Image Macro

Fig. 2-21   Outline of the 32bit microprocessor chip. (Reprinted from Erdelyi et al., 1984. Copyright © 1984, IEEE.)

Let us consider ratio circuits of NMOS gates. If the same load current is maintained while reducing the supply voltage, the gate capacitance of the active devices increases since the beta ratio must be increased to avoid noise margin degradation. Therefore, lower supply voltage offers a better power-performance tradeoff, as long as overall gate capacitance in the net is smaller than the wiring capacitance. The ratio of wiring to gate capacitance on a given net strongly depends on the layout style. When all circuits are closely packed by customizing shapes, the ratio is minimized. When a semi-custom approach, such as gate array is used, a penalty is paid with larger wiring nets. On the other hand, semicustom approach has an edge over the full custom approach for design efforts.

The following alternatives for random logic and data flow circuits were evaluated; gate array (with a 3.4 V circuit library) and full custom design. Table. 2-5 estimates chip size, performance, power and physical design time for gate array versus full custom design. For functions in the data flow, the hand-honed design approach proved to be a good tradeoff. If control logic were realized by the full custom approach, however, any logic change would have seriously impacted the schedule. Therefore, the total hand-honed approach should be rejected.

The best compromise was to design the control logic of the chip with the gate array technique by automatic placement and routing and the data flow portion by a hand-honed design. Figure 2-21 shows the chip floor plan when complex design approaches were used.

3. HIERARCHICAL LAYOUT DESIGN

3.1 Design Effort Reduction by Hierarchical Design

3.1.1 Needs for Hierarchical Design

According to G. Moore [37], the major hurdle faced in the construction of ever larger integrated systems is a complexity barrier. Large systems usually have a hierarchical structure which is broken down into such things as backboards, packages and ICs. Due to this physical hierarchy, system's design data are decomposed into an appropriate size of module which is easily manageable by a designer or a CAD program.

VLSIs fundamentally have no restriction in regard to partitioning or decomposition. A single chip can contain tens of thousands of circuit elements with no hierarchical structure. On the positive side, this freedom may be exploited for significant performance advantages. However, on the negative side, it may result in a hazardous situation where the complexity within a large, unstructured domain simply overwhelms the designer.[38] This is true not only for manual design, but also for CAD. In the following, we will examine the time complexity of layout CAD, and then prove the advantage of hierarchical design.

3.1.2 Time Complexity of Layout CAD

It is known that the time complexities of placement algorithms are in most cases between $O(n^2)$ and $O(n^4)$ according to their strategies (where n = number of placement modules). [39] For routing algorithms, the complexities are slightly smaller than that, still they may be $O(n\log n)$ – $O(n^2)$.

Figure 3-1 shows an example of growth rates of CPU time consumed by layout CAD programs compared to other CAD programs. PLASMA and COSMIC are a routing and a placement program, respectively.[19] TEGAS and SPICE are well known logic and circuit simulators [40, 41], and LOTAS is a timing simulator.[42] For all of these simulators, the growth rates are approximately $O(n)$. On the other hand, the growth rates for the layout programs are much higher than that. The time complexities of routing program PLASMA and placement program COSMIC are $O(n^{1.5}-n^{2.0})$ and $O(n^{2.0}-n^{2.6})$, respectively. This shows that when the circuit size (gate count) approaches 20,000, the CPU time required for placement will be more than 300 hrs; i.e., clearly out of a practical limit.

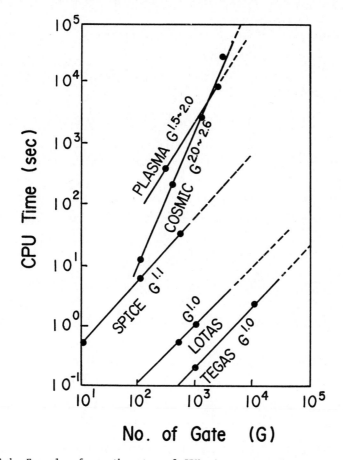

Fig. 3-1   Example of growth rates of CPU time consumed by layout CAD programs.

3.1.3 Effect of Hierarchical Design

This section examines the effect of a hierarchical design by using a simple model.
Consider a system composed of N gates as shown in Fig. 3-2(a). By decomposing the
system into M modules, a hierarchical structure system as shown in Fig. 3-2(b) can
be obtained. It is well known that the time complexities of layout problems are
$O(n\log n) - O(n^3)$ for most cases. Here, let it be $O(n^2)$, and S and S' be the
amount of layout design effort for the non-hierarchical and hierarchical
structures, respectively. Then, S and S' are expressed as

$$S = k \cdot N^2 \tag{3-1}$$

$$S' = k \left\{ M \left(\frac{N}{M} \right)^2 + \alpha \cdot M^2 \right\} \tag{3-2}$$

K. Ueda, R. Kasai and T. Sudo

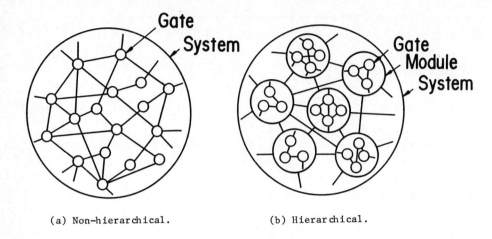

(a) Non-hierarchical.   (b) Hierarchical.

Fig. 3-2 Hierarchical design model.

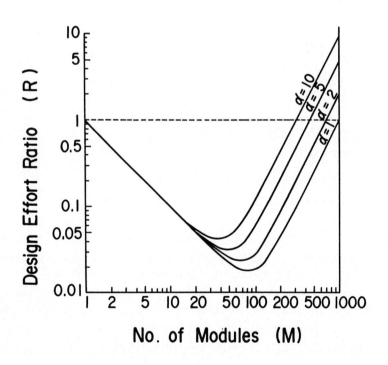

Fig. 3-3 Design effort ratio vs number of modules.

where k is a constant and α is a coefficient which may be a function of (number of terminals per module)/(number of terminals per gate).

From (3-1) and (3-2), ratio of the design efforts R is given by

$$R = \frac{S'}{S} = \frac{1}{M} + \alpha \left(\frac{M}{N}\right)^2 \qquad (3-3)$$

For a case where N = 1000, the value of R is changed as shown in Fig. 3-3 by altering M. From this, it is seen that the design efforts can be significantly reduced if an appropriate module size is chosen.

A hierarchical design has great advantages, such as design effort and turnaround time reduction. However, it also has some disadvantages, such as increases in physical and electrical performances. Therefore, these issues must be well considered before it is decided to what degree a hierarchical design method should be adopted. Table 3-1 summarizes the merits and demerits of hierarchical design.

Table 3-1 MERITS AND DEMERITS OF HIERARCHICAL DESIGN

Merit	Demerit
(1) Design effort reduction	(1) Hardware redundancy
(2) Ease of sharing design	(2) Electrical performance degradation
(3) Short design turnaround time	(3) Increase in production cost
(4) Ease of design modification	(4) Complicated support tools

3.2 Hierarchical Layout Design for 32 bit VLSI Processor [16]

As an example of logic VLSI chips designed using hierarchical layout approach, we will introduce a 32 bit CPU chip for a main frame computer developed by Atsugi ECL, NTT. The chip contains 17k gate random logic and a 2304 bit RAM. The function block diagram is shown in Fig. 3-4. The chip, fabricated with 2um rule silicon gate CMOS technology with double metal layers, was designed using the hierarchical automatic standard cell approach. A hierarchical and structural design automation (DA) system is used throughout the VLSI chip design. The DA system consists of a data base, data base manager and various kinds of CAD pro-

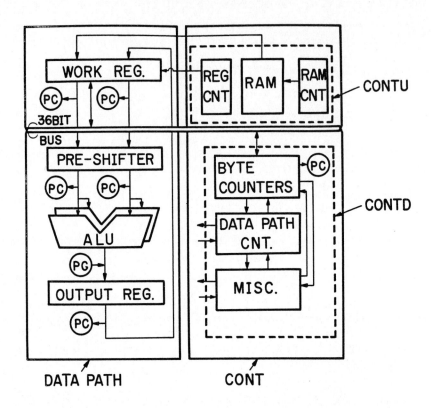

Fig. 3-4 CPU chip block diagram. (Copyright © 1982, IEEE.)

grams. A hierarchical specification language (HSL) is used as a common language, describing not only the structural design data on all hierarchical levels, but also information needed for data base management.

The hierarchy used in the layout design is shown in Fig. 3-5. Basically, the hierarchicy has four levels: cell, block, superblock and chip. Each block consists of 400 cells, on the average,; superblocks contain more than two blocks. The block is automatically laid out using the one-dimensional cell placement and the channel routing algorithm. In the superblock layout, once the intra-block connection is erased and inter-block and intra-block connections are routed simultaneously after the cell placements of all blocks inside the superblock. The overall cell placement is effected so that each cell row may be straight from end to end of the superblock. Efforts for cell placement and block terminal assignment are significantly reduced due to the aforementioned decomposition effect and their one-path routing over an entire superblock. Moreover, this method could improve the packing density by as much as 35%, in comparison with the conventional build-

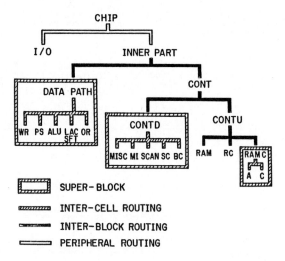

Fig. 3-5 Hierarchy used in the layout design. (Copyright © 1982, IEEE.)

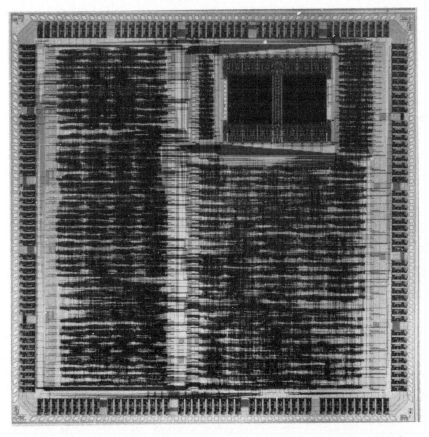

Fig. 3-6 Photomicrograph of CPU chip. (Copyright © 1982, IEEE.)

ing block approach when the super-block has less than ten thousand gates. The layout for a level higher than the superblock is performed in the conventional style of the building block approach.

Figure 3-6 shows a photomicrograph of the 32 bit CPU chip. The total transistor count, network count and average network length were 78k, 17125 and 0.59mm, respectively. The hierarchical automatic standard cell approach supported by a powerful DA system has resulted in a small amount of layout efforts of 2 man-months with a high packing density (542 transistors/ mm^2), even for hard-wired random logic VLSI.

4. LAYOUT CAD TOOLS

For the support of layout design, a variety of CAD tools including software programs as well as hardware equipment have been developed by many manufacturers and laboratories to date. In this chapter, layout CAD tools, which are categorized into three areas: software tools, interactive graphic systems and engineering work stations, are described. A summary of automatic layout systems is illustrated in Table 4-1.

4.1 Software Tools

4.1.1 PLA CAD Systems

APLAS [31]is an automatic PLA synthesis system which automatically generates a PLA for the control function of a design from a DDL-P description of a digital system. It can also minimize and partition the PLA to meet the design constraints. The system consists of three major programs. First, SALT (Stanford Automatic Logic Translator) translates the DDL-P description of a digital system into Boolean equations. Then, the second program, SPAM (Stanford Programmable Array Minimizer), minimizes and converts them into PLA format. Finally, PAPA (Programmable Array Partitioner) is used to partition a large PLA into smaller PLA's to meet the design constraints.

DDL-P is a Stanford version of DDL (Digital Design Language), a register transfer language. In a real circuit, a large PLA tends to be quite wasteful or not fast enough to support the other parts of the system. In this case, it can be split into several smaller PLA's to reduce the chip area and/or improve the speed. PAPA is a program which does this job. It also has a redundancy removal routine which detects input and output redundancy from a given PLA. The chip area and the overhead of using the many PLA's are the major factors considered in partitioning.

TABLE 4-1 SUMMARY OF AUTOMATIC LAYOUT SYSTEMS

System	Developer	Years	Style	Device	Algorithm	
					Routing	Placement
LILAC	Hitachi	1972	B.B.	MOS	Channel router 2 layer	Clustering
ROBIN	NEC	1974	B.B.	MOS	Channel router Global router	Manual placement
MP2D	RCA	1976	S.cell	MOS	Channel router	Two-dimensional placement
LTX	Bell Lab	1977	P.cell	MOS Bip.	Channel router	One-dimensional Two-dimensional
MARC	M.ECL N.T.T.	1978	G.A.	Bip.	Global router Line Search	AR method Iterative improve
MIRAGE	Mitsu-bishi	1979	H. S.cell	MOS	Global router Channel router	Manual placement
MASTER	NEC	1979	G.A.	Bip.	Global router Line search	Random Force-directed
----	Fujitsu	1980	G.A.	MOS	Global router Channel router	Min-cut One-dimensional
PLASMA/ COSMIC	M.ECL N.T.T.	1980	H. S.cell	MOS Bip.	Channel router	Linear placement Two-dimensional
APW	IBM	1981	G.A.	MOS Bip.	Global router Detailed router	Hierarchical Clustering
MARS-M3	Mitsu-bishi	1981	G.A.	ECL	Global router Channel router	Top-down Linear placement
MILD	Mitsu-bishi	1981	H. S.cell	MOS	Global router Channel router	Two-dimensional Linear placement
SHARPS	Sharp	1981	H. S.cell	MOS	Global router Channel router	Clustering Force-directed
ALEPH	Hitachi	1982	B.B. G.A.	MOS	Channel router Maze router	NetBalance,Zigzag Combine&Topdown
ALPHA	M.ECL N.T.T.	1982	H. S.cell	MOS	Global router Channel router	Linear placement Pairwise Exchange
LAMBDA	NEC	1982	G.A.		Global router Maze, Dynamic	Random, GFDR Hierarchical
VGAUA	RCA	1983	G.A.	MOS	Direct router Maze router	Pair interchange
GALA	Hughes	1984	G.A.	MOS	Global, Channel Maze router	Resistive network optimization
CHAMP/ ALPHA-I	A.ECL N.T.T.	1985	H. S.cell	MOS	Global router Channel router	Block packing Pairwise exchange

B.B.=Building Block, S.cell=Standard cell, P.cell=Poly cell, G.A.=Gate Array, H.S.cell=Hierarchical standard cell.

PLACAD [43] is a PLA design support system developed at Atsugi ECL, N.T.T. The
system's input data are described in either HSL (Hierarchical Specification
Language), TEGAS pit file, Boolean equations or state transition diagrams, and the
output is a PLA programming pattern which specifies AND and OR plane functions.
PLACAD performs four major functions: state assignment, transformation to product
term expression, conversion of Flip/Flop types, and minimization of product terms.
For the minimization of product terms, MINI's algorithm [44] is used. Table 4-2
illustrates the processing time for transformation to product term expression and
minimization of product terms.

Table 4-2 PROCESSING TIME FOR TRANSFORMATION TO PRODUCT TERM
EXPRESSION AND MINIMIZATION OF PRODUCT TERMS

Data	PLA Size		Transformation to Product Term Expression		Minimization of Product Terms	
	Input	Output	Product Terms	CPU Time (s)	Product Term	CPU Time (s)
1	25	15	87	58.1	75	1154.0
2	28	11	49	81.3	37	633.4
3	20	15	30	9.5	26	89.4
4	21	8	26	23.4	25	117.5
5	14	4	11	2.1	10	7.5

APSS [45] is integrated, fully automatic software that combines Boolean logic
translation, Boolean minimization, PLA folding, PLA topology generation, and
automatic PLA subchip interfacing to the MP2D standard-cell automatic placement
and routing program. The Boolean minimization routines operate upon the
two-level AND/OR description that was obtained either through the Boolean logic
translation or was accepted as a truth table. The objectives of the minimization
are 1) to minimize the number of products in the AND/OR description, thus,
minimizing PLA area, and 2) given the minimized PLA area, to minimize the number
of transistors in the PLA, thus reducing power dissipation and logical fanout,
possibly improving PLA speed. This minimization is approached in a heuristic
manner, in a method somewhat similar to that described by Hong, et al.[44]

The FSM design system [46] is a fully-automated finite state machine (FSM)
synthesis system developed at AT&T Bell laboratories. It permits the designer to
describe an FSM in a high level language and results in a mask level layout
description in PLA implementation.

The PLA synthesis system consists of several programs. A compiler translates the PLA description language, which is C-like in syntax, to an intermediate form. Minimization programs operate in the intermediate form to generate PLAs that are logically and topologically minimized. A set of audit programs gives valuable feedback to the designer about the PLA speed, area, testability and other connectivity information. An automatic layout program converts the intermediate file to layout information, which can be combined with similar information for the remaining parts of the chip, to produce a composite layout.

A designer can compile, minimize, and generate all modules for a reasonably sized PLA (around 200 product terms) in less than 10 minutes CPU time on a DEC VAX-11/780. The system's primary advantages are: first, the layout, circuit model, logic model, and functional model are all generated from the same high level description; second, the system generates predictable results when changes are made to a design.

4.1.2 Gate Array CAD Systems

Automatic Placement and Wiring program (APW) [47] is a layout subsystem of PHILO developed at IBM, which consist of three main sections: placement, global wiring, and detailed wiring.[48, 49] A key feature of the system is the ability to design chips with large macros (RAMs and PLAs). An example of a chip image is shown in Fig. 4-1. The key objective is to make the programs as technology independent as possible. This objective was completely met for the placement and global wiring programs. However, the detailed wiring programs are more technology dependent.

The placement program utilizes a hierarchical model. The chip is divided into large rectangles. In the first pass, the placement of the clusters in the large rectangles is optimized. Next, each of the clusters and rectangles is broken down into smaller clusters and rectangles. This process is repeated until the final placement is obtained. In the global wiring phase, the chip is first divided into a coarse grid of rows and columns. Wires are routed in these rows and columns but are not assigned specific wiring tracks. The objective is to minimize congestion. Then, the subsequent detailed wiring program completes the wiring of the chip by assigning wire segments to actual wiring tracks.

The PHILO system has been used in IBM's FSD (Federal Systems Division) [50]. Several chips have been designed using the system. Table 4-3 shows the statistics on three chips designed with 2.0 micron NMOS technology (A, B, C), and one prototype chip. The number of gates specified in Table 4-3 designates the number of equivalent three-way NOR gates.

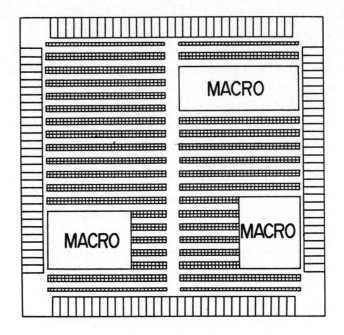

Fig. 4-1 Example of chip image. (Reprinted from Donze, et at., 1983. Copyright
          © 1983, IEEE.)

Table 4-3 CHIP STATISTICS

Part Number	A	B	C	D
# cells on chip	2432	2432	2432	8468
% cells populated	88	68	80	79
% cells occupied by macros	17	14	17	55
# gates	8276	7638	8928	36272
# gates/mm^2	197	182	214	735
# global nets	1016	997	1176	1991
% auto wired	99.8	99.7	99.9	98.4
CPU* place	6.5	6.0	6.5	10
CPU* wire	3	3	3	3.5

* IBM 3033 CPU Minutes

MARS-M3 [27] is a fully automatic chip layout design system for ECL and MOS gate array masterslices using two levels of metal and a standardized rectangular chip image.   In the placement subsystem, global optimization capability of top-down initial placement is combined with the local optimization capability provided by iterative placement improvement. The routing is also performed hierarchically by using two processes; channel assignment and track assignment. The CPU times for a typical gate array with 683 gates and 716 nets are 19.7 min for initial placement 45.7 min for placement improvement, and 31.8 min for routing by using MELCOM-COSMO 700.

LAMBDA [28] is an automatic/interactive layout design system for designing masterslice (or gate array) LSI chips. The system aims to achieve complete net connectivity in as short a design time as possible, and efficient automatic procedures as well as highly interactive functions are implemented. The system operates on an NEC/MS super minicomputer which has a 4M-byte memory capacity and a 200M-byte disk storage capacity, coupled with a 21" beam-directed refresh graphic display. The system is connected to NEC ACOS/700 and 900 general purpose, main frame computers via communication links to form the total custom LSI CAD system.

In the placement program, a two-level hierarchy placement algorithm is adopted to obtain good solutions in both the global and local meanings. One level provides the optimal positions for a set of blocks (group placement); and the other one is the detailed placement procedure to place all the individual blocks into cell arrays (individual block placement). For routing, three different routes are used; a global router, a detailed router and a dynamic router. The global router gives the topological route for each signal net. The detailed router, which adopts a modified maze search algorithm, decides the geometric path for each signal net. The dynamic router improves the connection rate by modifing paths. Some key features of LAMBDA are: 1) automatic and interactive functions are fully integrated. 2) The designer does not need the exact geometric or symbolic wiring patterns. A rough sketch of the wiring patterns as a guide is sufficient. 3) Blocking factors, signal nets or terminals are pictorially shown on the graphic display when a signal net cannot be connected.

Variable Geometry Automated Universal Array (VGAUA) [51] is a fully automatic physical layout system for LSI gate arrays developed by RCA with support from the United States Army's Electronics Research and Development Command (ERADCOM). The system accommodates multiple gate array sizes between 800 and 3,500 gates. For the placement algorithm, a pair-interchange technique is used, which is divided into two phases. In Phase I, the algorithm provides a Y-placement, or a cell to cell row assignment, such that a number of Y-distance criteria are minimized. In

Phase II, the placement algorithm provides an X-placement, or a cell to cell row
position assignment.

The routing function in VGAUA uses two types of routing algorithms: direct routing
and pathfinder algorithms. The direct routing algorithms are not designed to
complete 100% of the routing themselves; their goal is to efficiently route as
many wires as possible, with minimum impact on the remaining routes. The
pathfinder route is used to complete the remaining wires. The pathfinder
algorithm is a version of the maze runner algorithm that is optimized to the fixed
geometry patterns. One of the prominent features of the system is a critical path
or minimum delay option. The cells associated with these critical paths receive
the highest consideration in the placement function. An example of an 800 gate
design is shown Fig. 4-2. In Fig. 4-2(a), no nodes are specified as being
critical. In Fig. 4-2(b), the critical path is optimized.

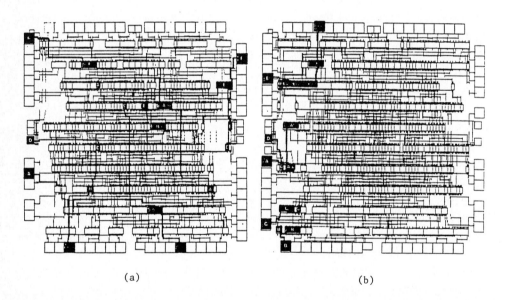

(a)   (b)

Fig. 4-2 Critical path optimization. (a) Critical path routing before
          optimization. (b) Optimized critical path routing. (Reprinted from
          Smith, et al., 1983. Copyright © 1983, IEEE.)

4.1.3 Building Block CAD Systems

ROBIN [21] is a building block LSI layout design system developed at NEC. The
chip structure used in the system is shown in Fig. 4-3. A group consisting of
cell rows is used as the layout unit. Besides the usual group, some large

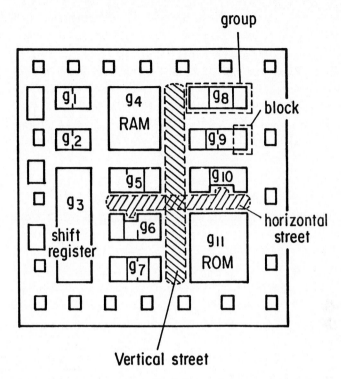

Fig. 4-3 Chip structure for ROBIN. (Reprinted from Kani, et al., 1976. Copyright © 1976, IEEE.)

predefined blocks, such as ROMs, RAMs, and PLAs, are also treated as a group. The determination of interconnection routes between these groups is performed according to the following procedure: 1) for a given topological layout, a mapping process sequence that can be applied to each street without contradiction is determined; 2) by a heuristic algorithm, the interconnection nets are assigned to streets aiming to minimize the chip area (routing process); and 3) using a heuristic approach, the physically precise positions of wire segments are also determined in order to minimize the width of streets (mapping process).

4.1.4 Standard-cell CAD Systems

LTX [52] is a modular semi-automatic layout system developed at the AT&T Bell laboratories. LTX provides both two-dimensional and one-dimensional placement capabilities. One-dimensional placement algorithms perform in-row placement optimization. A dog-leg channel router, which usually performs within ten percent of optimum, is used to make interconnections. Interactive editing of the routing can be performed on a graphics screen under program control. Output of LTX is a partial mask description in the XYMASK which is then used in the interactive graphics system to perform power bus routing, to add features required by the mask shop, and finally to create a magnetic tape for driving a pattern generator. A new version of the LTX system provides extensive support for hierarchical layout. Several new algorithms for placement, and signal and power bus routing, combined with bottom-up and top-down design capabilities and on-line rule checking make this system a powerful tool for VLSI layout. This system has been used to design a large number of custom MOS and bipolar chips including a complex custom CMOS chip containing about 32,000 transistors.

MILD [53] is a standard-cell based layout system for MOS LSI developed at Mitsubishi. It can handle macro blocks (or general cell) as well as standard cells. The chip structure for MILD is shown in Fig. 4-4. The standard cells are automatically placed, while for general cells, the designer must prepare exculsive benches for each of them as well provide for their orientations. (A 'bench' means an area in which cells are placed.) The placement algorithm employed in MILD is similar to that of LTX. The process consists of three steps: 1) assign standard cells to benches, 2) determine the standard cells positions in each bench, and 3) improve placement. The routing is performed by a global router and a gridless channel router.

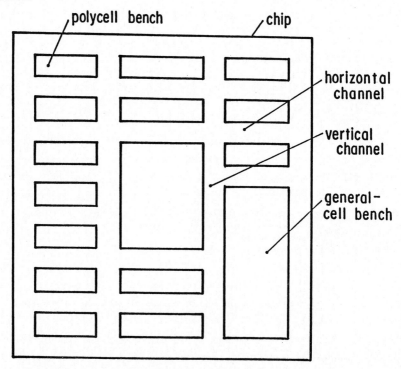

Fig. 4-4 Chip structure of MILD. (Reprinted from Sato, et al., 1981. Copyright
          © 1981, IEEE.)

SHARPS [54] is a hierarchical automatic placement and routing system for building-block and poly cell layout of custom LSIs developed at Sharp. The system is implemented through the use of a hierarchical structure such that each functional block is constructed of lower level blocks in a bottom-up manner. In the placement program, linear ordering and placement improvement by using a force-directed relaxation algorithm is performed. Some of the features of the router are: 1) rectanglular functional blocks of arbitrary size and shape can be designed, 2) loops on the channel order constraint can be automatically dissolved by a shrinking operation, and 3) bundles are constructed by putting together connection requirements and the global routing is achieved for bundles on a routing graph. The system is implemented on a DEC VAX 11/780 minicomputer coupled with several color graphic display terminals.

ALEPH [55] is a hierarchical layout program for both building block and masterslice LSIs developed at Hitachi. The system's features are: the design object can be partitioned into an arbitrary size, each partitioned circuit can be concurrently designed, automatic and manual design functional assignment is optimized, and verification programs, such as logical/physical/delay checkers, are prepared. The hierarchical structure and the layout model for a VLSI chip is shown in Fig. 4-5.

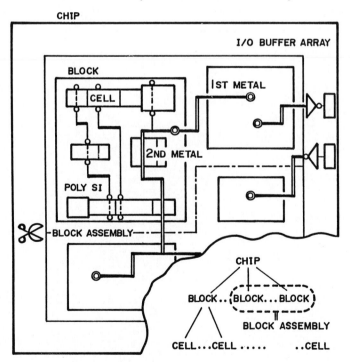

Fig. 4-5 Layout model for VLSI chip. (Reprinted from Terai, et al., 1982. Copyright © 1982, IEEE.)

The cell is a minimum logical unit consiting of 1 to 50 gates. The block is a
basic unit of logic design, and has about 500 to 1000 gates. The hierarchical
layout design is performed in bottom-up manner; first each block is designed, then
they are put together and interblock interconnections are routed. For routing, a
channel router as well as a maze router are used. Manual modification is also
performed if some portions can not be wired automatically. The system is modified
partially and can be applied to masterslice (or gate array) LSIs.

CHAMP/ALPHA [56, 18] is a hierarchical, highly automatic layout system developed
at Atsugi ECL, NTT, which semi-automatically performs a chip floor plan at the
block level and then automatically excutes the cell placement and routing for
custom VLSIs. The system can handle both standard cell blocks and macro-cell
blocks such as RAMs, ROMs or PLAs. The chip structure uses a hierarchical
methodology, but it is more or less different from a conventional hierarchical
structure; it has no inter-block channel region between blocks. Instead,
laterally-located blocks must have common cell rows through the whole chip area.
The layout system flow is shown in Fig. 4-6.

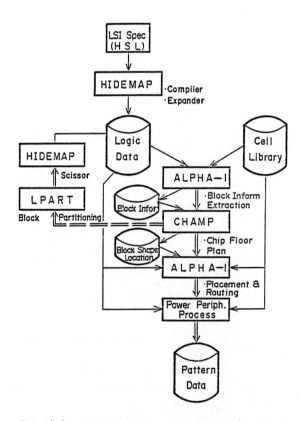

Fig. 4-6 Layout system flow of CHAMP/ALPHA.

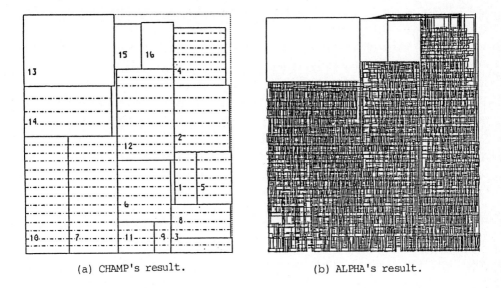

(a) CHAMP's result.   (b) ALPHA's result.

Fig. 4-7 Layout results for 16-bit CPU circuit using CHAMP/ALPHA.

The chip floor plan objectives are: to accommodate blocks within a nearly square-shaped chip boundary having as small an area as possible, to determine the height/width ratio (or the number of cell rows) for each variable-shaped (standard cell type) blocks, and the location for all blocks, and to estimate the chip area based on parameters associated with blocks. The chip floor plan processes with CHAMP are divided into two phases; the first is initial placement and the second is block packing. The initial placement is obtained manually or automatically by using the attractive and repulsive force method [57], which is a sort of gravitational placement method. An initial placement usually has some overlaps and/or dead spaces between blocks. So, in the subsequent block packing phase, such overlaps and dead spaces will be eliminated by moving and/or reshaping blocks with the chip boundary being shrunk. A chip area estimation, which is based on a calculation from empirically obtained equations, is also performed during the block packing phase. The chip area can be estimated well within ± 10 % accuracy.

In ALPHA, cell placement and routing is performed based on the floor plan, i.e., the block placement obtained by CHAMP. ALPHA consists of six major steps. First, in the block-level global routing, optimal block terminal directions are determined by obtaining optimal routes (quasi-Steiner trees) for interblock interconnections. Second, in the intra-block cell placement phase, cell locations are determined using the block terminal directions obtained in the block-level global routing process. Every block is merged together to form an entire chip, and then placement improvement is tried. In this placement improvement phase, cell exchanges are done beyond the block boundaries. However, it is limited to within the same row because of the amount of computing time. Next, the cell-level global router decides the optimal route on the entire chip in which channel congestions are taken into account. Finally, all the paths are determined by a channel router based on a left-edge algorithm. Fig. 4-7 shows the layout results for a 16-bit CPU circuit (5K-gate + 24K-bit ROM + 0.4K-bit RAM + 0.2K-bit RAM) obtained using CHAMP and ALPHA. Using CHAMP and ALPHA in combination, only a 7 man-day design effort is required for a 20K-gate VLSI layout design, provided that a standard cell library and macro-cells were pre-designed.

In a standard cell LSI layout, hundreds of standard cells must be configured into a cell library. This standard cell library concept is quite effective for the layout design of custom VLSIs in a short turnaround time with a small amount of effort. However, the cell library preparation itself usually requires several man-months of design effort. Further, a design rule change causes other work requiring almost the same amount of design effort. To cope with this problem, automatic cell generation systems have been developed.

CPGEN [58] is a program to design a polycell layout at the symbolic level. This symbolic code is independent of the design rules used in this technology. The symbolic code and technology file, in which layout rules are described, are used as inputs to CPGEN. The output of CPGEN is the layout description of the polycell(s) in the Bell lab mask description language XYMASK. The characterization of the designed polycells is performed by using a layout characterization and verification system.

Automatic Cell Generation system for CMOS (ACG) [17] is an automatic cell generation system for CMOS LSI. ACG can handle an arbitrary standard cell which is expressed by a circuit network instead of a Boolean equation. In this system, Building Pair Transistor (BPT) style is employed for the CMOS standard cell structure. First, transistor pairs are extracted from a circuit network and then they are placed in a one-dimensional array. In the BPT structure, it is empirically known that the unification of diffusion regions is effective in reducing cell area. So, in the cell placement phase, minimization of the number of diffusion regions is taken as the primary objective. In the subsequent routing phase, the nets are classified into five categories and routing is performed by using this net type information. A set of physical mask patterns, including P-diffusion, N-diffusion, well, channel stopper, polysilicon, contact hole, aluminum, etc., is generated from the placement and routing results. An example of a cell pattern automatically generated by ACG is shown in Fig. 4-8.

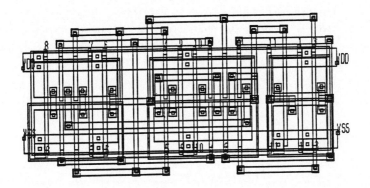

Fig. 4-8   Example of cell pattern generated by ACG.

4.1.5 Symbolic Layout Systems

MULGA [59] is a symbolic layout design system based on a "virtual grid" concept developed at Bell laboratories. The system uses a grid-based placement scheme as in coarse-grid layout methods but allows the final geometric spacing between grid lines to be determined by the density and interference of elements on neighboring grid locations. This virtual grid concept is illustrated by a simple example as shown in Fig. 4-9. The system allows the interactive editing, layout compaction, circuit connectivity extraction, parasitic audit, and timing simulation of MOS ICs

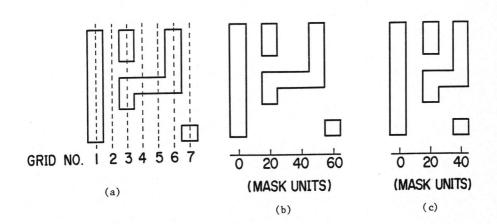

Fig. 4-9   Comparison of fixed-grid and virtual-grid layouts. (a) Circuit Specification based on a grid. (b) Fixed grid expansion. (c) Virtual-grid expansion. (Reprinted from Weste, 1981. Copyright © 1981, AT&T.)

within the symbolic domain. The programs make use of an intermediate circuit description language (ICDL), which captures both geometric placement and circuit connectivity. A convenient interface is provided to enable the procedural definition of symbolic layouts in the C programming language. A data path chip for a 16-bit CMOS processor was designed on the MULGA system in 8 man-weeks. It contains 5,000 transistors in a highly stylized layout modeled in gate-matrix style. In addition to the CMOS chip, two small NMOS chips have been designed, each comprising approximately 1,000 transistors. These chips were designed in 3 man-weeks.

SLIM [15] is a symbolic layout design system developed at Bell laboratories. The symbolic input is a loose topological description of the layout made up of single-connection-per-side symbols (transistors, interlayer contacts, etc.) and multiple-connection-per-side symbols (predefined RAMs, ROMs, flip-flops, etc). The output is a legal mask description and graphical displays. The initial placement procedure is the scheme for translating the relative location layout information obtained from the symbolic description layout information which is obtained from the symbolic description to a mask layout. The initial placement is not supposed to be a tight placement. On the contrary, it is supposed to be loose. In SLIM, several strategies such as critical path analysis, local compaction, automatic jog insertion and global compaction are used. Experiments have been performed with the SLIM symbolic circuit compactor. [60] The area comparison between SLIM vs Hand design is illustrated in Table 4-4.

TABLE 4-4   AREA REQUIRED BY SLIM VS HAND DESIGN

Cell	SLIM's Area*	Hand Area*	Overhead
Stack (no shift)	17574	10350	70 %
Twoport	52668	32578	60 %
Stack (shift)	34000	17990	90 %

* in square microns.

4.2 Interactive Graphic System

As one of the design tools of IC and LSI mask pattern design, interactive mask pattern design systems were first built at the end of the 1960's. Since then, they have been used by many semiconductor manufacturers and laboratories. At early dates, mask pattern designs were performed completely manually; draw layout patterns at a desk, cut Mylarsheets, and produce a set of mask patterns. Interactive graphic systems significantly reduced the amount of design effort. Especially for mass production LSIs such as microprocessors, microcalculators, and digital clock chips, a hand-crafted design using an interactive graphic system is suitable because the requirements for chip area reduction are severe for these full-custom LSIs. Memory LSIs, which have fundamentally an iterative layout structure, are also appropriate devices for hand-crafted design. Currently, semi-custom LSIs such as gate arrays and standard cell LSIs are usually designed automatically. However, some additional mask pattern data, which are needed for a final production mask set, are also designed by using these interactive graphic systems.

4.2.1 Major functions

(1) Interactive Mask Pattern Entry and Editing

Mask pattern data of circuit elements (transistors, resistors, etc.) and wiring are entered through a graphic display terminal and/or a digitizer. Rectangles or polygons are used as a basic pattern representing various circuit elements. For efficient entry, nestings of pattern data are usually allowed. Mask pattern data are edited by using a graphic display terminal. Edit commands usually include ADD, MOVE, DELETE, COPY, MAGNIFY, SHRINK, ROTATE, MIRROR, NEST, etc. Some systems offer edit commands which perform Boolean mask operations.

(2) Design Rule Checking

Designed mask pattern data is checked before it is handed over to a mask pattern generator because a manual design is prone to error. A small circuit such as a basic circuit can be checked by a DRC package on an interactive graphic system. However, for a whole LSI circuit, a main frame computer is usually used because of the amount of computing time required.

(3) Transformation to Mask Pattern Generation

Following the mask pattern design, the designed data are transfered to mask pattern generating systems such as a pattern generator (PG) or an Electron Beam Exposure System (EBES). The mask data transformation package is provided to offer an interface to such a system.

4.2.2 Hardware Configuration

Fig. 4-10 shows the system block diagram of CALMA's GDS-II as a typical interactive graphic system. A 16-bit ECLIPSE S/2 30 is used as a CPU, with

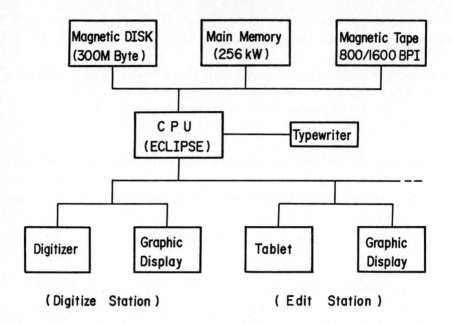

Fig. 4-10   System block diagram of CALMA's GDS-II.

256K-byte memory. During the graphic process, graphic information is stored onto a magnetic disk and retrieved from it. For input devices, a graphic display, tablet, keyboard, and digitizer are provided.

4.2.3 Trends

As the scale of design objectives is increasing, from LSIs to VLSIs, higher CPU capability is required. To satisfy this requirement, a high performance CPU is being introduced; shifting from 16-bit to 32-bit and from minicomputer to super-minicomputer. A more sophisticated software package including symbolic layout and compaction is becoming a standard option in many CAD systems.

4.3 Engineering Work Station

Engineering work stations (EWS) were first introduced at the beginning of the 1980's. The main purpose of these EWSs was to enhance engineering productivity which had been left at a relatively low level for a long time. For example, the layout design of a gate array can now be performed routinely by a batch process using an automatic placement and routing program. Whereas the logic design is usually performed on a cut and try basis. The primary purpose of EWSs is to offer an interactive design means allowing logic designers an efficient logic design environment. It includes a schematic entry and logic simulation function. However, its functions are being extended to cover circuit design, layout design, and further, testing and diagnosis. Table 4-5 illustrates a summary of Engineering Work Stations.

Table 4-5   SUMMARY OF ENGINEERING WORK STATIONS

System	Manufacturer	CPU	OS	Logic Sim.	Layout
CAE2000	CAE	Bit-slice	UNIX	IDEAL	
IDEA1000	Mentor	Bit-slice	AEGIS	ILOGS	GARDISYS
LOGICIAN	Daisy	80286	UNIXlike	DLS	GATEMASTER
λ750	Metheus	68000	UNIX		MERLYN-G
SCALD	Valid	68000	UNIX	SCALD	MERLYN-G
SL-2000	Silver-L.	Bit-slice	AEGIS	BIMOS	GARDS,CAL-MP
TEGAStation	Calma	Bit-slice	AEGIS	TEGAS	T-ARRAY
U-GAL	Sumitomo	68000	UNIX		GALS

4.3.1 Functions

A work station requires a high performance CPU as its controller since it must be interactive. Currently, many available work stations have CPU(s) in the 0.5 - 1 MIPS range. For main memory, 0.5 - 8M-byte capacity is provided with many systems. In line with the increase of CPU capability, the main memory is also increasing in its capacity. In many sytems, 800 x 1000 range high-resolution bit-map displays are used. Many systems support a multiple window function which enables several views of the same design to be displayed on the screen at once. By using the multiple window function in combination with a mouse (a pointing device), a designer can excute a highly interative design. For the link between work stations or host computer, local area network interfaces are used. The defacto standard in the industry, Ethernet supports many systems.

4.3.2 Trends

The main objective of these work stations is the design of semi-custom LSI, especially, gate array or standard cell LSIs. With the increase in capability of

EWS, the function and load sharing at terminal side will be increasing. The keys for increasing the number of work stations are: 1) improving of their functions and reduction of cost, 2) developing user-friendly, high-performance application software, 3) providing a standard interface to customer's systems.

5. LAYOUT INTERFACE

In the past few years, custom, particularly semi-custom, LSIs such as gate arrays have prevailed due to CAE/CAD developments and silicon foundry services. As the usage of CAE/CAD tools is spreading over a wide area, interface problems have increased. Layout design systems are involved with a variety of design data such as net lists, macros, cell libraries, symbolic-level layouts, mask-level layouts, etc. This diversity brings about some standardization difficulties.

In this Section, we will explain two typical interface formats. The CIF (Caltech Intermediate Form) and the EDIF (Electronic Design Interchange Format). They are the defacto standard for IC mask pattern interfaces and a forthcoming standard for electronic design interfaces including IC layout information, respectively.

5.1 Caltech Intermediate Form (CIF)[4]

CIF is the result of a standardizing effort which enables sharing mask pattern designs among several layout design groups. A CIF file consists of a sequence of characters containing a list of commands and delimiters described with the standard notation proposed by Wirth.[61] The main object of this format is to accurately describe the mask patterns of LSI circuits. Particularly, for a cooperative work among several design groups, ambiguity must be avoided. In designing the file format, many decisions were made to avoid ambiguity or troublesome errors. Floating point numbers and iterative constructs are avoided. Only a simple file format such as polygons, boxes, flashes, and wires are included.

For example, boxes are described as follows:

> Box Length 10 Width 30 Center 100, 50 Direction -10, 10;
> (or B 10 30 100, 50 -10, 10;)

The fields that define a box are shown in Fig. 5-1. Currently, many CAD systems support CIF form mask pattern output. A cell library described by using CIF has been published. [62]

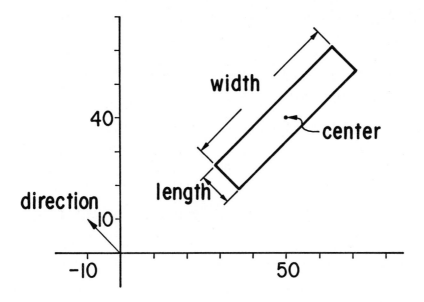

Fig. 5-1 Example of BOX.

5.2 Electronic Design Interchange Format (EDIF) [63]

EDIF was developed by a commitee consisting of the principal developers of the existing formats such as CIDF (Common Interchange Description Format), GAIL (Gate Array Interface Language), TDF, and TIDAL. It is aimed at a single, public domain, interchange format. In the development of the format, the following design data requirements are considered: the initial version can handle the data for exchanging information related to semi-custom LSI design. All information that a silicon foundry sends to customers and also that a customer sends to foundries must be supported. In the format design, several requirements are also considered, i.e. minimization of translator implementation difficulty and extensibility, etc.

The syntax of EDIF is based on that of the LISP programming language, in which all data are represented as symbolic expressions. This LISP-like syntax leads to ease of implementation and extensibility. Since the semantics is given by the

keywords, not by the syntax, the format has a large extensibility. An EDIF file need not be complete at any time; the file may contain only a partial description of a design. An EDIF description is expressed in the form of a hierarchy, abstract at its higher levels, and becoming progressively more detailed as it descends the hierarchy.

```
(Cell NAND2
  (View Topological Logic
    (Interface
        (Declare (INPUT PORT a b) (OUTPUT PORT z))
        (Body (Layer SchematicsSymbol
          (shape (20 0) (arc (40 0) (60 20) (40 40)) (20 40))
          (circle (60 21) (70 20))
             )
        )
        (PortImplementation a (Layer SchematicPort (Path (0 30) (20 30))))
        (PortImplementation b (Layer SchematicPort (Path (0 10) (20 10))))
        (PortImplementation z (Layer SchematicPort (Path (70 20) (90 20))))
        (Permutable a b)
     )

    (Contents
      (Instance Nand2Primitive)
     )
   )
  (View Physical
    (Interface
        (Declare (INPUT PORT a b) (OUTPUT PORT z))
        (Body (Layer MetalBlockage (Rectangle ....)))
        (PortImplementation b (Layer Poly ..... ))
             .
             .
             .
        Permutable a b)
     )
    (Contents
      (Layer Metal (Path ...) (Polygon ...) ...)
      (Layer Poly  (Path ...) (Path ...)..)
             .
             .
             .
     )
   )
 )
```

Fig. 5-2 Simple example of EDIF.

To avoid complicating the simple tasks, while still proving the power of procedural descriptions for those who need them, EDIF descriptions can be specified in three different levels of complexity. These are: Level 0, the basic level; Level 1, Level 0 with the addition of variables and expressions; Level 2, Level 1 with the addition of procedures and complex data types. EDIF covers the descriptions of cell libraries, netlists, mask layouts, symbolic layouts, logic models, functional tests and array-based design methods. A simple cell example is shown in Fig. 5-2. It is seen that EDIF supports multiple views of a cell such as its logic model and detailed layout.

6. CONCLUSION

VLSI layout design problems and their solutions have been reviewed. Until now, many layout styles and strategies have been proposed and many layout CAD tools have been developed and applied to practical use. Any layout style CAD tool more or less makes use of standardization at the cost of chip area, yield, electrical performance, etc. Although many kinds of CAD tools have already been developed, much wider spectrums of layout CAD tools will still be required as VLSIs become larger and more complex, and CAD users spread from device to system designers. New kinds of tools for VLSI layout design have appeared. The silicon compiler concept, which intends to make most of the design phases including layout into a 'black box', will offer one of the solutions to meet requirements, especially from system design people. As another promising solution, the expert or knowledge-based system approach will be exploited in the layout design field. By this approach, chip area, which has long been sacrificed by most conventional layout CAD tools, will be reduced as if by expert layout designers. Special hardware engine is regarded as another promising solution for the future VLSI layout problems. These hardware engines will bring a faster turnaround time and an easier implementation of interactive design to the layout design. As the LSI world is widening its boundary and also the population of CAD users, standardization of the interfaces between CAD tools becomes essential for the efficient design of VLSIs through the usage of many CAD resources.

REFERENCES

[1] Rice, R., VLSI: the coming revolution in applications and design, Compcon Spring '80, (Feb. 1980), 19-20.

[2] Murphy B.T., Cost-Size optima of monolithic integrated circuits, Proc.IEEE, 52, (1964) 1537-1545.

[3] Tansho, K., Kasai, R., Horiguchi, S. and Kitazawa, H., Optimum block partitioning in three wiring layers, National Convention Record IECE Japan, (1984), 2-183, (in Japanese).

[4] Mead, C. and Conway, L., Introduction to VLSI system, (Reading, MA, Addison-Wesley) 1980.

[5] Ahmed, H.M., Delosme, J.-M. and Morf, M., Highly concurrent computing structures for matrix arithmetic and signal processing, Computer, Vol.15, (1982), 65-82.

[6] Hariland, G.L. and Tuszynski, A.A., A CORDIC Arithmetic processor chip, IEEE Tras. Computers, Vol.C-29, (1980), 68-79.

[7] Patterson, D.A. and Sequin, C.H., A VLSI RISC, IEEE Trans. Computers vol. C-31 (1982), 8-21.

[8] Alexander. D, A Technology independent design rule checker, Proc. 3rd USA-JAPAN Comp. Conf., (1978), 412.

[9] Preas, B.T. et al., Automatic circuit analysis based on mask information, Proc. 13th DA conf., (1976), 309.

[10] Yamada, S. et al., A mask pattern analysis system for LSI (PAS-1), Proc. IEEE Conf. of ISCAS, (1979), 858.

[11] Fan, S.P. et al., MOTIS-C : A new circuit simulator for MOS LSI circuits, IEEE Proc. ISCAS, (1977) 700.

[12] Shima, M., Two versions of 16-bit chip span microprocessor, mini computer needs, Electronics, Dec.21, (1978), 81-88.

[13] Sudo, T., Nakashima, T., Aoki, M. and Kondo, T., An LSI adaptive array processor, Digest. ISSCC'82, (1982), 122-123.

[14] Williams, J.D., STICKS- A graphical compiler for high level LSI design, National Computer Conf., (1978), 289-295.

[15] Dunlop, A. E., SLIM- the translation of symbolic layout into mask data, 17th DA conf., (1980), 595-602.

[16] Horiguchi, S., Yoshimura, H., Kasai, R. and Sudo, T., An Automatically designed 32b CMOS VLSI Processor, Digest of ISSCC'82, (1982) 54-55.

[17] Miyashita, H. and Ueda, K., CMOS standard cell automatic generation systems ACG, Monograph of TG on design automation of IPS of Japan, (July 1984), (in Japanese).

[18] Adachi, T., Kitazawa, H., Nagatani, M. and Sudo, T., Hierarchical top-down layout design method for VLSI chip, Proc. 19th DA conf., (1982), 789-791.

[19] Nagatani, M., Miyashita, H., et al., An automated layout system for LSI functional blocks: PLASMA, Monograph of TG on Design Tech. of Electronics Equip. of IPS of Japan, (1980), 41.

[20] Hashimoto, A. and Stevens, J., Wire routing by optimal channel assignment within large apertures, Proc. 8th Design Automation Workshop, (1971), 158-169.

[21] Kani, K., Kawanishi, H. and Kishimoto, A., ROBIN : a building block LSI routing program, IEEE Proc. ISCAS (Apr. 1976) 658-661.

[22] Mikami, K. and Tabuchi K., A computer program for optimal routing of printed circuit conductors, Proc. IFIP Conf. 68, (1968), 1475-1478.

[23] Lee, C. Y., An algorithm for path connections and its applications, IRE Trans. Electron. Comput., vol. EC-10, no. 3, (1961), 346-364.

[24] Ohkura, I., Noguchi, T., Sakashita, K., Ishida, H., Ichiyama, T. and Enomoto, T., Gate isolation a novel basic cell configuration for CMOS gate arrays, Proc. CICC, (May 1982).

[25] Fukuda, H., Yoshimura, H. and Adachi, T., A CMOS pair transistor array masterslice, Symposium on VLSI Tech., (1982).

[26] Sano, T., Matsukuma, S., Hashimoto, K., Ohuchi, Y., Kudo, O. and Yamamoto, H., A 20ns CMOS functional gate array with a configurable memory, IEEE. Digest of ISSCC (1983) 146-147.

[27] Tanaka, C., Murai, S., Tsuji, H., et al., An integrated computer aided design system for gate array masterslice; part 2 the layout design system MARS-M3, Proc. 18th DA Conf., (June 1981), 812-819.

[28] Matsuda, T., Fujita, T., et al., LAMBDA: a quick low cost layout design system for master-slice LSIs, Proc. 19th DA conf. (June 1982), 802-808.

[29] Suzuki, M., Horiguchi, S. and Sudo, T., A 5k gate bipolar masterslice with 500ps loaded gate delay, IEEE J. Solid-State Circuits, vol. SC18, (1983), 585-591.

[30] Wood, R.A., A high density programmable logic array, IEEE TC, (Sep. 1979) 602-608.

[31] Kang, S. and vanCleemput, W. M., Automatic PLA synthesis from a DDL-P description, Proc. 18th DA Conf., (June 1981), 391-397.

[32] Takeda, T., Matsuhiro, K. and Suzuki, M., High performance bipolar FPLAs, Review of the ECL, vol 31, No.4, (1983) 566-575.

[33] Weinberger, A., Large-scale integration of MOS complex logic: a layout method, IEEE J. Solid-State Circuits, vol. SC-2, (dec. 1967), 182-190.

[34] Lopez, A.D., and Law, H.F.S., A dense gate matrix layout method for MOS VLSI, IEEE J. Solid-State Circuits, vol. SC-15, (1980), 736-740.

[35] Kasai, R., Fukami, K., Tansho, K. et al., An Integrated Modular and Standard Cell IC Design Method, Digest of ISSCC'84, (1984), 12-13.

[36] Erdelyi, C.K., Bechade, R.A., Concannon, M.P. and Hoffman, W.K., A Comparison of Mixed Gate Array and Custom IC Design Method, Digest of ISSCC'84 (1984), 14-15.

[37] Moore, G. E., Progress in digital integrated electronics, presented at the IEEE Int. Electron Devices Meet., Talk 1.3 (Dec. 1975).

[38] Sequin, C. H., Managing VLSI complexity: an outlook, Proc. IEEE, Vol. 71, No. 1 (Jan. 1983) 149-166.

[39] Breuer, M. A., Design automation of digital systems, (Prentice-Hall, 1972).

[40] Szygenda, S. A., and Thompson, E. W., Digital logic simulation in a time based table driven environment, Computer, (Mar. 1975).

[41] Nagel, L. W., and Pederson, C. O., Simulation program with integrated circuit emphasis, in Proc. 16th Midwest Symp. Circuit Theory (Waterloo, Ont. Canada, Apr. 1973).

[42] Miyahara, N., Endo, M., Tanabe, N., and Nakamura, H., Timing simulator (LOTAS), in Monograph of TG on Electron Devices of Inst. Electron. Commun. Eng. Japan, EDD82-13 (1982) 61-69 (in Japanese).

[43] Miyashita, H., Takeda, T., and Sugiyama, Y., PLA programming support system: PLACAD, ECL Technical Journal, Vol. 32, No. 6 (June 1983) 59-72 (in Japanese).

[44] Hong. S. J., Cain, R. G., and Ostapko, D. L., MINI: A heuristic approach for logic minimization, IBM J. Res. Develop., Vol. 18, No. 5 (1974) 443-458.

[45] Stebnisky, M. W., McGinnis, M. J., Werbikas, J. C., et al, APSS: an automatic PLA synthesis system, Proc. 20th DA Conf. (June 1981) 430-435.

[46] Meyer, M. J., Agrawal, P., and Pfister, R. G., A VLSI FSM design system, Proc. 21st DA Conf. (June 1983) 434-440.

[47] Donze, R., Sanders, J., Jenkins, M., and Sporzynski, G., PHILO - A VLSI design system, Proc. 19th DA Conf. (June 1983) 163-169.

[48] Lallier, K., Hickson, J., and Jackson, R., A system for automatic layout of gate array chips, European Conference on Electronic Design Automation, Univ. of Sussex, Brighton, England (1981) 54-58.

[49] Chen, K., Feuer, M., Khokhani, K., Nan, K., and Schmidt, S., The chip layout problem: an automatic wiring procedure, Proc. 14th DA Conf. (June 1977) 298-302.

[50] Ahdoot, K., Alvarodiaz, R., and Crawley, L., IBM FSD VLSI chip design methodology, Proc. 20th DA Conf. (June 1983) 39-45.

[51] Smith, D. C., Noto, R., Borgini, F., Sharma, S. S., and Werbikas, J. C., VGAUA: the variable geometry automated universal array layout system, Proc. 20th DA Conf. (June 1983) 425-429.

[52] Bose, A. K., Chawla, B. R., and Gummel, H. K., A VLSI design system, ISCAS83 (1983) 734-739.

[53] Sato, K., Nagai, T., Tachibana, M., Shimoyama, H., Ozaki, M., and Yahara, T., MILD-A cell based layout system for MOS LSI, Proc. 18th DA Conf. (June 1981) 828-836.

[54] Chiba, T., Okuda, N., Kambe, T., Nishioka, I., Inufushi, T., and Kimura, S., SHARPS: a hierarchical layout system for VLSI, Proc. 18th DA Conf. (June 1981) 820-827.

[55] Terai, H., Hayase, M., Ishii, T., Miura, C., Kozawa, T., Kishida, K., and Ngao, Y., Automatic placement and routing program for logic VLSI design based of hierarchical layout method, Proc. ICCC'82 (1982) 415-418.

[56] Ueda, K., Kitazawa, H., and Harada, I., CHAMP: chip floor plan for hierarchical VLSI layout design, IEEE Trans. on CAD/ICAS, Vol. 4, No. 1, (Jan. 1985) 12-22.

[57] Ueda, K., Placement algorithm for logic modules, Electron. Lett., 10, 10 (1974) 206-208.

[58] Lee, C. M., Chawla, B. R., and Just, S., Automatic generation and characterization of CMOS polycell, Proc. 18th DA Conf. (June 1981) 220-224.

[59] Weste, N. H. E., MULGA-an interactive symbolic layout system for the design of integrated circuits, BSTJ, Vol. 60, No. 6 (July-August 1981) 823-857.

[60] Mc Garity, R. C., and Siewiorek, D. P., Experiments with the SLIM circuit compactor, 20th DA Conf. (June 1983) 740-746.

[61] Wirth, N., What can we do about the unecessary diversity of notations for syntactic definitions?, Communications of the ACM (Nov. 1977).

[62] Newkirk, J., and Mathew, R., The VLSI designer's library, (Reading, MA: Addison-Wesley) 1983.

[63] Crawford, J. D., EDIF: a mechanism for the exchange of design information, Proc. CICC'84 (May 1984) 446-449.

LAYOUT DESIGN AND VERIFICATION
T. Ohtsuki (Editor)
© Elsevier Science Publishers B.V. (North-Holland), 1986

Chapter 2

PARTITIONING, ASSIGNMENT AND PLACEMENT

Satoshi GOTO, Tsuneo MATSUDA

C&C Systems Research Labs., NEC Corporation
Kawasaki 213, Japan

1.   INTRODUCTION

Layout design for an electronic equipment requires the partitioning and assignment of logic circuits to physical design modules, placement of these design units onto larger function modules and finding the routing patterns for interconnection. According to the physical hierarchy of hardware elements, modules correspond to function blocks, LSIs, printed circuit boards and backboards.

The current integrated circuit approach usually results in three-level backplanes. On the lowest level, sets of logic elements form modules. Interconnected sets of modules define boards, and interconnected sets of boards define the backplane.

In this VLSI era, modules correspond to VLSIs, ICs or discrete components. VLSIs have tremendous number of gates in themselves, and usually have hierarchical structures. On the lowest level, sets of logic elements form blocks. Interconnected sets of cells define macros, and interconnected sets of macros define a VLSI chip. See Fig. 1 and 2.

Although the problems involved in partioning, assignment, placement and routing are closely related, they have been treated separately because of the inherent computational complexity of the total problem. This chapter deals with partitioning, assignment and placement problems, since other chapters in this text deal with examing the routing problem.

The partition problem is viewed in this chapter as a design process which a module is constructed from submodules at a given hierarchy level. The assignment problem is to find one-to-one correspondence between logic circuit elements and placeable modules. The placement problem is to find appropriate locations for individual modules on the chip or boards.

These problems are considered to be hard combinatorial problems, or NP-complete problems from the computational complexity point of view, in the sense that the computation time required to obtain the real optimum solution increases in exponential order when the problem size increases. And also, actual problems have various kinds of object functions with a lot of restrictions.

Unfortunately, no methods exist which guarantee an optimum solution for the real-world large scale problems. Hence, algorithms based on heuristic rationales have been employed. All of these algorithms are either constructive or iterative. The constructive algorithms produce a solution using heuristic rules, often in sequential, deterministic manner. On the other hand, the iterative algorithms improve a solution by repeated modification of it. The automated layout design system usually has the above two algorithms, i.e., an initial solution is obtained through a constructive algorithm and the solution is improved gradually by an iterative algorithm.

Although many intensive efforts have been carried out in these years, the results obtained from the automated layout design systems are still quite discouraging, when compared to manually designed ones. This is because a human being or circuit designer can well understand the given electronic circuits, and he can find a good solution by exploiting his knowledge and intuition. Several approaches have been introduced to cope with this situation. The man-machine interaction system, high level function description or artificial intelligent approaches are notable approaches in this field.

This chapter is intended to give the current status of the partitioning, assignment and placement solving methods with tutorials. The emphasis is on rather new methods which seem to have a potential for solving some of the current problems in a practical manner.

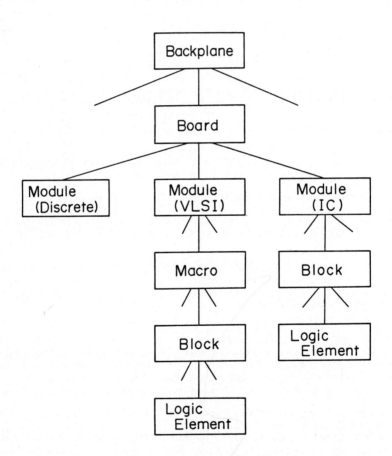

Fig. 1   Physical Hierarchy

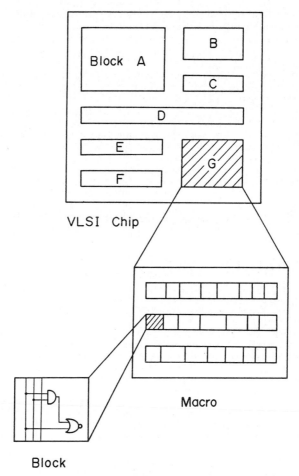

Fig. 2   VLSI Chip Hierarchy.

2.   PARTITIONING PROBLEM

2.1   Introduction

Partitioning is a process of creating the physical hierarchy of the hardware elements used to realize the design. It is essentially the subdivision of logic complexes into smaller subcomplexes. The physical structure has, in general, a hierarchical structure, with components such as LSIs, printed circuit boards, backboards and systems. The partitioning can define which LSIs contain what logic, and which boards contain what LSIs under several constraints. See Fig. 1 for the hierarchical structure.

There are two different applications for partitioning logic circuits. One is for circuit packaging, which divide a complex of logic circuit into subcomplexes, which can be fitted into the packaging structure used for the complexes. The other one involves a circuit layout for a VLSI chip. This partitioning is used in the layout of a complex, and is preparatory for the placement. Logic partitioning will find the global structure for a logic complex more easily than direct placement algorithms.

Parameters or constraints are associated with a partitioning problem which measure the solution of a partitioning problem. They are as follows.

(a) The size or area of each partition element is restricted.
(b) The number of external connections for each partition element is restricted.
(c) The maximum delay time through the circuits must not exceed a given time. Any external connection contributes a longer signal transmission delay time than an internal connection.
(d) A certain signal net must be externally available for testing purposes.

The above constraints are rather complicated and sometimes not well defined. In many practical cases, expert designers intuitively give good solutions based on their experience, taking the complicated situation into account. At this moment, very few automatic programs are practically used for partitioning because of the extreme complexities of the problems involved.

In the following sections, we look upon the partitioning problem from a mathematical point of view and give the formulation and its solving methods. The latest overview on this problem is given by Kodres [1], and we give considerably new results since then.

2.2   Mathematical Formuration

Any patitioning problem must consider two parameters, size and external connection requirements. See (a) and (b) in the previous Section. Neglecting, for the moment, any other parameters (c) and (d), we may define a partition problem as follows.

Let V be a set of nodes, and assign to each node $v \in V$ size $S(v)$, and the number of nets connected to v, $N(v)$, respectively. With each subset Vi of node set V, we define the number of nets, $I(v)$ for $v \in V_i$ such that nodes adjacent to v belong only to Vi. The number

$$E(v) = N(v) - I(v) \qquad (1)$$

denotes the number of external nets for a subset Vi. The size and external connection limits are denoted by S, and E, respectively. The partitioning problem is to find a family of subsets of V such that

$$\sum_{v \in V_i} S(v) \leqq S \qquad (2)$$

$$\sum_{v \in V_i} E(v) \leqq E \qquad (3)$$

where for $i = 1, 2, \cdots, n$,

$$\bigcup_{i=1}^{n} V_i = V \qquad (4)$$

$$V_i \cap V_j = \phi \qquad (5)$$

If the objective is to minimize the total number of external connections between partition sets, Eq. (3) is replaced by

$$T = \tfrac{1}{2} \sum_{i=1}^{n} \sum_{v \in V_i} E(v) \qquad (6)$$

$$T \rightarrow Min$$

It should be noted that, if a net common to more than two nodes exstis, the calculation of E(v) has to be modified. N(v) and I(v) have different values according to the way net connection patterns are set up.

Consider an example circuit shown in Fig. 3. with node set V={1, 2, · · · , 12}. We assume that S(v)=1 for v ∈ V, and want to partition the circuit into 3 sub-circuits with constraint S=4.

First, we want to have a partitioning for the limit of external connection E=4. A feasible partition is obtained by V1={1, 4, 6, 8}, V2={2, 3, 5, 12} and V3={7, 9, 10, 11}, shown in Fig. 4. However, this partition does not satisfy the minimum number of total external connections. On the other hand, the partitioning, V1={1, 2, 3, 12}, V2={5, 7, 10, 11} and V3={4, 6, 8, 9}, shown in Fig. 5, does satisfy it, and also satisfies the limit of external connection E=4. In many cases, the partition with the minimum number of total external connections between partition sets satisfies the limit of external connections for each partition set.

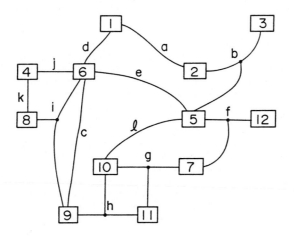

Fig. 3   An Example Network.

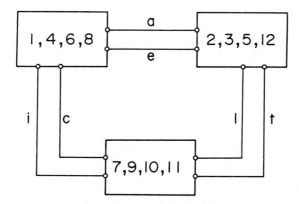

Fig. 4   Partitioning Result with External Connection E=4 Limit.

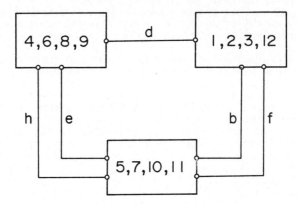

Fig. 5   Partitioning with the Minimum Number of Total External Connections, T=5.

2.3   Partitioning Algorithms [2-7]

There exist two kinds of heuristic algorithms, constructive and iterative improvement ones. The constructive algorithm produce a solution in a sequential and deterministic manner. On the other hand, the iterative one improves a solution by iteratively reducing the solution.

2.3.1 Constructive Algorithms

This method selects nodes, one at a time, based on an evaluation function, called IOC, which measures net connectivity to nets already or not yet selected, and then decides which partitioning set the selected node should belong to. The evaluation function is, in general, to be the number of nets connected to already selected nodes minus the number of nets connected to not yet selected nodes. The node with the highest value is the logical candidate to be selected. The selected node belongs to a partition set, which results in the minimum number of total external connections at this stage under the limit of external connection for each partition set.

This method starts with a seed or set of seeds for each partition set. A start with only one seed is called, "Max Conjunction-Min Disjunction Method," and a start with a set of seeds is called, "Clustering Method." However, these two methods do not involve appreciable differences between each other.

As an example, start with a seed, node 3 in Fig. 3, and obtain a partition in Fig.6 by applying the above algorithm to minimize the total number of external connections. Here, the number is 7 and, of course, it is not the optimum solution.

2.3.2 Iterative Improvement Algorithm

The method starts with an initially given partitioning result and improves it by a local transformation. The resultant partition is locally optimum for each initial partition in the sense that there does not exist a partitioning with a smaller cost for the given transformation, or modification rule and stopping rule.

We can generate a large number of locally optimum solution by generating initial partitions randomly, and thereby increase the probability that, at best, one of these solutions is close to the global optimum. In general, a more complicated transformation reaches one locally optimum with a better value in a greater computation time than a simpler one. In order to find a better solution within a given amount of computation time, the key problem is to find a suitable transformation.

For this type of problem, several discussions have been done as local optimization techniques which can apply to many other discrete optimization problems. We will discuss it in details in the placement section of Sec. 4. The result can be directly applied to a partitioning problem. Here, we describe a so-called, "Group Migration Method," proposed by Kernighan and Lin [5], which was developed in 1970 and is still one of the best methods in the past 15 years.

Interchange methods try to exchange nodes between partition sets. Pairwise interchange is applied to two nodes and triple interchange is to three nodes. An interchange operation is accepted if the resultant solution is a better one, and is not accepted if it is not. The new configuration is substituted for the old one.

Group Migration Method is used in an attempt to interchange more than two nodes simultaneously and improves a solution efficiently. The method is particularly well suited for bisection problem, where the circuit is divided into two subsets. Consider two initially constructed sets A and B, $A \cup B = V$ to start the process. If node $a \in A$ is exchanged with node $b \in B$, calculate the number of external connections gained by the interchange of a and b. This gain is denoted by g. If $g > 0$, then the interchange is beneficial, i.e., the total number of connections is reduced.

Let $a_1 \in A$ and $b_1 \in B$ be the nodes which result in the biggest gain , g., among all pair interchanges. Remove a_1 from A and b_1 from B, and repeat the process to find $a_2 \in A - a_1$ and $b_2 \in A - b_1$. By repeating this process until A and B are reduced to a null set, we get sequence $a_1, a_2, a_3, \cdots ; b_1, b_2, b_3, \cdots$, and $g_1, g_2, g_3 \cdots$. We find a value k such that

$$G = \sum_{i=1}^{k} g_i \qquad (7)$$

is maximum, then move $a_1, \cdots a_k$ to B and $b_1, \cdots, b_k$ to A. If G is positive and k is greater than zero, repeat the above process. Otherwise, stop the procedure.

For example, starting with a solution of Fig. 6, we obtain a solution of Fig. 4 by moving nodes 2, 7 and 8 simultaneously. However, in order to get a solution of Fig.5, we have to move nodes 4 and 6, 10 and 11, and 3 and 12 at the same time, which requires very complicated transformations.

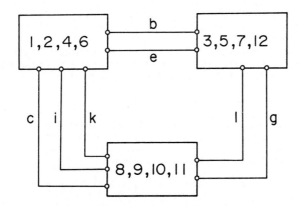

Fig. 6   A Partitioning Result using Constructive Algorithm, T=7.

2.3.3 Other partitioning methods

Several new approaches have been proposed up to now. Some approaches displayed remarkable results and some are now under investigation. Unfortunately, we cannot say clearly which method is better than the others because of the shortage of sufficient experimental results. We just show interesting and recently developed ones for the reader's reference.

(1)   Metric Allocation Method

This method attempts to find a metric allocation other than the graph structure on the nodes of the graph, which reflects the connectedness of the graph. As a metric, it uses an electrical analog for the network [8] or it calculates eigenvalue or eigenvector to obtain partitioning solutions [9].

(2)   Function Oriented Method

This method utilizes the information of logic circuits to determine a good partitioning. In [10], hierarchical design information is used to locate sets of identical logic, which can be realized as repeated physical entities to increase functional partitioning, and to divide the partitioning problem into multiple smaller problems. In [11], an algorithm is proposed for partitioning a behavioral hardware description written in the ISPS computer hardware description language. The partitioning is carried out before the actual registers, processing elements, and interconnections have been chosen, so that the partitioning information can be used to guide the design of the data path structure.

3.   Assignment Problem

An assignment is the phase of building a circuit of modules with size, shape and internal structure for a given logic circuit. The logic circuit is usually a schematic diagram where logic blocks, such as NAND, NOR gates or Inverters are connected by wires. The module circuit has to satisfy given circuit performance requirements such as delay time, power consumption or number of input-output terminals.

There exist different type assignment problems according to the different technologies. We will describe two typical problems [12], which plays important roles in the circuit layout.

3.1   Printed Circuit Board Problem

PCB design problem requires assigning abstract blocks to IC packages of TTL (Transistor-Transistor-Logic) circuits. Various kinds of IC packages are provided by IC makers, and can be used by selecting suitable ones. For example, IC package $\times \times \times 1$ contains four two-input NAND gates and $\times \times \times 2$ contains contains three two-input NOR buffers. If the design circuit includes 17 NAND gates and 20 NOR buffers, the board requires at least five $\times \times \times 1$ and seven $\times \times \times 2$ IC packages.

Some gates are left unused for later repairs or for circuit modifications. The assignment or selection rules are applied by taking into account easy routability and minimum signal transmission delay time.

The usual assignment algorithm is based on the following rules [12].

① Assign all ICs which contain only one logic block or gate.
② If an IC package has an available gate, use it for the gate which has the most connection in common with gates already assigned to this IC package.
③ If there is no partially used IC package, take the next unused package, and assign the first gate randomly.

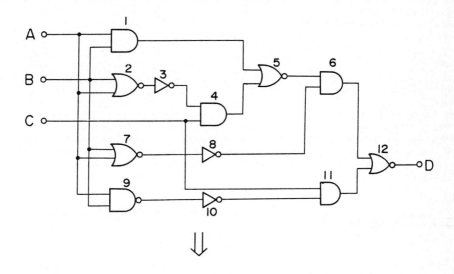

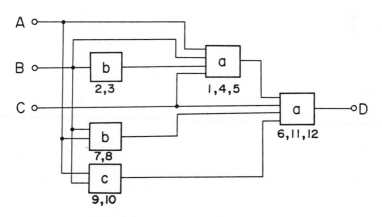

Fig. 7   Assignment from Logic Blocks to Circuit Modules.

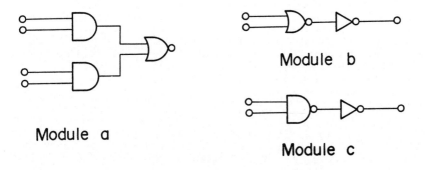

Fig. 8   Module Libraries Example.

3.2   LSI Layout Problem

In the gate-array or standard cell LSI design, fundamental logic elements, called circuit modules, such as NAND, NOR or FLIP-FLOP, are designed by giving interconnections between transistors or resistors in advance to meet the customer's LSI design requirements and stored in the library. The number of circuit modules in the library is usually between 100 and 200. Most logic blocks correspond to circuit modules in the library on a one-to-one basis, except that several circuit modules are available for each logic block. Such circuit modules have identical logic, but their output power and sizes are different. Also, certain circuit modules may be available in several shapes or terminal arrangements. Figure 7 shows an assignment example for use from logic blocks to circuit modules. The logic circuit with 12 blocks is converted to the module circuit with 5 modules by using 3 library modules, as shown in Fig. 8.

The assignment algorithm requires the following two steps.

  ① Find which part of logic circuits corresponds to library modules to have the same function.
  ② Determine which library module should be used to meet the required power, delay time and shape requirements.

The usual assignment algorithm is carried out in a straight forward way, picking up logic blocks one by one, according to the connectivities, and finding a suitable circuit module to meet function and performance requirements.

The above algorithms are based on constructive methods and are very simple to implement. There might be more complex ones, such as iterative improvement methods, which could be used to achieve better results, however no practically efficient algorithms have been found up to now, as far as the author knows. After the automatic assignment is accomplished by computer program, the results are usually improved on graphic stations by expert circuit designers.

3.3   Other Assignment Problems

In an automatic routing for a gate array LSI, a problem arises in regard to which one of the logically equivalent terminals is to be used to allow a signal net to be connected to any one of such terminals. There is also a problem regarding which vertical track in a feed-through cell is to be used for a signal net assigned to the feed-through cell. An efficient heuristic algorithm is proposed in [13], which tries to achieve 100% realization of the wiring requirements.

4.   PLACEMENT PROBLEM

4.1   Introduction

For each level of the hierarchical structure shown in Fig. 1, the placement problem has to be solved. Board placement has to be solved for backboard design. Module placement for board has to be determined, macro placement for VLSI must be settled, and block placement for macro or LSI must be accomplished.

The board placement, the module placement and the block placement problems can be treated in similar ways, since each component, in general, has the same size or the same height, even though components have different widths. Figure 9 shows a module placement example. Figure 10 and 11 are block placement examples, respectively. On the other hand, in the macro placement problem, individual components have various kinds of shapes or sizes, from larger to smaller, since it corresponds to RAM, ROM, PLA or a set of random logics, as shown in Fig. 2.

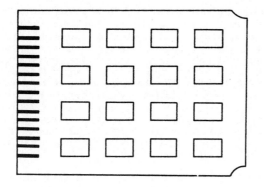

Fig. 9   Module Placement.

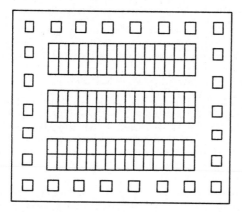

Fig. 10   Block Placement (Gate Array LSI).

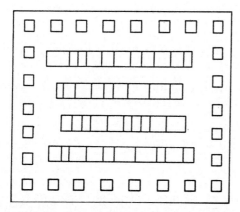

Fig. 11   Block Placement (Poly Cell LSI).

The macro placement problem has to be treated in a different manner from the others because of the irregular component shape. The module placement problem has been studied for many years to make the package design automatically, and its result can be applied directly to the block placement for macro or LSI. Particularly, the block placement problem in gate-array LSIs or poly-cell LSIs has been of great interest to LSI CAD engineers recently. The problem is of particular significance in the present day design of LSI chips, as a huge number of components has been mounted on one chip, i.e., 1000~10000 components. Therefore, efficient new algorithms have to be devised to meet such a situation.

The macro placement is a new problem which did not appear in the IC eras. The number of components, or gates per chip, has become too big to handle simultaneously, the hirarchical design methodology has been introduced.

For our convenience, we use, in the following section, the terminologies, blocks and a chip as the placement problem. Blocks are assumed to have various heights or sizes.

The layout problem is usually divided into two stages and solved as different problems, namely placement and routing. In the placement stage, locations of individual blocks, are decided on a chip to facilitate routing. In the routing stage, interconnections to external leads, called pins, are made in such a manner as to satisfy various physical constraints.

The true objective of the placement phase is to achieve 100% routing within a given area.
Such an objective is not mathematically well defined.

We are not able to know whether one placement result is good or not, until trying the actual routing. We replace the true goal by a simplified objective which is easy to calculate by computer. When we choose one objective, we hope that a solution to optimize the objective will lead to a high routability result. Until now, the following three objectives have been proposed to minimize:

(1) Total routing length.
(2) Maximum cut line.
(3) Maximum density.

For many years, the total routing length objective has been employed. However, this objective sometimes generates too crowded an unroutable routing area, and the second or the third objective has been introduced. In the following sections, the relations between these objectives are discussed and efficient algorithms for each objective are described.

Prior to the placement stage, block design and the assignment of logic gates into blocks have been accomplished. Logic elements, called blocks, such as AND, OR, or FLIP-FLOP, are designed by giving interconnections by conductors between transistors or resistors. The whole logic function of a given circuit can be realized by connecting between block terminals and external pads on the routing area. Therefore, circuit data is reduced to between-the-block connections, which are called a set of signal nets.

The latest overview on this problem is given by Hanan and Kurtzberg in [14]. They also presented valuable papers with many experimental results. In the following, considerably new results since [14] are described.

4.2   Interconnection Rules

In the placement stage, locations of individual blocks are decided on a chip to facilitate routing. Without trying the actual routing or geometric routing, it is necessary to decide whether one placement result is good or not at the placement stage. We need a simpler routing method at the placement stage, which can be incorporated into the routing program and reflect the actual routing with good approximaion. Sometimes, the technology used in the production dominates the interconnection rules, which must be satisfied in the actual routing. The followings are examples of interconnection rules.

(a) Minimum Steiner Tree
Minimum length tree, whose vertices include block terminals and junction points.

(b) Minimum Spanning Tree
Minimum length tree, whose vertices include only block terminals.

(c) Minimum Chain
Minimum length tree, which has at most two edges incident to any vertex.

(d) Minimum Source-to-Sink Connection
Minimum length tree, which has a connection from the source to each sink.

Figure 12 shows the examples. A vertex with ⊗ or × symbols represents the source terminal and sink terminal, respectively. A vertex with ○ symbol represents the junction point. The lengths of each tree are $l = 10$, 11, 12, and 15. In general, the shortest tree is the Minimum Steiner Tree, while the longest one is the Minimum Chain or the Minimum Source-to-Sink Connection. Except for the Minimum Steiner Tree, the optimum solution is obtained easily, i.e., in a polynomial order computation time. The problem of finding the Minimum Steiner Tree is proven to be on NP-complete problem, which possibly requires exponential computation time, when the problem size becomes bigger. A simple heuristic procedure is practically introduced to obtain the Minimum Steiner Tree.

Because the above four trees require some computation effort, even with simple heuristics, simpler approximation methods are introduced, called "Complete Graph Method or Half Perimeter Method." As far as the routing length is concerned, these simpler methods are considered sufficiently capable of reflecting achieving the goal.

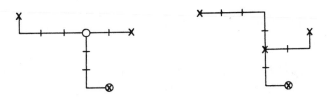

(a) Minimum Steiner Tree   (b) Minimum Spanning Tree
$l = 10$ $l = 11$

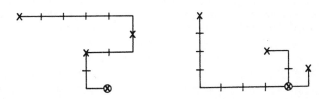

(c) Minimum Chain   (d) Minimum Source to Sink Connection
$l = 12$ $l = 15$

Fig. 12 Interconnection Rules.

(e) Complete Graph Method [15].
     Let a signal net be common to ρ blocks, the length of a signal net is calculated as

$$\frac{2}{\rho} \sum_{i=1}^{\frac{1}{2}\rho(\rho-1)} Length\ of\ between-the-block. \tag{8}$$

The intuitive rationale for this is that $2/\rho$ is the fraction of the $1/2\rho(\rho-1)$ edges needed to connect a signal net of size ρ. The length for the same example, in Fig. 12, is calculated as $2/4(8+3+4+5+3+6)=14.5$

(f)   Half Perimeter Method
      The length is defined as half-perimeter of the smallest rectangle which encloses the blocks in the signal net.

The length is calculated as $l=9$ for the same example, since the horizontal and vertical lengths are 4 and 5, respectively. These values are shrter than those of the Minimum Steiner Tree. However, in general, for a signal net with up to 3 blocks, the length is equal to the length of the Minimum Steiner Tree.

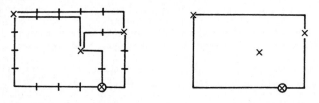

(d)Complete Graph Method   (e)Half Perimeter Method

l = 14.5 l = 9

Fig. 13   Simple Routing Length.

Figure 14 shows a comparison between Minimum Spanning Tree and Complete Graph Method. The comparison between Minimum Spanning Tree and Half Perimeter Method was made by calculating the total routing length of all signal nets in a logic circuit. Strong correlation exists between the two values. The stronger one is between Minimum Spanning Tree and Half Perimeter Method. Therfore, as far as the routing length is concerned, but not the routing pattern, we can adopt Half Perimeter Method because of its easier calculation.

4.3 Placement Goals

     In the placement stage, locations of individual blocks are decided on a chip to satisfy a number of constraints in actual cases.

·   The routing length of a signal net must be kept within tolerable bounds.
·   The heat dissipation or power dissipation level has to be preserved.
·   Signal cross-talk must be eliminated.

These constraints are interpreted in the placement stage as follows.

(1) A subset of blocks must be placed within some distance from each other.
(2) Some blocks must be placed in fixed positions on a chip.
(3) Some blocks must be placed in the next position for other specified blocks.

Under the above constraints, a good placement solution has to be obtained. We really don't know whether one placement solution is good or not without trying the actual routing. It takes a considerably large amount of computation time to try the actual routing for only one placement result in the real large-scale problems. In the process of obtaining a better placement solution through heuristic procedures, a large number of placement solutions will appear for the evaluation. If every placement solution is evaluated by the actual routing, the computation time would be tremendously long.

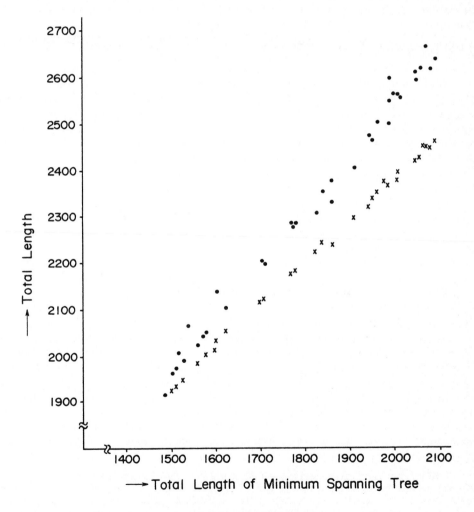

Fig. 14   Routing Length Comparisons.
     x :   Half Perimeter Method
     • :   Complete Graph Method

Therefore, the placement and routing problems are solved separately, not simultaneously. A simplified objective is introduced at the placement stage, which is easy to calculate and reflects the actual routing quite well. We will try to optimize a simplified objective, which hopefully leads to the true goal. Three simplified objectives have been proposed until now and are discussed in the followings.

4.3.1 Total Routing Length

The total routing length over all signal nets, T(P), for a placement P, is given by

$$T(P) = \sum_{i,j} W[i,j] \cdot d[P(i), P(i)], \tag{9}$$

if every signal net is common to only two blocks. Here, i and j designate block terminals, w[i, j] is a measure of the nets between the two block terminals i and j, and d[p(i), p(j)] is the distance between i and j. We want to find a placement which minimizes T(P) for all possible placements. If there exists a signal net with more than two blocks, distance, d, is replaced by the length of either one of trees described in Section 4.2.

The total routing length over all signal nets is the area size for the routing itself on a chip. If the routing area involves smaller size, the chip size becomes smaller for a dedicated custom LSI or the routability becomes higher for a fixed size chip LSI like gate-array LSI. The total routing length criterion neglects the interaction between signal nets, since the length is calculated independently for each signal net. This simplified calculation does not reflect the real routing area, but only realizes a rough approximation.

4.3.2 Maximum Cut Line

Maximum cut line, X(P) and Y(P), for placement P, is given by

$$X(P) = \underset{i}{Max}\, C_p(x_i) \tag{10}$$

$$Y(P) = \underset{i}{Max}\, C_p(y_i) \tag{11}$$

Here, Cp(xi) and Cp(yi) denote the number of signal nets which cross the vertical line X=xi and horizontal line Y=yi, respectively. A considerable number of vertical or horizontal lines is used to calculate the maximum cut line. We want to find a placement which minimizes X(P) or Y(P) for all possible placements. For a gate-array LSI, a line passing through boundary between cells is taken up as vertical or horizontal line. If we set the vertical or horizontal line for every grid (unit length), the total routing length is calculated as

$$T(P) = \sum_{i} C_p(x_i) + \sum_{i} C_p(y_i) , \tag{12}$$

which takes the total sum for every component instead of taking the maximum value only.

The difference value between the maximum cut line and the routing capacity of a chip represents the routability of signal nets. A greater positive value indicates a higher routability. If the maximum cut line is over the routing capacity on a fixed size chip, no routing procedure can achieve complete connection.

4.3.3 Maximum Density

Maximum density, D(P) for a placement P, is given by

$$D(P) = \underset{i}{M\ a\ x}\ d_p\ (e_i),\qquad\qquad(13)$$

$$where\quad d_p\ (e_i) = \frac{f_p(i)}{C_p(ei)}$$

Here, Cp(e$_i$) is the channel capacity for edge e$_i$ and fp(e$_i$) indicates the number of signal nets assigned to edge e$_i$.

In the case of the maximum density goal, the whole chip is divided into a rectangular array of blocks, called a portion, each of which contains some wiring grid lines and circuit pins. Global routing determines, for each signal net, a path of portions which the net will be routed through. Global routes for all signal nets are assigned in such a way as not to exceed the number of available grid lines, Cp(e$_i$), or to minimize the value fp(e$_i$)/cp(e$_i$) for all edges. The value fp(e$_i$) is the number of signal nets whose routes actually pass through edge ei.   This problem is similar to a multi-commodity flow problem, which is considered to be a hard combinatorial one, and the real optimum solution is not obtained within a reasonal computation time. Hence, heuristic algorithms are employed. In this situation, we calculate dp(e$_i$) for each edge, called density.

The maximum density criterion reflects the routability of signal nets more precisely than the maximum cut line criterion.   More detailed information regarding the routability inside the chip can be obtained by calculating the density. If the maximum density is greater than one for a placement, complete routing is almost impossible for the placement.

4.3.4 Comparison between goals

The above three objectives lead to different solutions for each other. In many cases, the optimum solution for one objective guarantees an optimum solution for another objective. A solution with the least total routing length quite often satisfies minimizing the maximum cut line or minimizing the maximum density. We don't know exactly how often it happens. We illustrlate solutions with a seven-block problem. The following are 5 signal nets.

$$S1 = \{A1,\ B1\}$$

$$S2 = \{A2,\ B2\}$$

$$S3 = \{A3,\ C1\}$$

$$S4 = \{C2, D1\}$$

$$S5 = \{E1, F1, G1\}$$

Let the chip have 3 rows and 3 columns. The distance between two adjacent cells is defined as, either vertical or horizontal, one unit length.   The capacity between two adjacent cells is equal to one unit. The problem is to place all blocks on the cells so that each objective is minimum.

Placement p1, shown in Fig. 15(a), realizes a solution with the minimum total routing length T(p1)=7. However, p1 has the maximum cut line, X(p1)=4 and Y(p1)=2, since

$$X(P1) = Max\{4, 0\}$$

$$Y(P1) = Max\{2, 1\}$$

Also, p1 has maximum density D(p1)=2, since two signal nets have to be assigned to at least one section edge by any routing under this placement,

$$D(P1) = Max\{2, 0, 1, 0, \cdots, \} .$$

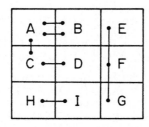

T (P₁) = 7
X (P₁) = 4, Y(P₁) = 2
D (P₁) = 2

(a)

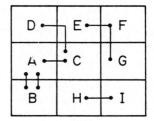

T (P₂) = 8
X (P₂) = 2, Y (P₂) = 2
D (P₂) = 2

(b)

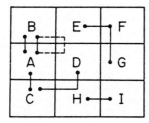

T (P₃) = 9
X (P₃) = 2, Y (P₃) = 3
D (P₃) = 1

(c)

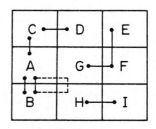

T (P₄) = 7
X (P₄) = 2, Y (P₄) = 2
D (P₄) = 1

(d)

Fig. 15   Placement Results and Goals.

For placement P2, each value is calculated and shown in Fig. 15(b). Placement P2 realizes a solution with minimum vertical and horizontal maximum cut line $X(P2)=2$, $Y(P2)=2$. However, the total routing length is longer than that for P1.

Although placement P3 does not realize a solution with either the minimum total length or the minimum maximum cut line, it achieved the minimum maximum density, $D(P3)=1$, shown in Fig. 15(C). One of the signal nets between block A and B can be routed as $A{\rightarrow}D{\rightarrow}E{\rightarrow}B$, and, at most, one net is assigned to each edge.

Each placement P1, P2 and P3 realize the optimum solution for one of the objectives, but not the optimum solution for the remaining two objectives. Therefore, we cannot say that the optimum solution for one objective does not guarantee the optimum solution for the other objectives. However, placement P4, shown in Fig. 15(d) realizes the optimum solution for three different objectives.

The problem of finding an actual optimum solution for even one of the objectives is considered to be hard combinatorial problem, which requires exponential order computation time, when the problem size become larger. Therefore, algorithms based on heuristic rationale have been employed. The problem of finding the real optimum solution for all objectives is a much harder combinatorial problem, and sometimes such a solution does not exist.

In the followings, we will discuss heuristic algorithms for either one of objectives, which have been proposed up to now.

4.4   Minimum Length Algorithms

There are, in general, two types of heuristic methods for this kind of problem. One is a constructive method, which obtains a solution using heuristic rules, often in sequential manner. The other one is an iterative improvement method, which improves a solution by means of local transformations. The algorithms described here are of these two methods, and published in [16]. The constructive ones here are rather simple algorithms and more complex ones are described in Sec. 4.8.

Consider a two-dimensional chip on which blocks are to be placed. The chip is characterized in terms of a finite array of cells. A matrix location, or cell may be represented by a point in an x-y coordinate system. Blocks are the entities which are to be assigned to cells on the chip. It is required that one block can occupy one and only one cell, and on each cell not more than one block is allowed.

Blocks contain pins for connection by phyical wires to form signal nets. In this study, the pins on the blocks are ignored and the distance is measured from the center of the block. Hence, a signal net becomes a subset of blocks, and a signal net specification defines the connection of all blocks on a specific chip.

Let the chip have m rows nd n columns. It may be assumed that the number of blocks is equal to $m{\times}n$ without loss of generality, because dummy blocks, which are not connected, can always be introduced. The distance between two adjacent cells is defined as, either vertical or horizontal, one unit length.

The routing length of a signal net is defined here as half perimeter of the smallest rectangle, which encloses the blocks in the signal set. The placement problem is now defined in the following.

Given a set of blocks with signal nets defined on subsets of these blocks and a set of cells, place all the blocks on the cells so that the total routing length over all signal nets is minimum.

4.4.1 Median of a block

Preceding a placement algorithm, it is necessary to introduce a concept, called median of a block, and present an algorithm to find it, since the present placement algorithm depends on it. Let us consider a chip on which every block is placed. Pick one block, denote it by M. Move only block M on the board, while the other blocks remain fixed. The routing length of a signal net does not change, as long as the signal net is not connected to block M. Therefore, consider the signal net connected to block M only and the sum of the routing length of these signal nets. This value is referred to as the routing lenght associated with block M.

Now, define the median of block M. Block M may be placed on $m \times n$ different positions. The block M median is defined as a position where the routing length associated with block M is minimum. Next, sort all the routing lengths associated with block M with respect to the block M position in ascending order. In this order, choose ε elements from the minimum one. The set of these ε position is defined as the ε-neighborhood for block M median.

Now consider how to find a median of a block. Let i $(i = 1, 2, \cdots, r)$ designate a signal net which is connected to block M. For each signal net i, consider the smallest rectangle which encloses the block in the signal net. Here, block M is excluded from the signal net when forming the rectangle. Let us denote the rectangle by I_i and its figure by parameters (X_i^a, Y_i^a) and (X_i^b, Y_i^b), where X_i^a and X_i^b are the minimum and maximum values in the x-direction on the rectangle, respectively. The same definitions are pertinent for Y_i^a and Y_i^b in the y-direction.

The routing length associated with block M, which is requied to place block M in position (x, y), is given by

$$F(x,y) = \sum_{i=1}^{r} (f_i(x) + f_i(y)) \tag{14}$$

where

$$f_i(x) = \begin{cases} x_i^a - x, & x < x_i^a \\ \\ 0, & x_i^a \leq x \leq x_i^b \\ \\ x - x_i^b, & x > x_i^b \end{cases} \tag{15}$$

$$f_i(y) = \begin{cases} y_i^a - y, & y < y_i^a \\ \\ 0, & y_i^a \leq y \leq y_i^b \\ \\ y - y_i^b, & y > y_i^b. \end{cases} \tag{16}$$

The problem is to find a pertinent position (x, y) on the chip, such that $F(x, y)$ is minimized. Since the function $F(x, y)$ has a separable form with respect to variables x and y, $F(x, y)$ can be calculated independently from each other for x and y. The y-component can be found in the same way as the x-component. Thus only the x-component will be discussed in the follwoing. Equation (15) is transformed as

$$f_i(x) = \tfrac{1}{2}\{|x - x_i^a| + |x - x_i^b| - (x_i^b - x_i^a)\}.$$ (17)

The problem is reduced to finding a position where

$$\sum_{i=1}^{r} (|x - x_i^a| + |x - x_i^b|)$$

is minimum, since $x_i^b - x_i^a$ is a constant value. The value $|x - x_i^a| + |x - x_i^b|$ indicates the sum of the distances from x to x_i^a and x to x_i^b, thus the problem is to find a point x such that the total sum of the distances from x to each point x_i^a, x_i^b (i = 1, 2, $\cdots$, r) is minimum.

In general, it is necessary to solve the following problem. There are a_i points on position i (i = 1, 2, $\cdots$, n) along the line. Find a position x on the line such that

$$\sum_{i=1}^{r} a_i |x - i|$$

is minimum. This problem is a particular case of finding an absolute median of a graph presented in [17]. The present problem, treating only a linear tree instead of a general graph, can be easily solved by using the following theorem.

Theorem I: Point q is the median if

$$\sum_{i=1}^{q-1} a_i < \frac{N}{2} < \sum_{i=1}^{n} a_i \quad holds,$$

where (18)

$$N = \sum_{i=1}^{n} a_i.$$

Fact 1
The total distance decreases monotonically from point 1 to a median. From a median to point n, it also increases monotonically. This fact is quite useful when more than one median are interested.

Fact 2
The x-component and y-component of a median can be calculated independently from each other. Each component has the characteristics mentioned in Fact 1. Therefore, the median on the two-dimensional chip can be found easily. When interest is in finding the kth minimum, instead of the first minimum, it is possible to use the efficient algorithm reported in [18]. Thus the ε-neighborhood of a block can be easily obtained.

4.4.2 Constructive Algorithms

This method selects blocks, one at a time, based on an evaluation function which measures signal net connectivity to blocks already or not yet selected, and then decides which cell the selected blocks will be placed on. Once a block is fixed into a position, it is not moved. The conventionl evaluation function, called IOC in Sec. 2.3, is adopted.

$$IOC = I - O$$

or

$$IOC = I.$$

Here, I is the sum of signal net connectivity to all placed blocks for each unplaced block. And O is the sum of signal net connectivity to all unplaced blocks for each placed block. The block with the first or the second highest IOC is the logical candidate to be selected. Each of them is seleced at random as the blocks to be put in place. The selected block is placed on the cell which yields the minimum total routing length among available cells. All available cells need not be examined here. Only a small part of them is examined, by calculating the ε-neighborhood of the mediam.

4.4.3 Iterative Improvement Algorithms

The purpose of this algorithm is to improve the placement by applying small local changes, such as pair-wise interchange of blocks. There is a large number of trials and it is essential that the total routing length is calculated on an incremental basis. There are many ways to interchange blocks efficiently, which are described in [14, 15]. We will just refer the names.

PI: Pairwise-Interchange
NI: Neighborhood-Interchange
FDI: Force-Directed-Interchange
FDR: Force Directed-Relaxation
FDPR: Force-Directed-Pairwise-Relaxation.

In this section, we will describe the details of GFDR (Generalized-Force-Directed-Relaxation) method. This method is considered to be a general one among relaxation methods and a most efficient method.

Let S be the set of all feasible solutions and let x be a feasible solution, $x \in S$. Consider the neighborhood of x, denoted by X(x), which is a subset of S. In the first step, x is set to a feasible solution and a search is made in X(x) for a better solution x' to replace x. This process, which is referred to hereafter as a local transformation, is repeated until no such x' can be found. A solution is said to be locally optimum if x is better than any other elements of X(x).

A lot of definitions may be considered for the neighborhood of a solution. In [19], the set of solutions transformable from x by exchanging not more than λ elements is regarded as the neighborhood of x. A solution x is said to be λ-optimum, if x is better than any other solutions in the neighborhood in this sense.

Although the λ-optimum solution gets better as λ increases, the computation time easily goes beyond the acceptable limit, when an exhaustive search is performed for large λ. The following method does not examine all the elements in the neighborhood, nor does it guarantee a λ-optimum solution. However, it is very efficient in the sense that it can be applied for a large value of λ with limited searches in the neighborhood.

The present search procedure is illustrated along with the search tree shown in Fig. 16, where each node represents a block and each edge represents a trial transformation. The root node of the tree A is a block chosen to initiate the trial interchange, it is referred to as the primary block. A path connecting node A and one of the other nodes defines a possible interchange. For example, the path A→B→E→O refers to the trial interchange of four blocks, as shown in Fig. 17. Here, block A is placed on the cell of B, B is placed on E, E on O, and O on A, in a round robin sequence. Although this transformaiton is a quadruple interchange, it includes a pairwise interchange as a special case, i. e., paths A→B, A→C, and A→D, as shown in Fig. 18. Value λ indicates the number of blocks to be interchanged.

The search tree is examined as follows. In this example, ε is fixed as 3. First, block A is interchanged with either one of the blocks on trial in the ε-neighborhood of A median ($\lambda=2$). The ε-neighborhood blocks are B, C, and D, thus pairwise interchanges between A and B, A and C, and A and D are performed (see Fig. 18). The trial interchange is accepted if it results in the reduction of the total routing length. If more than one reduction occurs in these transformations, the interchange with the greatest reduction is selected for

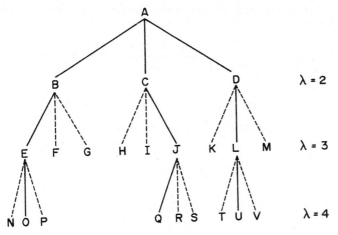

$\lambda = 2$

$\lambda = 3$

$\lambda = 4$

Fig. 16   Search Tree

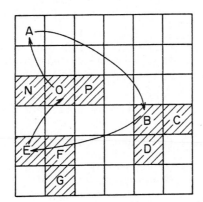

Fig. 17   Trial Interchange of Blocks (=4).

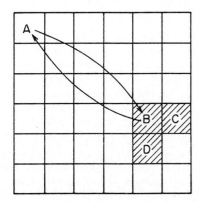

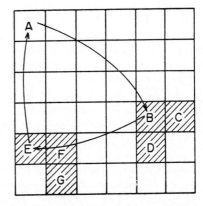

Fig. 18   Trial Interchange of Blocks ($\lambda = 2$).   Fig. 19   Trial Interchange of Blocks ($\lambda = 3$).

acceptance. If no interchange contributes to reducing the total routing length, the next step ($\lambda = 3$) is initiated.

Block A is placed on the cell of B. Then the median of B and its ε-neighborhood are calculated. In this case, the ε-neighbohood blocks are E, F, and G. Thus interchanges A→B→E, A→B→F, and A→B→C are tried (see Fig. 19).

These trial interchanges are accepted if one of them results in the reduction of the total routing length. Otherwise, consider the three interchanges of paths A→B→E, A→B→F, and A→B→G, and choose the best one (least total routing length) for the later tree search. Here, A→B→E is chosen, and A→B→F and A→B→G are omitted.

The solid lines in the tree search shown in Fig 16 indicate which searches are to be continued. Broken lines show the searches which are to be terminated. Terefore, no more search efforts are made along paths A→B→F and A→B→G. There is only one solid line under any node, except for root node A. Triple interchanges are performed for the other ε-neighborhood blocks, C and D, of root node A. Tree search will be continued following J or L, whereas no search will be accomplished through H, I, K, and M. The tree search is continued, i.e., a path from node A is extended as long as λ is no greater than λ^*, which is given as a parameter.

Each selection of a primary block is identified with an interchange cycle. Cycles are iterated until there is no reduction in the total routing length

The GFDR method is different from the FDPR method in [15] which tries to interchange only a pair of blocks. The GFDR, on the other hand, tries interchanges of more than two blocks at the same time. If the interchanges are performed randomly in the GFDR, like in the PI method of [5], the gain will not compensate for the comparative great increase in computation time. The GFDR method examines and performs trial interchanges for subsets of blocks which have a large possibility for improvement. This limited trial interchanges enable us to find a good locally optimum solution quickly.

Values ε and λ^* greatly affect the computation time and the total routing length. For a greater value of ε and λ^*, a better solution is obtained at the expense of computation time. In order to find a better solution within a given amount of computaiton time, the key problem is to determine how to set values ε and λ^*. In the following section, several experimental results are shown to point to the best values of ε and λ^*.

4.4.4 Efficiencies of Algorithms

In order to check and compare the results of the present algorithm with others, the algorithm was programmed and tested for five examples in a real problem. The program was written in FORTRAN and run on NEC ACOS-77/700.

Example logic graphs were obtained from [20]. Statistics for the example logic graphs and grid sizes are shown in Table I.

Experiment 1

The placement problem treated here has many locally optimum solutions. Thus different initial solutions lead to different solutions followed by an iterative improvement. It has been debated in the literatures whether it is better to use a random start or to use a constructive-initial solution, followed by an iterative-improvement. Experimental results in [15] and [21] showed that the constructive-initial start approach is superior to the random start in both solution value and computation time. Taking this fact into account, this constructive method generates good solutions randomly as initial solutions. In order to know the relation between the initial solution and its locally optimum solution, Example 5 was provided as a test with $\varepsilon = 4$ and $\lambda^* = 4$, and 25 different initial solutions were generated by SORG and improved by GFDR.

Table I Statistics for example graphs.

	Example 1	Example 2	Example 3	Example 4	Example 5
# blocks	67	108	116	136	151
# signal blocks	132	277	329	432	419
Av. # signal nets per blocks	6.32	6.41	7.30	7.86	5.98
Av. # blocks per signal net	3.43	2.78	2.66	2.73	2.35
Placement grid size	5×15	8×15	8×15	10×15	11×15

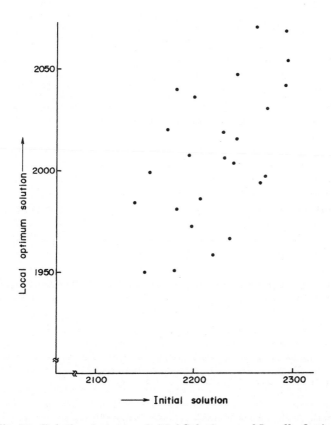

Fig. 20 Relation between Initial Solutions and Locally Optimum solutions.

Figure 20 shows the relation between initial solutions and their locally optimum solutions. A better initial solution does not always result in a better locally optimum solution. This result does not recommend generating many good random starts and following only the best one with iterative improvement. But it recommends repeating random generation of an initial solution and its improvement to obtain a set of localy optimum solutions. The best one among them is chosen as a final solution.

Experimnt 2

A simple transformation, such as pairwise interchange with smaller ε in general, does not yield better locally optimum solutions than more complicated ones with larger ε or λ^*. However, from the computational complexity point of view, the latter one takes much more time than the former one. In order to explore the influence of value ε or λ^* on computation time and the solution, the program was run by changing the value of ε or λ^*.

In Fig. 21, total routing length versus computation time curves are plotted for five values of ε operating on Example 5. Here, λ^* is fixed as 4. Each mark on the curves represents one cycle of GFDR method. The curve with smaller ε results in steeper descent, whereas it does not converge to a better solution. On the other hand, the curve with larger ε converges to a better solutions at the expense of computation time. A fairly small value of ε, i.e., $\varepsilon = 4$ or 5 is sufficient to lead to good solutions.

In Fig. 22, total routing length versus computation time curves are plotted for five values of λ^* operating also on Example 5, where $\varepsilon = 4$. The same discussion can be made for the value of λ^*, as for ε. The value $\lambda^* = 3$ or 4 seems to be enough to have good solutions.

The aim of the overall scheme is to generate as many locally optimum solutions as possible within some time interval. Then, the best one among them is chosen. Let P be the probability that the cost of the locally optimum solution is less than V and let T_0 be the running time to produce the local optimum solution. Then, the best locally optimum solution produced within time interval T has a probability

$$P = 1 - (1 - P)^{T/T_0} \tag{19}$$

of being better than V.

Random starts were repeated as many times as possible within $T = 30$ min.for a value of ε. Here, Example 5 was provided as a test and λ^* was fixed as 4. Let P'_ε be the probability of (19) associated with ε, and calculated P'_ε of being less than 2000. The experimental results show the $P'_2 = 0.68$, $P'_3 = 0.71$, $P'_4 = 0.95$, $P'_5 = 0.92$, $P'_8 = 0.70$. This procedure performs best at $\varepsilon = 4$.

The same analysis was done for finding the value of λ^*. Let P'_{λ^*} be the probability of (19) associated with λ^* and calculated P'_{λ^*} of being less than 2000. Here, ε was set to 4. We have the results that $P'_2{}^* = 0.74$, $P'_3{}^* = 0.93$, $P'_4{}^* = 0.98$, $P'_5{}^* = 0.89$, $P'_8{}^* = 0.75$. The best value was $\lambda^* = 4$. The same tendency was observed for the other four examples.

Note 1
GFDR method is considered to be a general relaxation one. The algorithm with $\lambda^* = 2$ and $\varepsilon = \infty$ corresponds to PI (Pairwise Interchange) method, and the algorithm with $\lambda^* =$ and $\varepsilon = 1$ corresponds to FDR (Force-Directed-Relaxation) method.

Note 2
The algorithm with $\lambda^* = 2$ did not result in good solutions in the meaning of both the appropriateness for solution value and computation time. This algorithm is called as the FDPR method in [15] and considered to be best among existing algorithms.

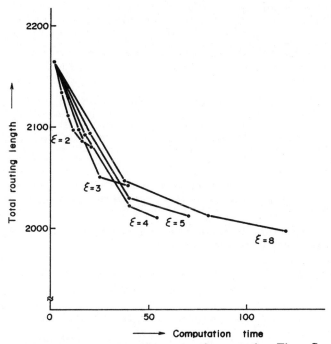

Fig. 21 Total Routing Length versus Computation Time Curve for Example 5, varing the Value of ε .

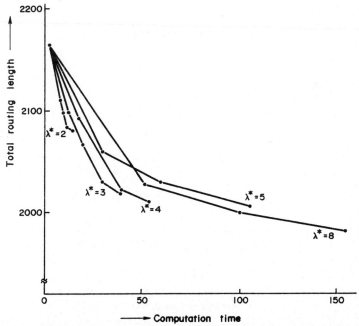

Fig. 22 Total Routing Length versus Computation time Curve for Example 5, varing the Value of λ^* .

Note. 3

The algorithm with larger ε, did not result in good solutions either. The method with the testing of all possible interchanges is exactly equal to the proposed algorithm with the largest value of ε. In this sense, the testing of all possible interchanges is said to be inferior to limiting the number of possible exchanges proposed here.

Experiment 3

In order to explore the influence of the number of blocks on the computation process, five examples were examined.

The graph in Fig. 23 shows computation time versus number of block for various values of ε. The computation time increases quite rapidly for $\varepsilon > 6$, when the number of blocks increases. Therfore, it seems infeasible to set $\varepsilon > 6$ for large scale problems.

The graph in Fig. 24 also shows computation time versus number of blocks, while varying the value of λ^*. From a practical point of view it may not be reasonable to set $\lambda^* > 5$ for large scale problems.

The computation time to obtain a locally optimum solution with $\varepsilon = 4$ and $\lambda^* = 4$ was linearly proportional to the number of blocks.

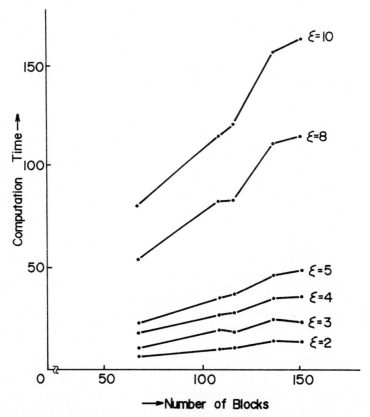

Fig. 23   Computation Time versus Number of Blocks, varing the Value of ε .

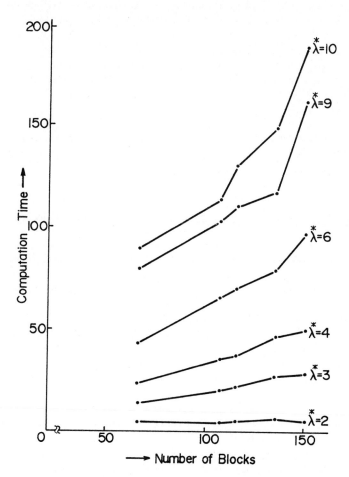

Fig. 24 Computation Time versus Number of Blocks, varing the Value of λ^*.

4.5 Minimum Cut Algorithms

The original idea for this algorithm was given by Breuer[22].

Let $C = \{C1, C2, \cdots Cr\}$ be a set of cut lines crossing the chip, either horizontally or vertically, and the value of Ci be the total number of signal nets crossing the cut line Ci, denoted by v(Ci). Objective function is to minimize some function F of the values of each cut line. This kind of placement algorithm is called minimum cut algorithm (or Min-cut algorithm).

An ideal objective function for this algorithm is

$$F1 = min \ (MAX \{v(c) \mid c \in C\}) \ . \tag{20}$$

However, this function is too difficult to satisfy. In general, following function is used:

$$F1 = min\ v(\ C_r\ )\ |\ min\ v(\ C_{r-1}\ )\ |\cdots|\ min\ v(C_1)$$

(21)

where C1, C2 $\cdots$, Cr is a given set of cut lines, and " $|$ " can be read as "subject to." Here (1, 2, $\cdots$, r) represents an ordered sequence of cut lines. This objective function is highly dependant on the locations of cut lines, as well as chip geometry.

Breuer's algorithm has been developed to handle regularly structured chips, such as gate-array or poly-cell LSIs. However, it is not easy to apply the algorithm to irregularly shaped macro placement. Therefore, modified algorithms have been presented which are suitable to macro placement.

4.5.1 Basic Algorithm[22]

The algorithm is an iterative improvement method, which starts with an initially given placement result and improves it by a local transformation.

It is assumed that a chip consists of cells and that every block has been assigned in a cell. Consider the chip shown in Fig. 25, where some arbitrary assignment of blocks to cells has occured. An assignment of blocks to cell locations is refered to as a temporary assignment, denoted by T-assignment. Now, line C divides are A into two areas A1 and A2.

The basic problem is to decide on a re-assignment of blocks in A to cells in either A1 or A2, so that the total number of signal nets crossing C is minimized, subject to the constraint that blocks not in A cannot be moved. New assignments of blocks in A1 and A2 are again considered to be T-assignments.

An area A, along with the cells and blocks within A is called a group denoted by g. (In [22], it is called a block. In this paper, it is called a group to avoid conflict.) A block is considered placed when the group it is in has only one cell.

Let G be a set of disjoint groups. Initially, let the entire chip be group g1, and set G={g1}. The basic algorithm is as follows.

1) Select a sequence for processing the cut lines.
2) Select next cut line C in sequence.
3) Assume C cuts across a subset of groups G'={gi1, gi2, $\cdots$, git}. Re-assign blocks within these groups, such that v(C) is minimized.
4) In a natural way, form two new groups from each of the groups cut by C.
5) If there are no more cut lines to process, or if every group contains at most one block then stop, or else return to 2).

The sequence in which cut lines are to be processed can be either fixed or adaptive.

Fig. 26 shows an example of this algorithm along with five cut lines which are processed in a sequence C1, C2, C3, C4, and C5. Starting with the initial group g1, consisting of the entire chip and all blocks, g1 is processed with respect to C1, hence minimizing v(C1). Group g1 is now divided into two new groups, denoted by g1 and g2. Note that once a block has been assigned to the left (the right) of C1, the block can never be moved to the other side of C1, no matter where subsequent cut lines occur. Now g1 and g2 are processed (simultaneously) with respect to C2, producing the four groups in Fig. 26(d). Next C3, which only intersects g3 and g4, is processed. Note that the blocks in g1 and g2 are not moved. Therefore, the locations of those blocks need not be known when calculating v(C3). Continuing these process, the resulting 12 groups shown in Fig. 26(g), are obtained.

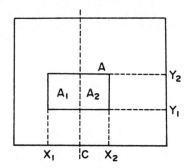

Fig. 25　Chip and Groups

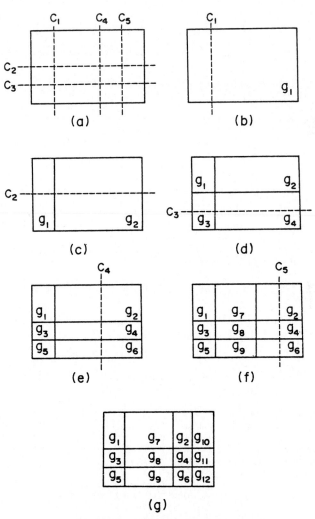

Fig. 26　Min-Cut Algorithm Example

In this algorithm, two key problems exist. One problem is to assign the block in a group g to two groups, g_1 and g_2, such that the number of signal nets crossing across a cut line C is minimized. This problem is a generalization of the following partitioning problems.

Given a graph G having n nodes, partition the set of nodes into two disjoint sets N_1 and N_2 of nodes having n_1 and n_2, respectively, where $n_1 + n_2 = n$, and such that the number of edges between N_1 and N_2 is minimal.
Heuristic algorithms for this problem have already been discribed in Sec. 2. 3.

Another key problem is to select cut lines and sequence in which cut lines are processed. The selection criteria is usually a function of the chip geometry and predicted routing density. For fixed ordering, it has been found that two types of cutlines are quite effective; they are refered to as a slice cut and a bisection cut. Slice cut C for a group g is a cut line which isolates a fixed number (K) of cells of g to one side of C and the remaining cells to the other side of C. A bisection cut C of group is a cut line which divides the blocks in g into either side of C as evenly as possible.

Two procedures are recommended by Breuer: Quadrature procedure and slice/bisection procedure. They differ in the order in which groups are processed and the type of cut lines employed.

In the quadrature procedure, the original group (entire chip) g_1 is first bisected by a vertical cutline producing two new groups g_1 and g_2. These two groups are then cut by horizontal bisection cut, thus producing four groups, which are next cut by vertical bisection cut. This process is repeated, until every block is placed.

This procedure tends to produce a placement which can be routed with a more uniform density. Therefore, it is suitable for chips having a high routing density in their center.

In slice/bisection procedure (see Fig. 27), initial set of n blocks is divided into a set of K blocks and (n-K) blocks, where $K>0$, such that v(C) is minimized. These K blocks represent the bottom row or slice of chip. This procedure is repeated on the remaing (n-K) blocks, again dividing them into a set of K blocks and (n-2K) blocks. This process is iterated until all blocks have been assigned to rows. The blocks are then assigned to columns via vertical bisection. This technique is best fit for chips where there is a high interconnected density at the external terminals.

These techniques have been widely used in designing gate-array and poly-cell LSIs.

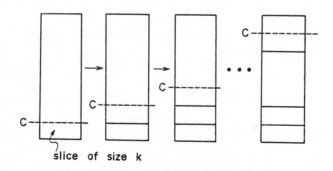

Fig. 27   Slice/Bisection Procedure

4.5.2 Min-cut algorithms for Macro Placement [24]

The basic algorithms described in the previous section are suitable for block placement, where each block has the same size or the same height even though it has different width. On the other hand, macro placement problem has to deal with irregular components which has various kind of shapes or sizes ranging from smaller ones to much larger ones. In addition,exact block location cannot be defined ,since chip shape is obtained after completing placement. Therefore, to apply the min-cut algoritm to macro placement, various block size must be taken into account and particular data structure has to be adopted, which represents the block locations.

Polargraph has been introduced to represent the macro placement by the first author, whose idea was implemented in the routing program ROBIN [23]. Since that time, several macro placement algorithms, based on graph representation, have been proposed [24] [25] [26]. Most of them are based on the min-cut algorithm with some modification or top-down approach.

Throughout the placement, macro placement is represented by a pair of mutually dual graphs $G_x = (V_x, E_x)$ and $G_y = (V_y, E_y)$. G_x and G_y are planar, acyclic directed graphs containing one source and one sink each; parallel edges are permitted. There is a one to one correspondence between the edges of G_x and G_y. Each pair of edges $(e_x{}^i, e_y{}^i)$ represents a rectangle Ai with x-dimension $\ell(e_x{}^i)$ and y-dimension $\ell(e_y{}^i)$, where $\ell(e)$ denotes the length associated with edge e. On the other hand, each pair of edges is defined to correspond to also a group gi; the area $\ell(e_x{}^i) \cdot \ell(e_y{}^i)$ equals the total area of the macros in the associated gi.

When the algorithm starts, the two graphs contain one edge. This pair of edges represents initial group g with all macros of the circuit. The area covered by group g is assumed to be square.

Therefore,

$$\ell(e_x^1) = \ell(e_y^1) = \sqrt{\sum_i a_x^i \cdot a_y^i}$$

where $a_x{}^i$ and $a_y{}^i$ are the dimensions of macro i (Fig. 28 a, b). Group g is now partitioned into two groups, g_1 and g_2, in such a way that the number of nets incident to macros in different groups is minimal and that the difference between total macro areas in the two groups does not exceed a predefined threshold value.

In the graph representation, this step corresponds to a splitting of the edge pair into two new edge pairs, each representing one of the two groups . The lengths of the edges in Gx are adjusted according to the total cell area in the respective group. Note that the two resulting rectangles in Fig. 28(d) have different areas.

In the next step, the partitioning procedure is applied to both of the groups. Now, however, the direction of the cutline is changed and the edge lengths in G_y are adjusted. The resulting situation is shown in Figs. 28(e) and (f). Group partitioning is carried out sequentially, alternating between the two groups, and iterating the process until the overall count of nets cut cannot be further reduced.

The described procedure is applied recursively to the new groups and terminates when each group contains one macro (Figs. 28(g) and (h)). Now, each edge-pair of the two graphs corresponds to one macro and the area 1 $(e_x{}^i) \cdot 1 (e_y{}^i)$ of the corresponding rectangle is equal to the area of the respective macro i, whereas the shape of the cell (length to width ratio) is not correctly represented.

It is easy to convert this representation to actual macro dimensions, while maintaining the local relations as derived by the algorithm.

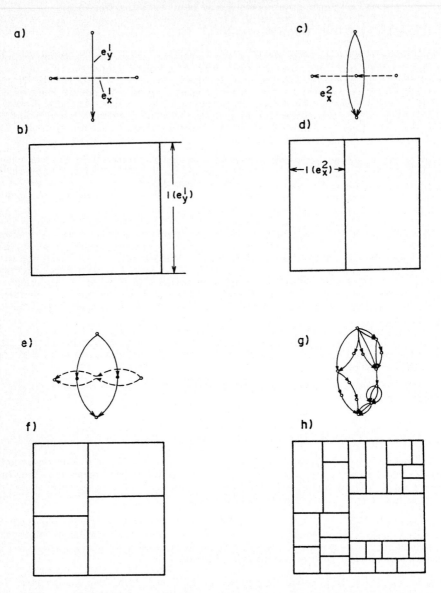

Fig. 28 Macro Placement and Associated Graphs

4.6 Minimum Density Algorithms

In the case of the minimum density goal, the whole chip is divided into a rectangular array of cells, each of which contains some grid lines and circuit pins. Global routing determines, for each signal net, a path of cells which the net will be routed through. The boundary between two cells is called a segment. To each sigment, the number of allowable signal nets, called capacity, is given. The routing path of a signal net is expressed as a sequence of segments. Thus, if the routing path for every signal net is given, the number of signal nets assigned to each segment can be calculated.

In this section, the same problem assumption as in Sec. 4.4 is adopted. We will rewrite Eq. (13) into a new form which is easy for the density calculation.

We now introduce the following symbols (see Fig. 29):

$e_{ij}{}^x$: X-directional segment, where $i=1, 2, \cdots m; j=1, 2, \cdots, n-1$.

$e_{ij}{}^y$: Y-directional segment, where $i=1, 2, \cdots m-1; j=1, 2, \cdots, n$.

$C_{ij}{}^x$: the number of signal nets segment $e_{ij}{}^x$ can accomodate

$C_{ij}{}^y$: the number of signal nets segment $e_{ij}{}^y$ can accomodate

X_{ij}: the number of signal nets assigned to segment $e_{ij}{}^x$

Y_{ij}: the number of signal nets assigned to segment $e_{ij}{}^y$

$f(X_{ij}, r)$: the cost of $e_{ij}{}^x$ when X_{ij} signal nets are assigned to segment $e_{ij}{}^x$, where r is a congestion parameter

$f(Y_{ij}, r)$: the cost of $e_{ij}{}^y$ when Y_{ij} signal nets are assigned to segment $e_{ij}{}^y$, where r is a congestion parameter

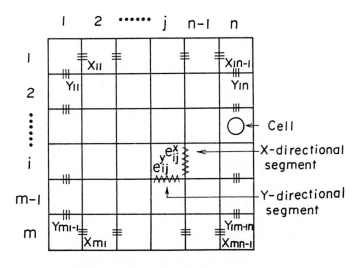

Fig. 29 A chip Structure.

$X_{ij}/C_{ij}{}^x$ and $Y_{ij}/C_{ij}{}^y$ are called segment densities. The cost function is assumed to satisfy the following equation;

$$\lim_{r \to \infty} f(X_{ij}, r) = \begin{cases} 0 & : X_{ij} \leq C^x_{ij} \\ \infty & : X_{ij} > C^x_{ij} \end{cases} \tag{22}$$

That is, when the segment density is greater than 1 (the number of signal nets assigned exceeds the number of signal nets that can be accomodated), the cost becomes quite high if r becomes large. When the segment density is less than 1 (the number of signal nets assigned is smaller than the number of signal nets that can be accommodated), the cost approaches 0 as r becomes large. There are a number of functions that satisfy Eq. (22). Here we use Eq. (23) (see Fig. 30).

$$f(X_{ij}, r) = \left(\frac{X_{ij}}{C_{ij}} \right)^r \tag{23}$$

Cost function $f(Y_{ij}, r)$ can be defined similarly. The cost Fr for the entire segment is defined by the following equation and we take as an objective function the minimization of this cost:

$$F_r = \left(\sum_{i=1}^{m} \sum_{j=1}^{n-1} f(X_{ij}, r) + \sum_{i=1}^{m-1} \sum_{j=1}^{n} f(Y_{ij}, r) \right)^{\frac{1}{r}} \tag{24}$$

$$= \left(\sum_{i=1}^{m} \sum_{j=1}^{n-1} \left(\frac{X_{ij}}{C_{ij}} \right)^r + \sum_{i=1}^{m-1} \sum_{j=1}^{n} \left(\frac{Y_{ij}}{C_{ij}} \right)^r \right)^{\frac{1}{r}}$$

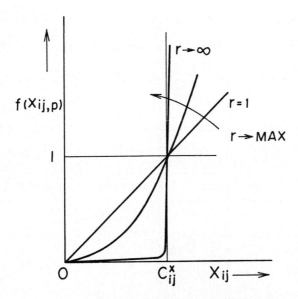

Fig. 30 Cost Function $f(X_{ij}, r)$.

The problem in this section is to place blocks on a chip so that F_r is minimized for a given value of r. By setting an objective function in this manner, we can evaluate various placements on a common ground. For example, if we let $r=1$ and $C_{ij}{}^x = C_{ij}{}^y = C$ in Eq. (24), then we have

$$F_1 = \frac{1}{C}\left(\sum_{i=1}^{m} \sum_{j=1}^{n-1} X_{ij} + \sum_{i=1}^{m-1} \sum_{j=1}^{n} Y_{ij} \right) \qquad (25)$$

Equation (25) represents the total routing length.

If we let $r \to \infty$, then

$$F_\infty = \underset{ij}{MAX}\left(\left(\frac{X_{ij}}{C_{ij}^x} \right), \left(\frac{Y_{ij}}{C_{ij}^y} \right) \right) \qquad (26)$$

Equation (26) gives the maximum segment density.

4.6.1 Calculation for density

Strictly speaking, the goodness of a placement should be judged by whether or not 100% wiring is possible when an ideal routing is taken. However, this makes the problem very complex. Thus, in practice, we introduce a set of criteria which approximately evaluate placements and isolate the placement problem from the routing problem. [27]

First, we calculate for each segment the number of signal nets which pass it. We make the following assumptions for routing since they produce routings close to actual cases using simpler calculations.

Assumption 1. The routing path of a signal can be determined independently of the routing paths of the other signal nets.

Assumption 2. The routing path of a signal is inside the smallest square containing the blocks connected by the signal net.

Assumption 3. The direction of a routing path is parallel to either the x-axis or the y-axis and it can make at most two turns.

Assumption 4. When more than one routing path is possible for one signal net, each of them has an equal probability of occurrence.

Complex processes and hence considerable computation time are required to consider the effects of signal nets on each other. Hence Assumption 1 is adopted. Also, most of the routings in practice satisfy Assumptions 2 and 3. Assumption 4 means that there are no priority orders among routing paths satisfying Assumptions 2 and 3. If priority needs to be given to a particular routing path, we can assign a higher probability to the routing path.

First we obtain the number of wires for the case of two-point wirings under Assumptions 2 and 3. Let A and B be blocks and let (x_A, y_A), (x_B, y_B) be the cells for them, respectively. There are four possible combinations of their positions, but we consider only the case $x_A \leq x_B$, $y_A \leq y_B$. The other three cases can be treated similarly. Let l_{AB} be the number of routing paths connecting A and B. Then

$$l_{AB} = \begin{cases} 1 : x_A = x_B \ \ or \ y_A = y_B \\ (x_B - x_A) + (y_B - y_A) : x_A = x_B \ \ and \ y_A = y_B \end{cases} \tag{27}$$

Let $l(e_{ij}x)$ be the number of signal nets passing through segment $e_{ij}x$. Then it can be obtained by the following equations:

(1) If $x_A = x_B$, then

$$l(e_{ij}^x) = 0 \tag{28}$$

(2) If $x_A \neq x_B$, and $y_A = y_B$, then

$$l(e_{ij}^x) = \begin{cases} 1 : x_A \leq i \leq x_B - 1 \ \ and \ j = y_A \\ 0 : otherwise \end{cases} \tag{27}$$

(3) If $x_A \neq x_B$, and $y_A \neq y_B$, then

$$l(e_{ij}^x) = \begin{cases} x_B - i : x_A \leq i \leq x_B - 1 \ \ and \ j = y_A \\ i - x_A + 1 : x_A \leq i \leq x_B - 1 \ \ and \ j = y_B \\ 1 : x_A \leq i \leq x_B - 1 \ \ and \ y_A < j < y_B \\ 0 : otherwise \end{cases} \tag{30}$$

The number of signal nets $l(e_{ij}y)$ passing through $e_{ij}y$ can be obtained similarly. Let X_{ij} and Y_{ij} be the number of signal nets assigned to segments $e_{ij}x$ and $e_{ij}y$, respectively, for a signal net connecting blocks A and B. Then from assumption 4 we can obtain

$$\left. \begin{aligned} X_{ij} &= l(e_{ij}^x) / l_{AB} \\ Y_{ij} &= l(e_{ij}^y) / l_{AB} \end{aligned} \right\} \tag{31}$$

Figure 31 enumerates routing paths between blocks A and B and $l_{AB} = 5$.

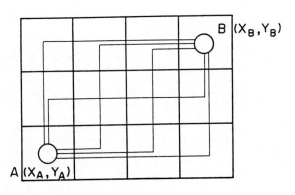

Fig. 31   Routing paths between Block A and B.

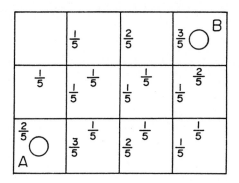

Fig. 32 Number of Signal Nets assigned to each Segment.

Figure 32 shows the number of signal nets that a routing path connecting blocks A and B assigns to each segment. If there are multipoint wirings among given signal nets, they are decomposed into two-point wiring by the method given in section 4.2. Then, the number of signal nets to be assigned to each segment is calculated by Eqs. (27) to (31) and the total number of signal nets for each segment is obtained. This value corresponds to X_{ij} and Y_{ij} and Eq. (24) can now be calculated.

4.6.2 Algorithms and experimental results

Since Eq. (26) is considered to be an extreme case of Eq. (24), GFDR method mentioned in Sec. 4.4 is efficiently applied. We will describe the results obtained in [27]. Here, Example 5 of Table I was provided as a test and $C_{ij}^x = C_{ij}^y = 10$ was used.

(1) The total routing length and the maximum density

We gave a value to the density parameter r, let $\lambda^* = 4$ and $\varepsilon = 4$. As a result, a placement of blocks was obtained and the total routing length and the maximum segment density were calculated for that placement. For the value of r, 1, 2, 4 and 8 were used. The total routing length and the vertical and horizontal maximum segment densities were obtained as shown in Fig. 33. The maximum density becomes smaller as the total routing length becomes longer. It has been found that for r, $2 \leq r \leq 4$, the total routing length is not large and neither is the maximum density.

(2) Distribution of segment density

The distribution of values of the segment densities (occurrence frequencies) against the change in the wire congestion parameter r is given in Fig. 34. From the figure we can see that for any value of r, the density exceeds 1 for about half the segments. Hence for $C_{ij}^x = C_{ij}^y = 10$, 100% wiring seems impossible. If we set the values of C_{ij}^x and C_{ij}^y at about 15 (in this case the distribution of segment density is obtained by dividing by 1.5 the values of the horizontal axis of Fig. 34), less than 10% of the segments have density greater than 1.0. Thus we can expect 100% wiring if a good routing algorithm is used. When r = 1 or 2, the wire congestion of segments has an average distribution. But it does not give good placements in the sense that the maximum density of segments is not minimized.

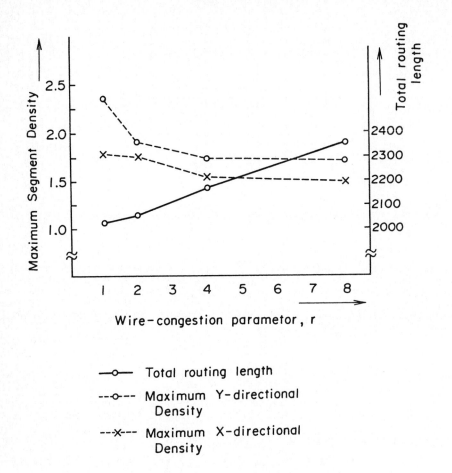

Fig. 33 Relation between the Maximum Segment Density
and the Total Routing Length.

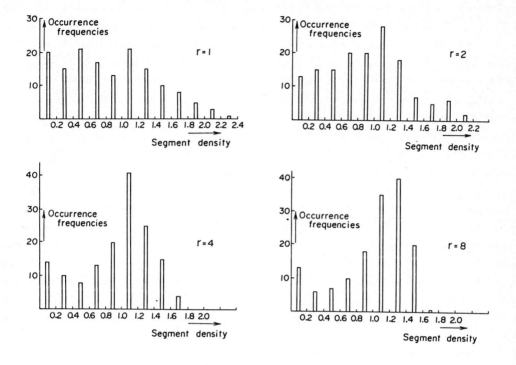

Fig. 34   Occurrence Frequencies of Segment Density

with regard to Wire Congestion Parameter r.

5. CONCLUSION

This chapter dealt with partitioning, assigment and placement problems and gave the current status of solving methods. Although those three problems are closely related, they were treated separately in practical cases because of its inherent computational complexities of the total problem. In this chapter, each problem was formulated as a mathematical problem and heuristic solving methods were given. Each problem is considered to be hard combinatorial problem (NP-Complete), and hense algorithms based on heuristic rationales have been employed.

Although these problems have been attached by many researchers in the past 20 years, the results obtained from heuristic algorithms are still not satisfactory, in comparison to an expert designer's result. This is because an expert designer can well understand really complicated problems with numerous constraints, and he can find a good solution by exploiting his knowledge and intuition.

Under this situation, the authors believe that the following approaches are indispensable.

(1)   The design problems themselves should be changed or simplified to make the problems easy solvable. One typical example is a gate array LSI designed with sufficiently large chip size for easy routing.

(2)   More research should be carried out to investigate various objective functions for the optimum layout. Three different objective functions are presented for the placement problem. However, more detail relations among them should be made clear. Altogether, the research should be forcused toward finding the most appropriate objective function which reflects the complete net connectivity.

(3)   If using algorithmic approaches make it essentially impossible to get better results, and real design would require much better result, a new approach should be divised, which would be substantially different from conventional algorithmic approaches. The authors believe that the knowledge based approach incorporated with an algorithmic approach represents the most hoepful from now on.

ACKNOWLEDGMENTS

The author would like to thank Prof. E. S. Kuh of University of California, Berkeley, Prof. T. Ohtsuki of Waseda University, Drs. Y. Kato, M. Naniwada and colleagues of NEC Corporation for their encouragement and useful suggestions.

REFERENCES

[1]   Kodres, U. R., "Partitioning and Card Selection," Design Automation of Digital Systems (Ed. by M. A. Breuer), Prentice - Hall (1972).

[2]   Lawler, E. L., "Electrical Assemblies With a Minimum Number of Interconnections," IEEE Trans. on Electronic Computers (Correspondence), Vol. EC-11 (February, 1962), pp. 86-88.

[3]   Lawler, E. L. et al., "Module Clustering to Minimize Delay in Digital Networks," IEEE Trans. on Computers, Vol. C-18 pp. 47-57 (January, 1969).

[4]   Luccio, F. and M. Sami, "On the Decomposition of Networks in Minimally Interconnected Subnetworks." IEEE Trans. on Circuit Theory, Vol. CT-16, No. 2, May, 1969.

[5]   Kernighan, B. W. and S. Lin, "An Efficient Heuristic Procedure for Partitioning Graphs." Bell Sys. Tech. J., Vol. 49, pp. 291-307 (1970).

[6]   Russo, R. L. "A Heuristic Procedure for the Partitioning and Mapping of Computer Logic Gates," IEEE Trans. Comput., Vol. C-20, No. 12, pp. 1455-1462 (1971).

[7]   Schweikert, D. G. and B. W. Kernighan, "A Proper Model for the Partitioning of Electrical Circuits," Proc. of 9th DA Workshop, pp. 57-62 (1972).

[8]   Charney, H. R. and D. L. Plato, "Efficient Partitioning of Components," Proc. of 5th DA Workshop, pp. 1-21 (1968).

[9]   Donath, W. E. and A. J. Hoffman, "Lower Bounds for Partitioning of Graphs," IBM Journal of Res. and Dev. 17, pp. 420-425 (1973).

[10]   Payne, T. S. and W. M. vanCleemput, "Automated Partitioning of Hierarchical Specified Digital System, Proc.19th Design Automation Conf., pp. 183-192 (1982).

[11]   McFarland, M. C., "Computer-Aided Partitioning of Behavioral Hardware Descriptions, Proc. 20th Design Automation Conf., pp. 472-478 (1983).

[12]   Soukup, J., "Circuit Layout," Proc. of IEEE, Vol. 69, No. 10, pp. 1281-1304 (1981).

[13]   Tsukiyama, S., M. Fukui and I. Shirakawa, "A Heuristic Algorithm for a Pin Assignment Problem of Gate Array LSI's," Proc. of ISCAS 84, pp. 465-469 (1984).

[14]   Hanan, M. and J. M. Kurtzberg, "Placement Techniques," Chap. 5 in Design Automation of Digital Systems, 1 (Ed. by Breuer, M. A.), pp. 213-282, Prentice Hall (1972).

[15]   Hanan, M., P. K. Wolff, Sr. and B. J. Agule, "Some Experimentl Results on Placement Techniques," Proc. of 13th DA Conference, pp. 214-224 (1973).

[16]   Goto, S., "An Efficient Algorithm for the Two-Dimensional Placement Problem in Circuit Layout," IEEE Trans. Circuits & Syst., CAS-28, 1, pp. 12-18 (1981).

[17]   Hakimi, S. L., "Optimum Locations of Switching Centers and the Absolute Centers and Medians of a Graph," Oper. Res., 12 pp. 450-459 (1964).

[18]   Johnson, D. B. and T. Mizoguchi, "Selecting the kth Element in $X + Y$ and $X_1 + X_2 + \cdots + X_m$, SIAM J. Computing, 7, 2, pp.141-143 (1978).

[19]   Lin, S. and B. Kernighan, "An Effective Algorithm for Travelling-Salesman Problem," Oper. Res., 11, pp. 498-516 (1973).

[20]   Stevens, J. E., "Fast Heuristic Techniques for Placing and Wiring Printed Circuit Boards," Ph. D. Dissertation Comp. Sci., Univ. of Illinois (1972).

[21]   Goto, S. and E. S. kuh, "An Approach to the Two-Dimensional Placement Problem in Circuit Layout," IEEE Trans. Circuits Syst., CAS-25, pp. 208-214 1978).

[22]   Breuer, M. A., "A class of Min-cut Placement Algorithms," Proc. 14th DA Conference, pp. 284-290 (1977).

[23]   Kani, K., H. Kawanishi, and A. Kishimoto, "ROBIN: A Building Block LSI Routing Program," Proc. IEEE ISCAS, pp. 658-660 (1976).

[24]   Lauther, U., "A Min-cut Placement Algorithm for General Cell Assemblies Based on a Graph Representation," Proc. 16th DA Conference, pp. 1-10 (1979).

[25]   Preas, B. T. and C. W. Gwyn, "Methods for Hierachical Automatic Layout of Custom LSI Circuits Masks," Proc. 15th DA Conf., pp. 206-212 (1978).

[26]   Sato, K., T. Nagai, H. Shimoyama, and T. Yahara, "MIRAGE-A Simple-Model Routing Program for the Hierachical Layout Design of IC Masks," Proc.16th DA Conf., pp. 297-304 (1979).

[27]   Jung, J., S. Goto and H. Hirayama, "A New Approach to the Two-Dimensional Placement Problem of Wire Congestion in Master-Slice LSI Layout Design, Trans. Inst. of Electronics and Communications Engineers of Japan, Vol. J64-A, No. 1, pp. 55-62 (1981).

LAYOUT DESIGN AND VERIFICATION
T. Ohtsuki (Editor)
© Elsevier Science Publishers B.V. (North-Holland), 1986

Chapter 3

MAZE-RUNNING AND LINE-SEARCH ALGORITHMS

Tatsuo OHTSUKI

*Department of Electronics and Communication Engineering
Waseda University, Tokyo, Japan*

1. Introduction

This chapter is devoted to review the maze-running and line-search algorithms and their variations. They are the most classic, but still up-to-date, algorithms for detailed routing.

As far as the author knows, the Lee algorithm [1] is the first one aimed at automated wire routing. It is actually an application of the shortest path algorithm presented by Moore [2] to a grid structure representing wiring space. The original Lee algorithm was followed by a large number of contributions in order to extend it to variations of path selection criteria. These algorithms are often referred to as grid expansion or maze (running) algorithms due to their similarity to finding an entrance-to-exit route in a grid-structured maze.

Even though the original Lee's idea was created in a pre-IC period and has been applied to printed circuit board design, it is still alive in the current VLSI design as the Lee algorithm with its extensions is very general and guarantees finding a path if one exists. Because of these attributes, many existing routers use the Lee algorithm exclusively, or initially use some other algorithms to rapidly interconnect most of the nets, and then employ it in order to interconnect the remaining nets.

The major disadvantage of the maze algorithms lies in that they need a great deal of memory and running time for finding a connecting path. The negative feature inherent in maze algorithms has given rise to another class of routing tools which are referred to as line search algorithms [3], [4]. Unlike maze algorithms in which a path is represented by a sequence of grid points, the line search algorithms search a path as a sequence of line segments. Consequently, they save memory and quickly find a simple-shaped path. On the other hand, it is generally understood that this approach does not guarantee finding a path even if one exists.

In Section 2, the original Lee algorithm dealing with the simplest single-layer interconnections is introduced. This is followed by Sections 3 and 4 in which its extensions to more general routing problem with various constraints and path selection criteria are reviewed.   In Section 5,   the line-search algorithms, together with the line-expansion algorithm (known as a recent improvement of them),   are described and compared with the maze-running algorithms in terms of speed and memory requirement. Finally, the controversial topic on the ordering of nets to be routed is discussed and then recent rip-up and reroute techniques aiming at 100% routing completion are described in Section 6.

2. The Original Lee Algorithm

The major constraints imposed on the dimensions of nets to be wired are characterized by two values: a minimum width w and a minimum clearance c,   which must be maintained between wires.   It is assumed for standardization that these two values are fixed throughout all the nets,   and then combined into a single center-to-center constraint $\triangle = w + c$.   This leads to a uniform rectangular grid having an incremental spacing of $\triangle$.   Then,   by requiring that all the nets are routed following the lines of this grid,   the width and clearance constraints will be satisfied automatically.   It is also assumed that the pins (terminals) of nets are located on grid intersection points.   The use of such a grid automatically precludes any curved or diagonal runs for the nets which are typically seen in manual layouts.   However,   this loss of generality makes the routing problem well tractable by automation.

The original Lee algorithm was dealing with single layer interconnections on a planar rectangular grid.   It is sometimes referred to as a <u>wave propagation</u> <u>method</u> due to its similarity to the wave created by dropping a pebble into a pool of water. We shall see this interpretation by taking an example of Fig. 1.   Let A and B be a pair of terminals on a net.   We want to know whether there exists a path connecting them and,   if it is the case,   to find the shortest one.   At this

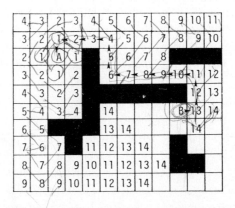

Fig. 1 Lee algorithm.   Target B is labeled with "14".

point, it is convenient to displace the grid intersection points by square cells so that the numbers representing labels can be entered into the "empty" cells. Here the "empty" cells indicate the grid points available for routing, whereas the "occupied" cells, blacked out in this figure, indicate those occupied by obstructions or already-routed nets.

The algorithm may be decomposed into three distinct phases. The first phase, called wave propagation, begins with selecting either pin, say A, as the "source"; the other pin B is regarded as the "target". Then we simulate the wave created at the source A propagating. Initially, a "1" is entered in every empty cell immediately adjacent to the cell A. Next, a "2" is entered in every cell adjacent to those labeled "1"; and so on. This process is continued until one of the following two results occurs. Either in the k th step no empty cell adjacent to those labeled "k-1" exists, or the target B is reached. In the former case, we know that there is no path between A and B. In the latter case, as in the example of Fig. 1, we know two facts: There is a path from A to B, and the length of the shortest one is equal to the number entered in the target cell, which in this case is 14.

A key property of the above procedure is that the set of same numbered cells forms a wavefront and the number (label) entered in each cell represents the distance from the source. This process can be viewed as a breadth-first-search [5] in graph theory, as the nodes (cells) closer to the starting point (source) are searched with higher preference.

The next phase of the algorithm, called backtrace, is to actually find a shortest path. Since B was reached on the 14th step in this example, it follows that there must be a cell with "13" adjacent to B. Likewise, such a cell with "13" must be adjacent to a cell with "12", etc. By tracing the numbered cells in descending order from B to A in this way, it is easy to find a desired shortest path. In the backtrace phase, there is generally a choice of cells when there are two or more cells with "k-1" adjacent to a cell with "k". Theoretically any of these cells may be chosen and a shortest path will still be found. However, a practical guideline is "not to change direction" unless one has to do. Thus, in the example of Fig. 1, the path indicated by arrows is found. This rule tends to minimize the number of bends among the shortest paths. Once a desired path has been found, those cells used for the route connecting A and B are regarded as "occupied" for subsequent interconnections.

The Lee algorithm for routing a single net is completed by the last phase, called label clearance, in which all the labeled cells, except those used for the path just found, are cleared to "empty" for subsequent interconnections.

As is clear from the above arguments, the Lee algorithm requires $O(N^2)$ memory for an N X N grid plane. In addition, some non-trivial amount of temporary storage is needed to keep the positions of the cells currently on the wavefront.

It also requires $O(N^2)$ running time in the worst case. To be more precise, $O(L^2)$ time is needed to find a path of length L in the wave propagation and label clearance phases; the backtrace phase runs in $O(L)$ time.

3. Extensions of Lee Algorithm

There are a large number of variations [6]-[13] possible on the original Lee algorithm. Among various extensions, some basic ones are described in this section according to the author's selection, rather than a comprehensive compilation.

3.1 Some speed-up techniques

A straight-forward idea for time-saving implementation of the Lee algorithm is to reduce the number of cells being searched in the wave propagation phase, since the running time is proportional to this number. The following three speed-up techniques, based on this idea, are commonly used and can be easily implemented with a slight modification of the original algorithm [8].

(1) Starting point selection. For a given pair of terminals, start the wave propagation on the one farther from the center of the grid plane. If the source thus selected is close enough to the frame, the area of wave propagation is bounded by it.

(2) Double fan-out. Generate waves at both of the terminals and continue labeling until a point of contact is reached. Usually, this technique halves the area of wave propagation.

(3) Framing. An artificial rectangular boundary is imposed about the terminal pair, and no labeling is allowed beyond this boundary. Typically this frame is 10 to 20% larger than the rectangle defined by the terminal pair. This approach usually speeds up the search considerably. Of course, if no path is found, the frame should be removed so as to continue the search.

These techniques save the running time of the Lee algorithm to some degree. But more essential speed-up techniques will be described in Section 4.

3.2 Connecting a multi-pin net

When a net with three or more terminals (called a multi-pin net) is to be interconnected, we encounter the problem: How should these terminal points be interconnected with a minimum total wire length ? This problem is called Steiner tree problem in graph theory and known to be NP-hard [14]. However, it is not the ultimate purpose in practice to achieve the strict minimum but to limit the total length less than some prescribed value. In this sense, the Lee algorithm can be used to get a sub-optimal interconnection in the following way [8].

Consider the five terminals shown in Fig. 2. In ▓▓▓▓ly one terminal, say A, is chosen as a source. Then a wave is propagated starting from A with the other four terminals as targets. It continues until the one of the targets (in this case, B) is reached. Then the path thus obtained (route 1 in Fig. 2(a)) is laid out. Now all cells in this route are marked as source points with the other three terminals remaining as targets. Again the wave is propagated starting from the path A-B until the first target point (C) is reached. All cells in the path A-B-C wire now become sources, and this process continues in this way. Fig. 2(a) shows the final pattern of the interconnection with the numbers indicating the order in which the paths were laid out.

The interconnection obtained by this process is not always (even in this example) of minimum length. There is, however, a simple technique which will often allow a shorter interconnection to be found. It is well known in graph theory that removing any segment from a tree will result in two sub-trees. Then the shortest path between these sub-trees can be found by applying the Lee algorithm again with all cells in one serve as sources and all cells in the other serve as targets. If the shortest path is shorter than the length of deleted segment, then by inserting this path, a new shorter length interconnection is obtained. Applying this technique to the segment between sub-tree A-E and sub-tree B-C-D (Fig. 2(a)), a shorter interconnection as shown in Fig. 2(b) is found.

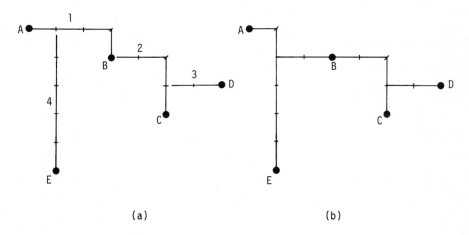

(a)   (b)

Fig. 2 Connecting a multi-pin net.

3.3 Multi-layer interconnection

One of the most useful variations of the Lee algorithm involves its extension to three-dimensional grid --- in particular, to a multi-layer interconnection through via holes. A straight-forward way for routing on a three-dimensional grid is to consider a cellular array consisting of unit cubes as shown in Fig. 3, where

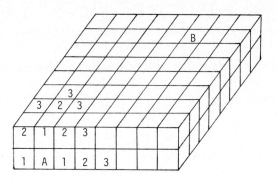

Fig. 3 Cellular array for two-layer routing.

the two-layer case is illustrated [15].   Note that the unit squares representing
grid points in the single-layer case in Section 2 were replaced by the unit cubes.
As before a pair of source and target cells is given.   To find the desired path
using the Lee algorithm, we follow the same wave propagation as in Section 2,
except that all adjacent "empty" cells including those accessible from above or
below must be labeled at each step (see Fig. 3).   This process leads to a path
which occupies a minimum number of cells.   This implies that an inter-layer
connection through a via-hole is assumed to have the same weight as the unit
length on one layer.

The three-dimensional cellular array involves considerable overhead in memory
and running time, although it is very general.   A simpler extension of Lee
algorithm to two-layer routing is to use a cell with a couple of superposing
squares at each grid point on the plane, and to neglect the weight for inter-layer
connection through via-holes [7].   Moreover, we assume that pins are accessible
from both of the layers.   Then it can be easily proved that any coupled two grid
points, one on the first layer and the other on the second layer, have the same
rectilinear distance from a source point.

An example of applying the two-layer Lee algorithm based on this model is
illustrated in Fig. 4(a), (b), in which two planar cellular arrays are used.   The
process of wave propagation is same as the single-layer Lee algorithm except that
(1) wavefronts are advanced on both of the layers simultaneously and (2), whenever
an "empty" square is labeled, the same number must be entered in the corresponding
square on the other layer unless it is "occupied".   The shortest path obtained in
this way is shown in Fig. 4(c).

The two-layer Lee algorithm outlined above is quite useful and easily coded.
One disadvantage is that any path entering a cell on one layer and returning the
same cell on a different layer (see Fig. 5) will not be found.   Note that such a
path makes sense when a via cannot be placed at immediate neighbor of occupied

cells due to some minimum spacing rule. Even in this case, such a path rarely exists in practice.

Years of manual experience with printed circuit board routing problems have shown that the most efficient approach is to route most of horizontal runs on one layer and most of vertical runs on the other. Two layer routing following the horizontal-vertical rule strictly can be achieved by blocking vertical runs on one layer and horizontal runs on the other at the wave propagation phase of the Lee algorithm (see Fig. 7(a)). A practically attractive way of implementation is to connect most of the nets following this horizontal-vertical rule and, after congested, to connect the remaining nets by means of the general two-layer Lee router.

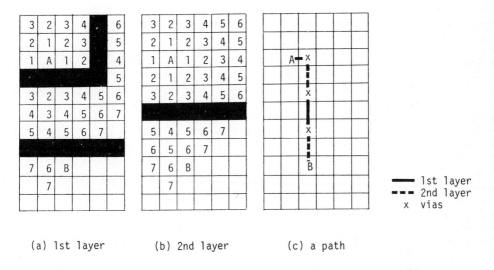

(a) 1st layer   (b) 2nd layer   (c) a path

Fig. 4 Two-layer Lee algorithm.

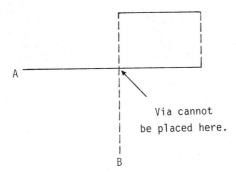

Fig. 5 A path not obtained by Geyer's router.

3.4 Coding schemes for labeling

A nontrivial storage problem lying in the Lee algorithm is that a unit of memory space is needed for every cell (grid point). This problem is serious for large-scale grids. Suppose a 1000 X 1000 grid plane (a million cells) per layer is given for routing. Then, if wire lengths up to 1000 are expected, each cell may be labeled by a number as large as 1000 during the wave propagation phase. This implies that at least 10 bits must be allocated to each unit of memory. However the following observation leads to more efficient storage schemes for labeling information.

Consider the labeling process illustrated in Fig. 1 again. Then we see that, for each cell (labeled "k"), all adjacent cells are labeled either "k-1" or "k+1". Thus it is only necessary in the backtrace phase to distinguish the predecessor cells from the successor cells. The following coding schemes based on this idea are widely used.

(1) At each step of wave propagation, the adjacent "empty" cells are labeled with   ←(left), →(right), ↑(up), or ↓(down), which indicates the direction of wave propagation. If a cell is reached from two or more directions, one of them is entered.

(2) A sequence: 1,2,3,1,2,3,... is exploited for labeling. The example of Fig. 1 labeled in this sequence is shown in Fig. 6(a).

(3) A sequence: 1,1,2,2,1,1,2,2,... is exploited for labeling. The example of Fig. 1 labeled in this sequence is shown in Fig. 6(b). Note that, in this example, target B has been reached with a "1" preceded by another "1". Therefore the path must be backtraced in the sequence: 1,1,2,2,1,1,.....

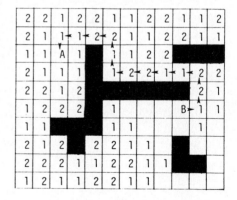

(a) 1,2,3 sequence. B is
labeled with "2".

(b) 1,1,2,2 sequence. B is
labeled with "1" preceded
by "1".

Fig. 6 Efficient labeling sequences.

Among these coding schemes the last one proposed by Akers [6] is most economic. Since each cell will be in one of just four states: "empty", "occupied", or labeled with "1" or "2", two bits will suffice to proceed the single-layer Lee algorithm. The first two coding schemes need three bits for the same information, but are almost equally efficient for some variations of Lee algorithms.

In designing practical routers, the coding scheme should be determined according to a trade-off between the number of bits per cell needed and the efficiency when the algorithm proceeds. For example, consider what we have to do in the label clearance phase. Namely, searching all the labeled cells would be equally involved as the wave propagation phase. However, a coding scheme with few additional bits allowing redundant information will simplify programming and also speed up the label clearance phase.

A suggested coding scheme for the two-layer router outlined in Section 3.3 is to use 5-bit storage per cell in the following way. A one-bit field is used to indicate via availability at each cell. A two-bit field is used to record for each layer whether a cell is "empty" or "occupied". Another two-bit field is used to record numbers for the 1,2,3,... or 1,1,2,2,... sequence plus an additional state to denote an unlabeled cells (including "occupied" ones). Now the last field is the essential part for simplifying the label clearance phase. Because what is needed after generation of a path is to clear (to the "unlabeled" state) this field of all cells within some specified rectangle including the source and the target. This coding scheme requires only (M+3) bits per cell for M layer routing. But the Geyer's router proposed in [7] uses (2M+3) bits per cell in order to speed up the backtrace phase.

3.5 Finding more desirable paths

The original version of the Lee algorithm is designed to route a net so as to minimize the total number of grid cells it traverses. On the other hand, practical situations often require a more desirable path. Perhaps the main reason for the widespread use of the Lee algorithm is the ease with which it can be modified to accommodate various constraints and path selection criteria. Many of such modifications preclude the simple coding schemes described in Section 3.4, but usually a similar coding scheme can be devised. (The basic requirement on the labeling sequence is that the desired path be unambiguously backtraced.)

As typical useful modifications, few additional bits of memory per cell may control the router to reserve wide area for via holes or power and ground connections, and to preclude long parallel runs. Likewise the Lee router is easily modified to generate a path with partial diagonal runs. Two-layer interconnection subject to the horizontal-vertical rule (Section 3.3) can be

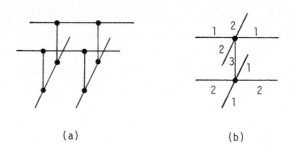

Fig. 7 Two-layer grids considering the horizontal-vertical rule.

achieved by considering a grid as shown in Fig. 7(a).   A more flexible way of
considering this rule is to vary the incremental weights by which the labels are
increased at each step of the wave propagation, so that the desired directions are
given higher priority.   Fig. 7(b) shows how this can be done.   In this case, an
incremental weight "1" is given to the desired direction and "2" to the other
direction.   Likewise, if a via is permitted but not preferred, a large weight ("3"
in this example) would be given.   This method with a pertinent set of incremental
weights allows the router to generate a desired path with given priority.

The simplest versions of the Lee algorithm tend to generate a path traversing
minimum number of grid cells, vias, bends, etc. These path selection criteria are
mathematically interesting as they match graph-theoretical approaches, and such a
path tends to have a small resistance value.   But resistivity of wires usually
varies depending on layers as in the current metal-polysilicon or double-metal
wiring technology,   likewise an interlayer connection has a high value of
resistivity.   Consequently, a pertinent set of weights should be given to grid
cells on both of the layers and to grid segments for interlayer connections.   Then
a path with minimum weighted sum of the used grid cells and grid segments fits the
minimum resistance criterion. Yet, it might be argued that minimization of such a
weighted sum is really not important in practice.   As long as there are feasible
paths in terms of signal delay, power dissipation, current flow capacity, etc., a
more preferable path would be such a path that will cause the least difficulty for
subsequent path connections.   Akers [6] has proposed an interesting modification
of Lee algorithm to this direction.

The Akers approach is based on the idea that a path running along
obstructions would leave more room for subsequent interconnections.   Suppose, for
example, that a net x has already been routed on a grid as shown in Fig. 8, and
that a pin A is now to be connected to another pin B.   The normal labeling
procedure tends to generate a shortest path like z, but the longer path y would be
more superior in the above sense.   The essential point here is that those grid

segments incident with "occupied" grid points are not usable any more, and that path y eliminates less usable grid segments than path z. Such path selection can be accomplished by preparing the weighted cellular array as shown in Fig. 9(a), where each grid point in Fig. 8 is replaced by a square. Note that the number of usable grid segments minus 1 is entered in each cell. Then the desired path may be generated by routing a net so as to minimize the total weight of used cells. In this example, paths y and z have total weights 13 and 15, respectively.

The wave propagation process for a weighted grid involves more careful considerations than the original Lee algorithm described in Section 2. Let us examine the point by considering the weight array illustrated in Fig. 9(a). What will be described here is not restricted to the grid segment routing, but equally applicable to any router which intends to minimize the total weighted sum of grid points and/or grid segments.

If pin A (with a weight of 0) is chosen as a source, the process begins by assigning each empty cell adjacent to A a value equal to its weight. Each such cell is marked as a "latest" cell. At each step from here on, each latest cell C (with value V) is examined, and every empty cell (with weight W) adjacent to C receives a value V+W and itself serves as a latest cell for the next step. If a cell adjacent to C already has a value greater than V+W, then the old value is replaced by this value and serves as a new latest cell. Otherwise the value of the adjacent cell is left as it stands. It should be noted here that some old latest cells may become new latest cells again if their values are decreased.

Fig. 9(b) shows the generalized labeling procedure, where the latest cells at the end of 3rd and 7th step are indicated by circled numbers. At the 7th step, the target has been reached and received a value "15". Now a careful consideration is needed here, as it does not mean that a desired path has been found. In general the minimum value among the latest cells (except target) is less than the value of some old latest cell, then it is possible that the old

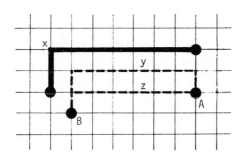

Fig. 8 Grid segment routing.

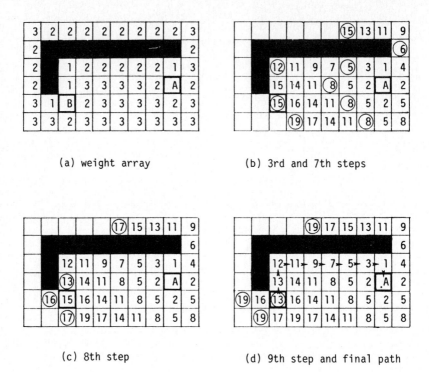

(a) weight array   (b) 3rd and 7th steps

(c) 8th step   (d) 9th step and final path

Fig. 9   Wave expansion on a weighted grid.

latest cell will receive a smaller value later.   This implies that an already
generated path may be overridden by a new one with a less total weight.   As it
really happens, the value "15" of the cell just above the target is renewed to
"13" in the 8th step and it again serves as a latest cell as shown in Fig. 9(c).
In the next (9th) step, the value of target cell is renewed to "13", which is less
than the minimum value of the other latest cells.   This implies that the desired
path, which is indicated by arrows in Fig. 9(d), has been found.

As has been mentioned, the original Lee algorithm runs in $O(N^2)$ time in the
worst case for an N X N uniform grid. However, the generalized labeling procedure
does not run as fast in the worst case, as each cell may become a latest cell
several times.   As far as the author knows, the best way of implementation
considering worst-case running time is to use a priority queue for storing the set
of latest cells, which results in $O(N^2 \log N)$ running time for an N X N weighted
grid. Another way of implementation is not to explicitly obtain the minimum value
of latest cells at each step but to continue the labeling until no cell receives a
smaller value.   This way requires $O(N^4)$ worst-case running time, but is more
efficient than the former according to the author's experience unless cell weights
are distributed in a wide range.

4. Enhanced Search Toward the Target

As has been mentioned, the Lee algorithm is the special case, called the breadth-first-search, of the "search" techniques in graph theory. The search techniques, in general, are capable of finding a path no matter in which sequence the cells are searched, and their worst-case running time is of $O(N^2)$ for an N X N grid. On the other hand, the running time for a particular instance of source-target pairs is proportional to the number of cells being searched until the target is reached. In Section 3.1 we have already seen a few speed-up techniques with intension to reducing the number of cells being searched. In this section, we describe two approaches which achieve more dramatic speed improvement. The common idea behind them is to advance wavefronts with higher priority to the target direction.

4.1 Hadlock's minimum detour algorithm

Hadlock [12] proposed a shortest path algorithm based on a new measure for labeling, named <u>detour number</u>. For a specified pair of source A and target B, we denote by M(A,B) the Manhattan distance between A and B. Let P be a path connecting A and B, then the <u>detour number</u> d(P) of path P is defined as the total number of cells (grid points) directed away from its target B. Now it is easily proved that the length $\ell(P)$ of a path P is given by

$$\ell(P) = M(A,B) + 2d(P)$$

It is also clear that P is a shortest path if and only if d(P) is minimized among all the paths connecting A and B. It should be noted that M(A,B) is fixed independent of the paths being selected. Fig. 10 illustrates an example of how path length is represented by the detour number.

Based on the above idea, the wave propagation phase of the Lee algorithm can be modified in such a way that (1) detour numbers with respect to a specified target, rather than the distances from the source, are entered in searched empty cells and that (2) those cells with less detour numbers are expanded with higher

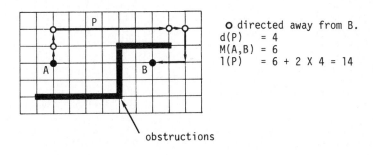

o directed away from B.
d(P)   = 4
M(A,B) = 6
1(P)   = 6 + 2 X 4 = 14

obstructions

Fig. 10 Path length and detour number.

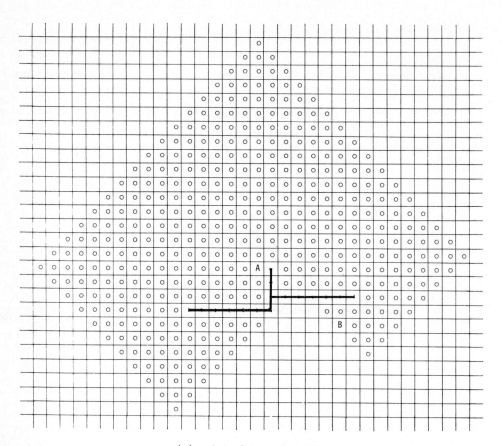

(a) original Lee algorithm

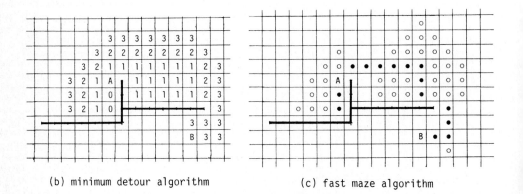

(b) minimum detour algorithm   (c) fast maze algorithm

Fig. 11 Cells searched before target is reached.

priority.   Fig. 11 shows a sample problem of finding a path around a simple combination of obstructions.   In Fig. 11(b), the searched cells are marked with their detour numbers.   In comparing this with Fig. 11(a), in which the searched cells by the original Lee algorithm are marked with circles, we shall notice the remarkable speed improvement by means of the detour number algorithm.

The Hadlock's minimum detour algorithm, with this modification, still guarantees to find a shortest path.   The running time for an N X N grid plane is ranging from $O(N)$ to $O(N^2)$ depending on the positions of source-target pairs and distributions of obstructions. To implement this algorithm for a general graph, a specific data structure, e.g., a priority queue, is needed to manage the set of latest cells with their detour numbers.   If a given graph is a grid with uniform weight, however, the algorithm can be implemented by a simple scheme.   Hadlock devised a scheme to run the minimum detour algorithm using two stacks, without explicitly enter the detour number to each cell.

4.2 Soukup's fast maze algorithm

The fast maze router proposed by Soukup [13] can be viewed as a combination of two methods: the Lee algorithm and the line-search algorithm (which will be described in the next section). Graph-theoretically, it is actually a combination of breadth-first-search and depth-first-search [5].   In this algorithm, a line segment starting from the source is initially extended toward the target, and the cells on the line segment are searched first.   The line segment is extended without changing directions unless it is necessary.   When the line hits an obstacle, the Lee-type wave propagation is used to search around the obstacle. Once a cell approaching the target is found, another line segment starting from there is extended again toward the target.   Fig. 11(c) shows the set of searched cells by an implementation of this idea for the same sample problem as before, where the blacked circles indicate the cells directed toward the target.   Again, we shall notice from this figure speed improvement achieved by this algorithm.

This algorithm again guarantees finding a path whenever it exists, but the generated path is not always shortest.   Usually it generates a suboptimal path in terms of the length and/or the number of bends.   Apart from the minor disadvantage, it runs extremely fast when the routing space is not very congested. Soukup claims that it is 10-50 times faster than the Lee algorithm on typical two-layer routing problems.

Rubin's router [10] is also based on the same idea.   It achieves the similar effect by sorting the cells on the wavefront using the key representing an estimated distance to the target.   On the other hand, Soukup's router uses only two stacks without involving the sorting procedure.   This simplifies coding and saves memory.

5. Line-Search Algorithms

The class of detailed routing algorithms, referred to as "line-search", was first proposed by Mikami-Tabuchi [3] and Hightower [4] independently, aimed at reducing memory space required for implementing the maze-running algorithms. The line-search algorithms also proceed to find a path running on a grid. However, the essential difference lies in that the routing space and paths are represented by a set of line segments, unlike the Lee algorithm or its extensions in which a unit of memory space is allocated for each grid point. In this sense, the line-search algorithms can be viewed as proceeding on an "imaginary" grid. This feature makes it possible to reduce necessary memory space and running time, provided that the routing area is not very congested. In this section, the line-search procedure for finding a path is reviewed based on the original papers [3], [4]. Next a recent improvement of the line-search algorithms, called line-expansion algorithm [16] is introduced. Finally, the running time and memory required for the above algorithms are discussed.

5.1 Line-search procedures

The first report dealing with the line-search algorithms was presented by Mikami and Tabuchi [3]. Their algorithm aims at single-layer routing or two-layer routing with the horizontal-vertical rule. We only describe here point-to-point interconnections. Note that this process can be extended with obvious modifications to more general cases with multiple source/target points.

Let A and B be a pair of terminals of a net located on some intersection point of an imaginary grid. The initial stage of the line-search is to generate lines starting with A and B. There are four such lines, either horizontal or vertical, and they are extended until they hit obstructions or the external frame of the routing space. If a line originated from A intersects another originated from B, a connecting path without bend or with one bend has been found. Otherwise, these four line segments are identified as trial lines of level (0) and stored in a temporary storage. Then at each iteration step, say i th step, the following operations are done.

(1) Pick up trial lines of level (i) one after another from the temporary storage. Along each such trial line, trace all its grid points (called base points). Starting from these base points, generate new line segments perpendicular to the trial line. Then the generated line segments are identified as trial lines of level (i+1).

(2) If a trial line of level (i+1) intersects a trial line originated from the other terminal point, then backtrace from the point of intersection to both of the terminals to generate a path.

(3) Otherwise, all trial lines of level (i+1) are stored in the temporary storage.

Fig. 12 illustrates an example of finding a path by means of the above procedure. In this example, a trial line of level (1) originated from B intersects that of level (1) originated from A at the grid point marked with a circle, and the generated path is indicated by bold lines. In general, if a level (i) trial line originated from A intersects a level (j) trial line originated from B, a path with (i+j+1) bends (vias) is generated.

The Mikami-Tabuchi line-search algorithm guarantees to find a path if one exists, provided that all possible trial lines up to the deepest possible level are examined. Moreover, a path with minimum number of bends (or via holes for two-layer routing with the horizontal-vertical rule) can be found by means of an appropriate level scheduling of the trial lines.

The Hightower line-search algorithm [4], invented independently and published soon later, is very similar to the Mikami-Tabuchi algorithm. Instead of generating all line segments perpendicular to a trial line, the Hightower algorithm only consider those extendable beyond the obstacle which has blocked the preceding trial lines. Such trial lines are called <u>escape lines</u> in his paper. The Hightower router, for each line, generates the longest perpendicular escape line. If there is a multiple choice, the escape line nearest to the current base point is taken. The set of escape lines generated by the Hightower router for the same example as in Fig. 12 is shown in Fig. 13. As can be observed in this figure, a smaller stack of escape lines (trial lines) is needed and therefore finds a simple-shaped path faster compared with the Mikami-Tabuchi algorithm. On

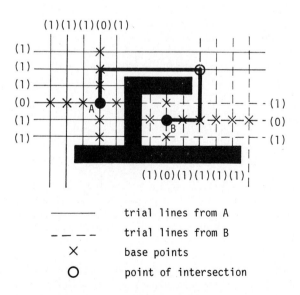

——————	trial lines from A
_ _ _ _	trial lines from B
×	base points
O	point of intersection

Fig. 12 Mikami-Tabuchi line-search algorithm.

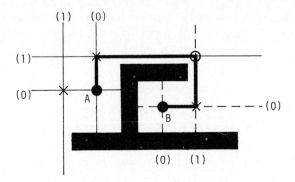

Fig. 13 Hightower line-search algorithm.

the other hand, it does not guarantee connection even if one exists.   It finds a
path with minimum number of bends in most cases, but not always.

As has been mentioned, the line-search algorithms need less running time and
memory as compared with the Lee algorithm.   This is because, if the routing space
is not very congested, a connecting path without detour or with a small number of
bends is likely to exist.   However, for complicated mazes, it runs slower and
needs more memory than the maze-running algorithms. (This point will be clarified
in Section 5.3.)   This observation has given rise to another class of routers,
referred to as <u>pattern routers</u>.   The philosophy behind these routers is to limit
patterns or number of bends of paths to be generated and to speed up the process
to find such a path.   Fig.   14 shows an example of the class of paths being

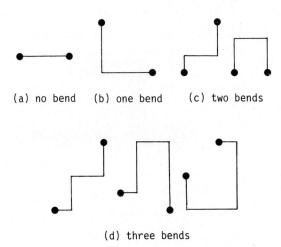

Fig. 14 Paths generated by a pattern router.

generated by a pattern router. In summary, a practical router would be to use a line-search or pattern router in the beginning followed by a maze-running router once the routing space has become congested.

5.2 The line-expansion algorithm

The line-expansion algorithm, recently proposed by Heyns et al. [16], is a new algorithm which combines the maze-running algorithms and the line-search algorithms. It is based on expanding a line in its perpendicular direction. For every grid point of the line it is investigated how far it can be expanded, i.e., until a line through this grid point is blocked by an obstacle. The expansion zone then can be defined as the zone consisting of all grid points that can be reached by a line beginning on the expanded line and perpendicular to it. Fig. 15 illustrates what the expansion zone of a line segment looks like. The algorithm searches all grid points in the expansion zone like the Lee algorithm. However, the grid points are not kept in memory; only its boundary segments, called active lines, are pushed onto a stack. In the next step, the generated active lines are expanded outside the zone for further search.

The above procedure is initiated from both of the terminals (A and B in Fig. 16) by entering them as starting active lines into the stack. Wavefronts, which do not advance point by point but zone by zone, are then generated starting from both of the terminals as shown in Fig. 16(a). A connection is found when an active line reaches the wavefront advanced from the other terminal. In the example of Fig. 16(a), this occurs at the shaded area, called a solution zone, in which the two wavefronts are freely interconnected. The rest of the algorithm is to backtrace from the solution zone towards the terminal points, generating a connecting path as shown in Fig. 16(b). Another important consideration in the

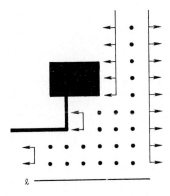

Fig. 15 Expansion of a line ℓ in the upward direction. The arrows indicate the direction in which the active lines are being expanded.

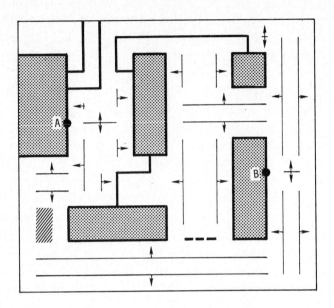

▨ solution zone   --- stop line

(a) active lines generated during the search

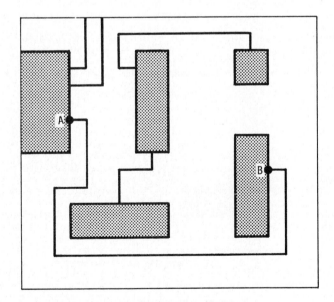

(b) generated path between A and B

Fig. 16 Line-expansion algorithm.

line-expansion procedure is to generate a <u>stop line</u> (see Fig. 16(a)) when an active line meets the wavefront advanced from the same terminal. The stop line is imposed to prevent duplicated search of a zone, thus to speed up the algorithm.

The line-expansion algorithm compares favorably with the Mikami-Tabuchi line-search algorithm in that it guarantees connection if possible and that it can be readily extended to more general cases with multiple terminals, source/target points, layers, etc. It has more subtle way of preventing duplicated search than the Mikami-Tabuchi algorithm. A trivial disadvantage is that it, unlike the latter, does not always find a minimum bend path.

5.3 Computational complexity analysis

It is generally considered that the line-search algorithms do not guarantee finding a path even if one exists but find a simple-shaped path faster with less memory than the maze-running algorithms. This statement, however, is misleading in two ways. First, the Mikami-Tabuchi line-search algorithm and the line-expansion algorithm do guarantee finding a path provided that we do not mind considerable amount of time and memory. Secondly, the speed improvement is not so dramatic. Some engineers argue that the line-search algorithms, in many cases, are not as fast as the fast maze-running algorithms described in Section 4. The readers might have been impressed by the preceding descriptions that the running time is proportional to the number of generated lines (trial lines, escape lines, or active lines) independent of the path length. However, how do we extend a line until it is blocked by an obstacle? And, how do we investigate whether a grid point on the extended line is empty or occupied? In order to answer the questions, the data structure for managing the set of lines must be specified.

Unfortunately, the original papers [3], [4], [16] include no rigorous analysis of computational complexity, and data structures are only ambiguously described. Therefore we shall analyze these algorithms based on the following orthodox data structure of managing the lines by means of linked lists. Obstructions imposed by circuit components and already-routed nets are represented by a series of horizontal and vertical line segments. Each line segment is defined by three integers to specify the x, y coordinates of its two end points. The lines are sorted according to their x, y coordinates, and the co-horizontal vertical (co-vertical horizontal) lines are grouped together. For each y (x) coordinate there is a pointer to the first horizontal (vertical) line on that coordinate and for every line there is a pointer to the line that comes next. The lines generated while searching a connection can also been stored in a temporary storage with the similar data structure. With this data structure, it is clear that the required memory space is proportional to the number n of the lines. For an N X N grid plane, n could be as large as $O(N^2)$, but usually $n \ll N^2$. Therefore the line-search or line-expansion algorithms substituted to the maze-running algorithms saves memory.

Next we shall analyze the time complexity of these algorithms based on the best way of implementation that the author could work out subject to the above data structure.   The Mikami-Tabuchi algorithm consists of unit operations to generate new trial lines perpendicular to each old trial line.   Provided that there are n lines representing obstructions and n' trial lines are temporarily generated to find a path, this unit operation implies to search a total of $O(n + n')$ lines.   Since the unit operation is repeated $O(n')$ times, the total running time will be of $O(n'(n + n'))$.   Note that, if a solution must be guaranteed, the number n' of generated active lines are often as large $O(N^2)$ even if the routing space is not so congested.   The line-expansion algorithms can also be analyzed in the similar way.   However, it does not generate lines carrying redundant information, but only $O(n)$ active lines. Hence the algorithm runs in $O(n^2)$ time. The Hightower algorithm generates only one escape line at each level of the line-search.   But many other lines could be investigated in order to choose an optimal escape line.   To be more specific, it requires as before $O(n^2)$ time in the worst case.   If the algorithm is modified to choose a sub-optimal escape line which can go further beyond the current obstruction, it will run in $O(n\ell)$ time to find a path with ℓ line segments, i.e. $(\ell-1)$ bends.

The above argument implies the very conservative estimation of $O(N^4)$ running time for all of the three algorithms since the values of n, n' and ℓ could be of $O(N^2)$ in a complicated maze.   When the routing area is not congested, these algorithms are expected to run much faster.   Particularly, the Hightower algorithm, in such a case, is expected to run in time proportional to the number of bends of a path.   In conclusion, the line-search algorithms do not improve speed so dramatically in contrast to its memory saving; they run faster than the maze-running algorithms only when there are few obstructions.

In the final part of this section, it should be stressed that an improved line-search algorithm is available to always find a minimum bend path in $O(n \log^2 n)$ time with $O(n)$ memory.   This improvement is due to a more subtle data structure which enables us to find a desired line without the exhaustive search as above.   The improved line-search algorithm will be described in Chapter 9 as a computational geometry algorithm.

6. Aiming at 100% Connection Ratio

The final routing (or elsewhere referred to as detailed routing) of connections is a most exciting game, as its result is so vividly physical and convincing. There are two different goals of routers, depending on the situations they are used.   One goal is to route given nets in as small area as possible.

This is typical in the channel routing problem (Chapter 4) which encounters in standard-cell or building-block LSI design. The other goal is to complete as many connections as possible (or preferably all the necessary connections) within fixed routing space. This goal matches the switch-box routing problem (Chapter 4) or final routing procedures of gate-array LSI layout. The maze-running or line-search algorithms are typically applied to the routing problems with the second goal in such a way that the nets are connected one at a time.

The goal of 100% routing completion is often difficult or impossible when poor placement is given. Even if the placement is reasonably good, results obtained from fully automated routers are still discouraging compared with manual designs. A common way that the engineers in industries are doing is to use an auto-router to connect most of the nets (typically 99% of the nets), followed by manual rework of the result along with its plotted image in order to complete the remaining connections. This way of manual intervention (for connecting just 1% of the nets) occupies the major portion of the time spent for routing. This includes the time for correcting errors caused by manual intervention. Therefore more advanced design methodologies are still needed to cope with the difficult routing problems. This section discusses some recent results on the rip-up and reroute techniques which aims at completing the few remaining nets.

6.1 Is net ordering a critical aspect ?

The auto-routers which route one net at a time suffer from a fundamental shortcoming in that they provide no feedback or anticipation to avoid conflict between nets or to assure that some early connection will not prevent successive connections. Because of this blindness, it has been claimed that the order in which a set of nets is routed is of crucial importance to the successful routing completion. However, the topic of connection ordering is controversial.

It is true that one particular ordering leads to a higher connection ratio than the others. For example, the four nets given in Fig. 17 can be successfully

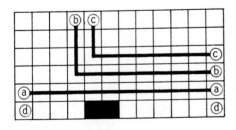
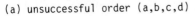
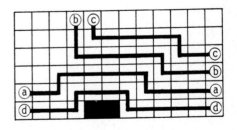

(a) unsuccessful order (a,b,c,d)   (b) successful order (d,a,b,c)

Fig. 17 An example in which a good ordering is successful.

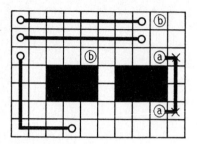
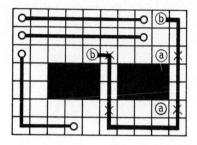

(a) connection a-a blocks   (b) connection b-b blocks
    connection b-b   connection a-a

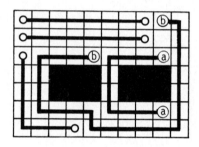

(c) perfect solution

Fig. 18 An example in which net ordering is of no help.

connected by a particular ordering whereas some other orderings are not successful.

On the other hand, there are several arguments against the importance of investigating an optimal ordering. First, there are n! possible orderings for connecting n nets, and it is impossible to try all orderings in the real life design task. Even heuristics aiming at obtaining a sub-optimal ordering are disappointing, as no effective means is known to compare the goodness of orderings. Secondly, an experimental result [9] has been presented to show that the performance of a router, when measured in terms of the total of the ideal lengths of connections successfully completed, is almost independent of the order in which connections are attempted.

The other argument is that, even if all the orderings have been tried, it does not always achieve the best available. Fig. 18 illustrates an example in which net ordering is of no help. Here we assume that the router generates a shortest path or a minimum bend path. If net a is connected first as shown in Fig. 18(a), it blocks net b. Conversely, if net b is connected first as shown in

Fig. 18(b), it again blocks net a. The perfect solution is given in Fig. 18(c), in which both nets a and b are not routed by optimal paths. This example suggests that what is more critical than net ordering is to specify those cells which must be or must not be used to lay out particular nets. This point will be made clear in the subsequent discussions.

6.2 Simulating the human way of routing

With regard to the question of the degree of difficulty of the routing problem, it has been proved that the disjoint connecting path problem (a simplified version of the routing problem) is NP-complete [14]. Aside from the theoretical point of view, how difficult problems do we encounter in the real life design ? The degree of difficulty seems to range within some bounds. The so-called difficult problems are those in which the random order sequential routing often fails to achieve 100% connection ratio. Otherwise the users of a router would input more difficult problems. On the other hand, the problems should be as easy as those in which an expert engineer can complete the few remaining connections. An auto-router may be criticized only when it fails to complete the connections in which the human rework is successful. Therefore, the goal of advanced routers would be to achieve the equally good performance as manual design with short turn-around time and without error.

Several attempts simulating the human way of routing have been reported [17]-[20], intended to provided some help for the time-consuming final cleanup in which expert engineers make the last few connections abandoned by an auto-router. The principle of these attempts is to detect connections that cause "blockages" for other nets, and to rip up and reroute the connections causing blockages so as to provide room for other nets. The papers proposing this principle, although some performance improvement is reported, only ambiguously describe the essential points: which already-routed nets or which parts of them really cause the blockages and how to find them. In the following subsections, the points will be made clear based on more recent contributions [21], [22].

6.3 Minimal separator --- a concept for connection blockage

When a particular net cannot be connected by means of the Lee algorithm, it implies that some already-routed nets have blocked its terminal pair. Such a blockage can be characterized by the notion of minimal separator in graph theory. For two non-adjacent nodes A and B in a graph, an A,B separator is a set S of nodes such that any path connecting A and B includes at least one node in S. A minimal A,B separator is an A,B separator such that its no proper subset is an A,B separator. The example shown in Fig. 19 will help understanding this notion.

For our purpose, we shall associate a cell of the cellular array (grid plane) with each node of a graph. Now minimal separators of our interest are those

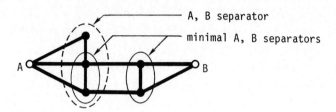

Fig. 19 (Minimal) A, B separators.

consisting only of the cells occupied by already-routed nets.   Among such minimal
separators, two particular ones, S_A (reachable from one of the terminals A) and S_B
(reachable from the other terminal B), can be found by means of the Lee-type wave
propagation, which consists of the following four phases.

Phase 1. A wave is propagated starting from A, going through empty cells, until
it hits occupied cells, followed by marking them with "a".

Phase 2. Another wave is propagated starting from B in the same way as Phase 1,
marking the reached occupied cells with "b".   If both "a" and " b" are entered
into an occupied cell, they are replaced by a new mark "γ".

Phase 3.   The wavefront at the end of Phase 1 is further advanced starting from
the cells marked with "a", going through both empty and occupied cells, until it
hits the cells marked with "b", followed by replacing their marks by "β".

Phase 4.   The wavefront at the end of Phase 2 is further advanced starting from
the cells marked with "b" in the same way as Phase 3 until it hits the cells
marked with "a", followed by replacing their marks by "α".

   Once the above four phases of the wave propagation are completed, the set of
occupied cells marked with "α" or "γ" and those marked with "β" or "γ" constitute
the desired minimal A,B separators S_A and S_B, respectively.   Fig. 20 illustrates
an example of such minimal separators.   If should be observed from this figure
that the minimal separators indicate the essential parts of the connection
blockage.

6.4 Reroute effect index

   Based on the minimal separators S_A and S_B, a measure is derived which
indicates the likelihood of rerouting particular nets to succeed.   For each pre-
assigned net W, let $N_A(W)$, resp. $N_B(W)$, be the number of cells of W contained in
S_A, resp. S_B.   Then

   $P(W) = \min \{N_A(W) - L_A(W), N_B(W) - L_B(W)\}$

is called <u>reroute effect index</u> for W, where $L_A(W) = 1$, resp. $L_B(W) = 1$, if the
terminals of W lies in the opposite sides with respect to S_A, resp. S_B, otherwise

it is equal to 0.   It should be noted here that a necessary condition for rerouting W to succeed is that $1 \leq P(W)$.   Moreover, the larger the value $P(W)$ is, more likely rerouting W is successful.

Let us examine how the concept of reroute effect index works along with the example of Fig. 20.   The terminals A and B are blocked by five nets $W_1 - W_5$.   By inspection, their reroute effect indices are obtained as follows,

$$P(W_1) = \min \{7 - 0, 0 - 0\} = 0,$$
$$P(W_2) = \min \{2 - 1, 4 - 0\} = 1,$$
$$P(W_3) = \min \{3 - 1, 2 - 1\} = 1,$$
$$P(W_4) = \min \{3 - 0, 3 - 0\} = 3,$$
$$P(W_5) = \min \{2 - 0, 3 - 0\} = 2.$$

From this result, we see that rerouting W_1 is not useless, whereas rerouting the other nets is worth trying.   In this example, pertinent rerouting of any of W_2, W_3, W_4, and W_5 is successful.

It is clear that rerouting a pre-assigned net does not always succeed even if its reroute effect index is greater than 0.   This case occurs when removal of a blockage still cannot create a path between the terminal pair under consideration because of other pre-assigned nets.   When rerouting a single net is not successful, our next trial would be to rip up multiple nets simultaneously and to reroute them.   For this purpose, the reroute effect index for multiple pre-assigned nets W_1, W_2, ..., W_k is defined by

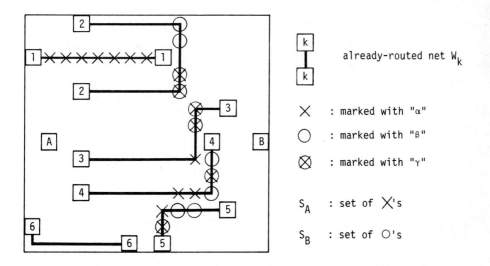

Fig. 20   Two minimal separators S_A and S_B obtained by Lee-typer wave propagation.

$$P(W_1,\ldots,W_k) = \min \{ \sum_{i=1}^{k} (N_A(W_i) - L_A(W_i)), \quad \sum_{i=1}^{k} (N_B(W_i) - L_B(W_i)) \}.$$

Algorithms for obtaining minimal separators and reroute effect indices are readily implemented. Minimal separators can be found by means of the modification of the Lee algorithm as described in Section 6.3. The reroute effect index of a pre-assigned net can be obtained by searching and counting the related cells along its route. However, a few more bits per grid point is needed in the algorithms in order to record the labels for the modified Lee algorithm and the directions of pre-assigned nets running.

6.5 Interactive rip-up and reroute

The rip-up and reroute techniques for the detailed routing is more suitable to implementation in a stand-alone interactive system than batched-mode use of a main-frame. The former has more flexibility to exploit expert designers' control based on their experience and intuition in such a way that they can make right choice at each critical branch of the routing procedure. On the other hand, the interactive system must be intelligent enough to preclude the tedious manual job of drawing a line one at a time on the graphic display and any error caused by human intervention. Moreover it must accommodate a set of high-level commands by which basic unit operations can be done automatically, e.g., connecting a specified pin pair and deleting a connection. Such interactive layout systems have been reported already, e.g. [18].

In order to realize an effective means for the rip-up and reroute techniques based on the concepts of minimal separators and reroute effect index, it is suggested that an interactive routing system accommodates the following commands backuped by the host computer.

(1) By means of a Lee-type algorithm, connect the given set of nets in a given order.

(2) Find the minimal separators S_A and S_B for an unconnected pin-pair (A,B).

(3) List all the pre-assigned nets with their reroute effect indices, excluding those with the value zero which are hopeless.

(4) Evaluate the reroute effect index for a specified set of pre-assigned nets.

(5) Delete a specified set of pre-assigned nets.

(6) Connect a pin-pair traversing specified cells.

(7) For specified cells, connect a pin-pair traversing as few those cells as possible.

By means of a pertinent combination of the above commands, the user at the interactive terminal will be able to try several ways to decrease unfinished nets. The simplest way is to temporarily remove a pre-assigned net, say W_1, which is blocking another net, say W_0; then W_0 is routed first and W_1 is rerouted after that. This way is successful for rerouting nets W_2, W_3 and W_5 of Fig. 20 if the pin pair (A,B) is connected by a shortest path. When rerouting one net at a time fails, the user can of course try rerouting multiple nets. The pre-assigned nets being rerouted can be tried with higher priority to those having larger values of reroute effect indices. Or particular pre-assigned nets can be tried first due to the user's decision.

Another way worth trying, when the above way of changing the order of net connections has failed, is to reroute a pre-assigned net so as to traverse as few cells in the minimal separators as possible, creating room for connecting the unfinished nets. In the example of Fig. 18, if nets are routed in an random order and net a blocks the connection of net b as in Fig. 18(a), then net a can be rerouted traversing fewer minimal separator cells (marked with X), creating a path for net b as shown in Fig. 18(c). This way also works when net b has been connected first as shown in Fig. 18(b).

The rip-up and reroute techniques in no way preclude the use of expert engineers' experience and intuition. By observing the graphically displayed image of the layout pattern with few nets remaining unfinished, the engineer can choose pre-assigned nets to be ripped up and specify pertinent routes on which they are laid out. But the minimal separators and the reroute effect indices obtained by the extensive use of the Lee algorithm will provide great help to such manual jobs. The author and his colleagues are developing an interactive router which provides the rip-up and reroute feature backuped by a hardware engine; an intermediate report on the prototype machine has been published [23].

6.6 Further topics on detailed routing

There still remain several topics which need further research in order to realize a routing tool with higher performance. Some of them are listed below.

(1) Two disjoint path algorithm. Polynomial-time algorithms have been discovered to find two vertex-disjoint paths for specified two node pairs in a graph [24] - [26]. Such an algorithm would be useful to find connections in a complicated maze as shown in Fig. 21, and to provide more successful rip-up and reroute techniques. Note that the human way of finding an optimal rerouting path by inspection in multi-layer cases is not as useful as the single-layer case. However these algorithms so far have too much overhead to be efficiently implemented for large-scale grid structures. Moreover, no polynomial-time algorithm is known for finding three or more disjoint paths.

(2) Bottle-neck analysis. Some experienced engineers argue that a few last connections with a really tangled routing are worse than more unfinished connections with more space to handle them. A way to cope with this argument is

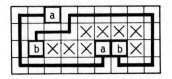

$\times$: occupied cell

Fig. 21 Two disjoint paths in a complicated maze.

halt the sequential routing in some intermediate stage and to find a bottle-neck
for connecting the remaining nets, even if a few more nets can be connected. Once
the bottle-neck turns out to be critical, the pre-assigned nets causing it had
better be rerouted in this stage. Examples of such bottle-necks are shown in Fig.
22.   It seems to be worth investigating how the maze-running algorithms can be
extended to finding bottle-necks.

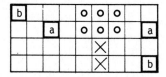

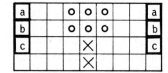

(a) bottle-neck cells (marked   (b) bottle-neck cells make
    with O) must be reserved   impossible to connect
    for connections a-a and b-b.   remaining nets.

Fig. 22 Bottle-necks.

(3) Hardware engine.   The feature of an interactive terminal of key importance
is the quick response time;   preferably a few seconds and at most a minute per
command.   Hardware routers are particularly useful to speed up the exhaustive
search for the rip-up and reroute techniques, in which the fast maze algorithms as
described in Section 4 are of no help.   Although several parallel processing
architectures have been presented to physically implement the Lee algorithm [27] -
[31], more economic and compact ones are really needed.

(4) Gridless routing.   One of the reasons why expert designers can complete
unfinished connections abandoned by auto-routers is because they flexibly remove
some standardization rules for automation, allowing diagonal runs, variable width
and spacing of wires, etc.   Although, the maze-running algorithms can be

theoretically extended to the cases with more complicated constraints, it involves a too precise grid to be handled in primary memory. Therefore gridless maze-running algorithms which really save memory are needed to cope with complicated design rules.

7. Conclusion

In this chapter, various extensions of the maze-running and line-search algorithms have been throughly reviewed, followed by introduction of recent rip-up and reroute techniques aiming at 100% connection ratio. Even though the recent trend of routing VLSI chips is to perform global routing followed by channel routing, the Lee algorithm and its extensions, because of their flexibility and generality, are powerful for dealing with difficult routing problems which encounter in the final cleanup phase. The research target in this area, of course, is to improve these algorithms for higher performance and less memory requirement.

References

[1] C.Y. Lee, "An Algorithm for Path Connections and its Application," IRE Trans. on Electronic Computers, vol. EC-10, pp. 346-365, 1961.

[2] E.F. Moore, "The Shortest Path through a Maze," Annals of the Harvard Computation Laboratory, vol. 30, Pt. II, pp. 185-292, 1959.

[3] K. Mikami and K. Tabuchi, "A Computer Program for Optimal Routing of Printed Circuit Connectors," IFIPS Proc., vol. H47, pp. 1475-1478, 1968.

[4] D.W. Hightower, "A Solution to Line-Routing Problem on the Continous Plane," Proc. 6th Design Automation Workshop, pp. 1-24, 1969.

[5] A.V. Aho, J.E. Hopcroft, and J.D. Ullman, The Design and Analysis of Computer Algorithms, Addison-Wesley, Reading, Mass., 1974.

[6] S.B. Akers, "A Modification of Lee's Path Connection Algorithms," IEEE Trans. on Electronic Computers (Short Notes), vol. EC-16, pp. 97-98, 1967.

[7] J.M. Geyer, "Connection Routing Algorithm for Printed Circuit Boards," IEEE Trans on Circuit Theory, vol. CT-18, pp. 95-100, 1971.

[8] S.B. Akers, "Routing," in Design Automation of Digital Systems, vol. 1, Breuer, M.A. (Ed).

[9] L.C. Abel, "On the Ordering of Connections for Automatic Wire Routing," IEEE Trans. on Comput., vol. C-21, pp. 1227-1233, 1972.

[10] F. Rubin, "The Lee Path Connection Algorithm," IEEE Trans. on Comput., vol. C-23, pp. 907-914, 1974.

[11] J.H. Hoel, "Some Variations of Lee's Algorithm," IEEE Trans. on Comput., vol. C-25, pp. 19-24, 1976.

[12] F.O. Hadlock, "A Shortest Path Algorithm for Grid Graphs," Netsworks, vol. 7, pp. 323-334, 1977.

[13] J. Soukup, "Fast Maze Router," Proc. 15th Design Automation Conf., pp. 100-102, 1978.

[14] M.R. Garey and D.S. Johnson, Computers and Intractability: A Guide to the Theory of NP-Completeness, W.H. Freeman and Company, San Francisco, 1979.

[15] M. Gardner, "Mathematical Games," Scientific American, 1963.

[16] W. Heyns, W. Sansen, and H. Beke, "A Line-Expansion Algorithm for the General Routing Problem with a Guaranteed Solution," Proc. 17th Design Automation Conf., pp. 243-249, 1980.

[17] M. Bollinger, "A Mature DA Systme for PC Layout," Proc. Internat. on Printed Circuits Conf., pp. 85-99, 1979.

[18] H. Mori, et. al., "BRAIN: An Advanced Interactive Layout Design System for Printed Wiring Board," Proc. Internat. Conf. on Circuits and Computers, pp. 754-757, 1980.

[19] W.A. Dees and R.J. Smith, "Performance of Interconnection Rip-up and Reroute Strategies," Proc. 18th Design Automation Conf., pp. 382-390, 1981.

[20] W.A. Dees and P.G. Karger, "Automated Rip-up and Reroute Techniques," Proc. 19th Design Automation Conf., pp. 432-439, 1982.

[21] S. Suzuki and T. Ohtsuki, "Some Considerations on Data Structure for Implementing Dynamic Routers (in Japanese)," IECE Monograph, CAS 82-153, pp. 85-92, 1983.

[22] S. Shimano and T. Ohtsuki, "Some Considerations on Two-Layer Dynamic Router (in Japanese)," ibid., CAS 84-130, pp. 15-22, 1984.

[23] T. Ohtsuki, M. Tachibana, and K. Suzuki, "A Hardware Maze Router with Rip-up and Reroute Support," Proc. Internat. Conf. on Computer Aided Design, 1985.

[24] Y. Perl and Y. Shiloach, "Finding Two Disjoint Paths between Two Pairs of Vertices in a Graph," J. ACM, vol. 25, pp. 1-9, 1978.

[25] Y. Shiloach, "A Polynomial Solution to the Undirected Two Paths Problem," ibid., vol. 27, pp. 445-456, 1980.

[26] T. Ohtsuki, "The Two Disjoint Path Problem and Wire Routing Design," Lecture Note in Computer Science. 108, Graph Theory and Algorithms, Springer-Verlag, pp. 207-216, 1980.

[27] A. Isoupouicz, "Design of an Iterative Array Maze Router," Proc. Internat. Conf. on Circuits and Computers, pp. 908-911, 1980.

[28] M.A. Breuer and K. Shamsa, "A Hardware Router," J. Digital Systems, vol. 4, pp. 393-408, 1981.

[29] T. Blank, M. Stfik, and W. vanCleempt, "A Parallel Bit Map Architecture for DA Algorithms," Proc. 18th Design Automation Conf., pp. 836-845, 1981.

[30] S.J. Hong, R. Nair, and E. Shapiro, "A Physical Design Machine," in <u>VLSI 81</u>. London: Academic Press, pp. 257-266, 1981.

[31] M. Tachibana, K. Suzuki, and T. Ohtsuki, "A Hardware Engine Architecture for Interactive Routing Design," <u>Proc. Internat. Symp. on Circuits and Systems</u>, pp. 209-212, 1985.

LAYOUT DESIGN AND VERIFICATION
T. Ohtsuki (Editor)
© Elsevier Science Publishers B.V. (North-Holland), 1986

Chapter 4

CHANNEL ROUTING

Michael BURSTEIN

IBM T.J. Watson Research Center
Yorktown Heights, NY 10598, U.S.A.

Channel Routing is one of the most important and probably one of the most popular phases of physical design of LSI and VLSI chips as well as PC boards. As a separate wire routing method it was first proposed by Akihiro Hashimoto and James Stevens in 1971. Since then the method was extensively studied and applied to many different technologies. In many instances it has undisputed advantages over other methods. In this chapter we review the most significant achievements in channel routing theory and algorithms. Although we intend to bring the reader to the state of the art stage in the field, this objective is almost unrealistic because of the explosion of new ideas and techniques, which is going on right now.

0. INTRODUCTION

The problem of channel routing is a special case of the wire routing problem when interconnections have to be performed within a rectangular strip having no obstructions inside, between terminals located on opposite sides of the rectangle. This problem arises in many different methodologies of physical layout.

Originally the feasibility of channel router [1] was demonstrated on actual design specifications for ILLIAC IV Control Unit boards. In the design of gate arrays, after the placement and global routing phases, the channel router is usually invoked to perform final interconnections within wiring bays (Figure 1; see [2, 3]). Similarly channel routers are used to complete the interconnections of polycell (standard cell) chips (Figure 2; see [4]). In custom layout of VLSI chips channel router completes interconnections between the "macros" (Figure 3; see [5]).

Since channel routers perform detailed (final) interconnections in layout design the general routing strategies and actual algorithms are very much dependent on technology restrictions. Different technologies introduce different instances of the problem. In this chapter we mainly concentrate on what is currently considered a classical model for routing: net terminals are located on vertical grid lines, two wiring layers are available for interconnections - one layer is used exclusively for vertical segments, another for horizontal and vias are introduced for each layer change. Multi-terminal nets are allowed. Some authors refer to this model as to "Manhattan Routing".

Typical Gate-Array

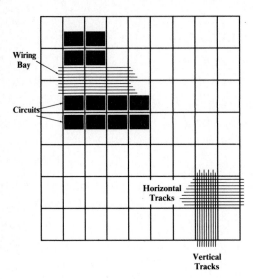

A Fragment of Polycell Layout

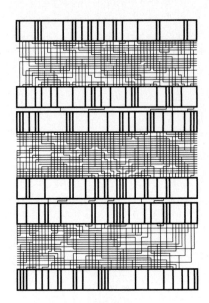

Figure 1.

Figure 2.

Custom Layout

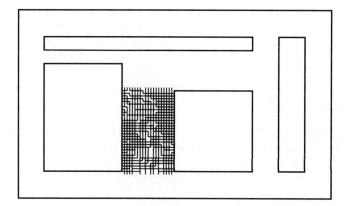

Figure 3.

We discuss however several aspects of switchbox routing problem since many ideas for switchbox routing were inspired by channel routers. The last section is devoted to a brief overview of the most important variations of routing models, such that three layer routing, routing in both directions on both layers, "knock knee" model.

1. THE PROBLEM

To address the problem more precisely we need the following definitions. A ***channel*** is a pair of vectors of nonnegative integers - *TOP* and *BOT* - of the same dimension

$$TOP = t(1), t(2), \dots , t(n) \text{ and } BOT = b(1), b(2), \dots , b(n)$$

with the condition that any positive integer having an entry in one of them $-TOP$ or $BOT-$ has at least one other entry, i.e. every positive integer represented in these vectors is represented at least twice. We assume that these numbers are the labels of grid points located along the top and bottom edge of a rectangle in a rectilinear grid (see Figure 4). Points having the same positive label have to be interconnected, i.e. they define nets. Zeros in *TOP* or *BOT* mean that no connection has to be made to the corresponding point.

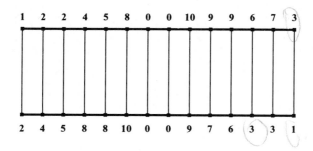

Figure 4.

A solution to a channel routing problem specifies for each net, a set of connected horizontal and vertical "wire" segments whose endpoints are gridpoints located on or between *TOP* and *BOT* lines. Two segments in the same direction are on the same layer, so they may not touch if they are for differenet nets. Two segments for the same net in different directions that touch at a grid point are said to be connected by a "via" at that point.

The thickness of this rectangle must be determined by the router, i.e. we are allowed to add horizontal tracks to the rectangle, but vertical columns must remain intact. The objective is of course to minimize the number of tracks, in other words to route within a channel of minimal thickness. Without loss of generality we assume that the set of labels presented in *TOP* and *BOT* is [0, *N*], i.e. all the integers from 0 to *N*, where *N* is the number of nets. All vertical columns are numbered from 1 to *n*, where *n* denotes the length of the channel. It is worth mentioning that certain channels are impossible to wire with even an arbitrarily large number of tracks if there are no free columns. For example, the channel

$$TOP = 1, 2 \text{ and } BOT = 2, 1$$

is not routable. But addition of one vertical column results in channel

$$TOP = 0, 1, 2 \quad \text{and} \quad BOT = 0, 2, 1$$

which can be routed in three tracks. We further assume that any number of free columns can be added to either end of the channel.

Two graphs are associated with the channel and are important for routing. Both have the set of nets $-\{1, 2,..., N\}-$ for the set of vertices. Every column $i(i = 1, 2,..., n)$ such that $t(i)$ and $b(i)$ are not zeros introduces a directed edge from the node $t(i)$ to the node $b(i)$ in the Vertical constraints Graph VG. So VG is the directed graph associated with the channel. We assume that edges of this graph are labeled by the corresponding column numbers (see Figure 5 for the channel of Figure 4).

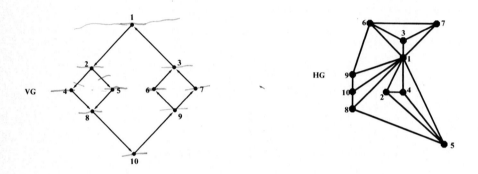

Figure 5.   Figure 6.

The horizontal constraints graph HG is constructed as follows. With every net i we associate an interval $I(i)$, where left point of $I(i)$ is the minimal column number k such that $t(k)$ or $b(k)$ equals i, and the right point of $I(i)$ is the maximal such column number. HG is simply an intersection graph of this system of intervals. So it is by construction an interval graph. We refer to monograph [6] for definitions of all graph theoretical notions. Figure 6 illustrates HG for the example presented in Figure 4.

The density of HG, or clique number, which is the maximal number of intervals crossing the same vertical line, presents a lower bound for channel thickness, i.e. the lower bound for the number of horizontal tracks:

$$den(C) = \omega(HG(C)) \le t(C),$$

where C is any channel, ω denotes clique number of a graph and $t(C)$ minimal number of horizontal tracks required for the channel. This lower bound, although trivial, is very important and for some time was the only lower bound available.

2. RESTRICTIVE ROUTING

These graphs $-VG$ and $HG-$ play very significant roles in the case of ***restrictive*** channel routing problem, when the number of horizontal tracks on which any net can be positioned is limited to one. In this case a wire geometry is very simple: any net is implemented as a single horizontal segment with vertical branches connecting it to the pins (Figure 7). The problem of restrictive routing is to determine a horizontal track number for every net. Restrictive routing has an important advantage besides the simplicity: it uses minimal number of vias.

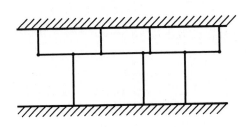

Figure 7.

Initially [1, 7] only restrictive routing was considered as acceptable solution. However, the restrictive solution may not exist. This is the case when the Vertical constraints Graph has directed cycles. Presence of an edge (a, b) in VG indicates that the horizontal segment of the net a must be positioned above the horizontal segment of the net b. Thus the cycle in VG cannot be resolved. But even if VG is acyclic the problem is very far from simple. We discuss that at the end of this section. However the problem can be easily and optimally solved in the case when there are no vertical constraints at all: the graph VG is empty.

CASE OF NO VERTICAL CONSTRAINTS

Note, that in the restrictive routing horizontal wire segments positioned on the same track do not intersect. This means that assignment of tracks to wire segments (i.e. restrictive routing) corresponds to proper coloring of the horizontal constraints graph HG, and vice versa. HG is by definition an interval graph, so it belongs to a very special category of graphs - perfect graphs [6]. Perfect graphs have several remarkable properties which make the problem of coloring relatively simple. We present here two simple algorithms for coloring of interval graphs. Optimality of these algorithms is a simple consequence of these properties of perfect graphs.

Line Packing Algorithm. Select a set of pairwise disjoint intervals as follows: Let I_1 be the interval with the minimum left end. Let I_2 be the interval with the minimum left larger than the right end of I_1. And so on. Let I_k be the interval with the minimum left larger than the right end of I_{k-1}. The selection process terminates when proper interval cannot be found. The intervals $I_1, I_2, \ldots, I_k$ are pairwise disjoint. They can be safely packed on one track. Corresponding nodes of HG can be colored using the same color. The important thing about this system of intervals is that the set of remaining intervals has smaller density. In fact, suppose that d is the density of the original channel and yet there are d intervals among remaining set of intervals sharing common point x. Then this point x must not belong to any of the intervals $I_1, I_2, \ldots, I_k$, because otherwise the density of the whole system would have been $d + 1$. Thus x falls between two consequtive intervals, say between I_{i-1} and I_i. Some of the d intervals containing x do not intersect I_{i-1} (because of the same density argument). Thus we can find an interval with the left end smaller than the left end of I_i and yet larger than the right end of I_{i-1} which contradicts the selection of I_i.

So, by packing the set $I_1, I_2, \ldots, I_k$ on one track we reduce the density of remaining system. We now can select a similar set from the remainder and pack it on the next track, and so on. We can be assured that no more then d tracks will be used. In the case of no vertical constraints this line packing algorithm is equivalent to restrictive routing. This algorithm is in fact the original router of Hashimoto and Stevens [1].

Left Edge Algorithm. Sort the set of intervals in ascending order of abscissas of their left ends. Let $I^1, I^2, \ldots, I^n$ denote this sorted set. Let $t_1, t_2, \ldots, t_d$ denote the routing tracks. Assign I^1 to t_1. Assuming that $I^1, I^2, \ldots, I^{i-1}$ are already assigned to tracks, assign I^i to an arbitrary available track. The availability of the track is guaranteed by the following argument. If I^i cannot be assigned to any track, the left edge of I^i intersects with previously assigned interval at every track $t_1, t_2, \ldots, t_d$. This brings the total density to $d + 1$.

After careful examination of both of these procedures one can realize that Line Packing Algorithm and Left Edge Algorithm are essentially the same algorithms. They produce identical solutions. For practical implementation, however, we prefer the Left Edge, because when properly implemented it runs in almost linear time (actually the running time of the Left Edge Algorithm is proportional to $n \cdot \log (d)$, where d is the channel density). It of course involves sorting, but since the elements that are sorted are integers from $[0, n]$ a radix sort technique may be applied.

VERTICAL CONSTRAINTS HANDLING

The length of the longest directed path in a vertical constraints graph VG (which must be acyclic) presents another lower bound for channel width which may easily be higher than density. It is often possible to use the Left Edge technique and/or its heuristic modifications and satisfy vertical constraints (so called **constrained** Left Edge or Line Packing algorithms). But in general only suboptimal solutions may be obtained. The authors of [7] correctly guessed that the restrictive routing problem, with presence of acyclic vertical constraints, is NP-complete. The fact was subsequently proved by A. LaPaugh [8]. A. LaPaugh reduced the known NP-complete problem of coloring of circular arc graphs to restricted channel routing, thus demonstrating that if there was an optimal restrictive channel routing algorithm running in polynomial time, than there would be a polynomial time algorithm for optimal coloring of circular arc graphs, and hence - for optimal solution of all NP-complete problems. This means that a computational complexity of any algorithm that guarantees an optimal solution to our problem is most likely exponential. And [7] suggests the algorithm performing actually an exaustive search based on branch and bound technique for track assignment.

In principal an exhaustive search routine has to enumerate all possible assignments of net segments to tracks satisfying vertical constraints. In the process of this enumeration an upper bound for channel width can be obtained by precedence: a width of any feasible solution obtained on the way presents an upper bound ub of the minimal channel width. On the other hand, suppose that we are examining the possibility of assignment of a current net segment I to track t_k. Suppose also that there is a directed path in VG starting at the vertex corresponding to I and having a length of l. This means that there will be at least l segments positioned one below another and below E, so that the channel width will become at least $l + k$. If $ub \le d + k$ that we can reject the possibility of assigning E to t_k right at this stage. This is the bounding condition of branch and bound search. If $ub > d + k$ then the assignment is accepted. We may arrive at another feasible solution with the width $w < ub$. In this case we reset $ub = w$. And the search continues.

There are certainly instances of restrictive channel routing problem that are easily solvable even with a non empty VG. Suppose that VG is a simple directed path. Then the solution is obvious (Figure 8). This fact inspires an idea of trying to reconstruct the original problem in such a way that VG will become a simple directed path. This idea belongs to T. Yoshimura and E. Kuh [9] : "Merging of nets". Consider two nets of the channel, I_1 and I_2, such that the corresponding nodes of the HG are not adjacent (i. e. the intervals do not overlap) and the corresponding nodes of VG cannot be connected by a directed path. In thit case the nets I_1 and I_2 can be merged into one. The corresponding operation in VG is simple node contraction. The implication on HG is more severe: if intervals of I_1 and I_2 were far apart than any net in between is adjacent to the new "merged" node in the new HG. But the new VG still has no directed cycles (remember that if there are cycles in vertical constraints graph, then restrictive solution does not exist). And VG is in a sense closer to a simple directed path. One can proceed with merging of nets until either VG is a simple directed path or any pair of nodes not connected by directed path in VG overlap (are connected in HG).

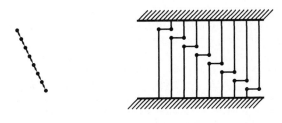

Figure 8.

This reduction process supposedly results in channel routing problem of much smaller size. T. Yoshimura [9, 10] suggests special heuristics for merging nets with minimal or no increase in the length of maximal directed path in *VG*. But the merging process may increase the density. Merging of nets is a heavy restriction: the merged nets must be assigned to the same track. However for many practical problems T. Yoshimura and E. Kuh were able to obtain best restrictive solutions much faster than exhaustive branch and bound search.

This basically concludes our discussion of the restrictive channel routing problem. We further demonstrate that allowing wires to occupy any number of tracks may lead to significantly better results if we stay within the bounds of classical (Manhattan) model. But the simplicity and effectiveness of Left Edge and Line Pack Algorithms may inspire completely different ideas for elimination of vertical constraints. One of them may be the introduction of the third layer for interconnections and realization the vertical segments connecting to the top row of the channel on the first layer, vertical segments connecting to the bottom row - on the third layer and using the second (intermediate) layer for horizontal segments. Future technological advances may very well allow something of this kind.

3. DOGLEG ROUTER

In this section we describe a dogleg router of D. Deutsch [11]. The idea is to remove the heavy constraint of restrictive routing on wire geometry, i.e. to allow net splitting between different tracks. In that case we must add vertical wire segments connecting horizontal pieces. These are called doglegs. Introduction of doglegs may enable us to decrease channel width significantly in the cases of long vertical constraints path in *VG* (Figure 9). They may also enable us to resolve cyclical constraints in *VG* (Figure 10). However, there are certain penalties we have to pay for those gains. From the electrical standpoint, each dogleg adds one or two additional vias and this increases the capacitance. Also, if a dogleg occurs at a position where local density is already equal to channel density, the required number of horizontal tracks increases by one and the problem cannot be completed in channel density. So it is also desirable to limit the number of doglegs. D. Deutsch [11] suggested a simple and very efficient way of introduction of doglegs: for a given net, doglegs are allowed only in those columns, where the net has terminals (except leftmost and rightmost columns). The following is the excerpt from [11] :

"By studying the output of non-dogleg routers, it was found that when they required many tracks in excess of the channel density, the usual cause was quite evident. There was a long constrain chain with a few crucial nets (typically clock lines) that were heavily connected to both sides of the channel. If these multi-terminal nets could be doglegged efficiently, a far more compact routing would be achieved. Based upon this observation, it was decided that doglegs would only be introduced at a terminal position for a net. This decision severely constraints the potential number of doglegs in that no 2-terminal net is ever doglegged, 3 terminal nets have only one potential site for doglegs, etc. Note that this rule is independent of net length (distance between leftmost and rightmost terminals). The number of added contacts is also minimized by introducing doglegs only at a position where the net already has a terminal."

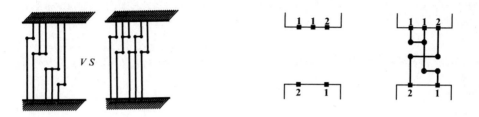

Figure 9.   **Figure 10.**

The idea of the Dogleg Router is to split every net into 2-terminal subnets as follows: if a net E has n terminals $t_1, t_2, \ldots, t_n$ which are sorted by ascending order of their abscissas: $x(t_1) \leq x(t_2) \leq \ldots \leq x(t_n)$ then E is substituted by $(n - 1)$ "sub"nets: $E_1, E_2, \ldots, E_{n-1}$ such that E_i connects t_i with t_{i+1}, $i = 1, 2, \ldots, n - 1$. However vertical constraints graph is not constructed in a straight forward manner for new set of nets: in fact, no vertical constraints are introduced between E_i and E_{i+1}, $i = 1, 2, \ldots, n - 1$. Note that this idea of splitting nets is quite opposite to merging process of Yoshimura and Kuh [9, 10]. The constrained Left Edge (or Line Packing) algorithm is then applied to the resulting set of nets with one modification: when a subnet ends, the next subnet of the same net can be placed in the same track (sharing a terminal with the previous subnet). Horizontal segments of subnets of the same net are finally connected by doglegs.

The above described Dogleg Router works reasonably well but it often adds many more doglegs than it is necessary. D. Deutsch [11] introduced a controlling parameter - "range " - in order to minimize this undesirable result. Range is the minimum number of consecutive subnets that must be assigned to the current track. As the range gets larger, fewer doglegs will be introduced (maybe at the expense of the width). The only additional rule is that a sequence of subnets shorter than range will also be accepted if this will terminate the processing of the original net. Without such a rule, 2-terminal nets would never get placed.

Introduction of Dogleg Router by D. Deutsch was at the time the breakthrough achievement and inspired further development of powerful heuristics for the general channel routing problem. Although the Dogleg Router still limits the possible wire geometry in a sense that no vertical line crosses the same wire twice, the term "dogleg" routing very often means general routing. Introduction of doglegs may decrease the number of required horizontal tracks significantly. It was demonstrated in [11] on several examples, one of which became a benchmark test case for numerous channel routers being developed ("Difficult" Channel). Since we will refer to this particular example often we will describe it in the remainder of this section. But before we proceed it is appropriate to mention that the general channel routing problem is also

NP-complete. This fact is not implied by the NP-completeness of the restricted problem and was established by T. Szymanski [12, 13] and in certain sence justifies the development of heuristic solutions.

We describe the "Difficult" Channel by presenting *TOP* and *BOT* vectors. Channel length is equal to 175.

TOP =

```
 1  3  5  7  9  5 12 14 15  7 12 14  7  4 13  8  6 15 18 14  8  6 11 22 21
 0 18 16 18 16  0  8  6 26 11  0 24 23 25 20  1 29  0 22  3 22  3  0  0  9
 2  9  2  0 32 23 33 19  6  8 30 27 34 35 36 37 39 31 39 35 38 31  8 30 37
41 19  6 44 45  0 33 31 33 31  0 27 35 36 48 49 31 39 46 47 50 52 20 53 24
 0 47 39  0 24 51 20 52 20 52 23  8 30 50 56  0  0 57 49 19  6  6 19 49 59
 0  0 61 50 30  8 55  0 24 64 20 52  0 67 68 63 55 24 52 20 69 24  0 46 62
63 68  0 24 65 20 52  0 70 60 62 54 63  0 24 71 20 52 67  0  0  0  0  0  0
```

BOT =

```
 0  2  4  6  8 10 11 13  3  9 16  5 17 11  5 14 14  7 12 17 19  1 20 21 23
24  0 16 10  3 11 25  0 26 11 26 11  0 27 28 11  3  9 16 30 27  5 31  1  5
 1 20 32 23 24  0  9  1 20 29 23 24  0  3  8 30 38 28 19  6 40 27 35 41 42
 6 19 34 43 30  8 31 43 39 46 36 46 47 48 31  0 24 23 45 20  1 51  0 40 39
40 39  0  8 30 50 54  0  0 55 49 19  6  0 47 42 47 42  0 53 58  6 19 49 50
30  8 60 62 59 54 55 54 56 63 55 65  0 66 68 66 68  0 60 68  0 46 44 46 44
 0 69  0 55 58 55 58  0 64 71  0 72 63 72 63  0 57 62 54 70 67 55 61 63 68
```

The density of this channel equals to 19. There are five columns where the maximal density of 19 is achieved: 75, 80, 81, 82, 83. Vertical constraints graph of this example has no cycles, but it has a directed path of the length of 28 (there was a typographic error in [11] stating that the number was 26), thus establishing the 28 as the lower bound of channel width for restrictive problem. The branch and bound search [7] results in 28 track solution (requires hours of computation time). The Dogleg Router [11] produces a solution requiring 21 tracks. Doglegging technique can be combined with net merging of Yoshimura and Kuh [9] yielding a 20 track routing of "Difficult" Channel (and in general slightly better results).

Remark. In our tables describing "Difficult" Channel the first column (1,0) was added to the original data of D. Deutsch (it was not there, but there was a requirement that net #1 should enter from the left side of the channel).

Dogleg Router along with routers of [9, 10] give usually superior results over Left Edge Router, very often requiring just a few tracks over the channel density, but there is a major drawback: they do not operate with cyclical vertical constraints. Certainly one may argue that directed cycles in *VG* are rare and can often be avoided by replacement and pin swapping, still their presence may be annoying; especially in cases when the channel router is a part of a silicon compiler, and no manual intervention is permitted. In this respect the development of general channel router capable of cyclic constraints handling becomes very important.

4. "GREEDY" ROUTER

If we try to extend the doglegging idea to permit   doglegging in any column not necessarily containing terminal pin of the doglegged   net, we notice the following interesting phenomenon. Instead of splitting every net E into "sub"nets like in Dogleg Router, let us split every net into "sub"nets connecting every two consecutive columns of the net span. We do not arrive at any meaningful channel routing problem as a result of this splitting, but if we assume that there are no vertical constraints between subnets of the same net and try to apply the Left Edge Algorithm we will notice that it will be performing   left-to-right column-by-column scan of the entire channel. And this is an underlying idea of Greedy Router of R. Rivest and C. Fiduccia [14].

The Greedy Router scans the channel in a left-to-right, column-by-column manner, completing the wiring within a given column before proceeding to the next. In each column the router tries to optimize the utilization of wiring tracks in a "greedy" fashion. It may place a net on more than one track and have a vertical line crossing more than one horizontal segment of the same net. The router always completes the routing, even in the presence of cyclical vertical constraints, often using no more than one track above the density. We will describe the routing algorithm pretty much in the same way it was originally presented in [14].

When routing a given column, the router classifies each net having a pin to the right as either *rising* , *falling* or *steady*. A net is *rising* if its next pin after the current column will be on the top of the channel (say in column k), and the net has no pin on the bottom of the channel before column $k + stc$, where stc is a steady-net-constant - a nonnegative integer controlling parameter. *Falling* nets are defined similarly. *Steady* nets are the remaining nets. The fundamental data structure for the router is the set $Y(E)$ for each net E of y -coordinates of "tracks currently occupied" by net E. If $Y(E) = \emptyset$, the net is not currently being routed. Otherwise, for each $y \in Y(E)$ (assuming the router is working at the column i) the point (i, y) is a "dangling end" of some wiring segment already placed for net E. Exactly one such "dangling end" is listed in $Y(E)$ for each connected piece of wiring already placed for net E. Eventually these "dangling ends" will be connected together. When extending the routing from column i to column $i + 1$, horizontal wiring will be used in every track $y \in Y(E)$ for some E.

A net E is said to be *split* or *collapsible* at the given time if $| Y(E) | > 1$. Following [14] we illustrate the steps of the routing algrithm using a set of "before-after" pictures (Figure 11). Nets entering the column from the previous column are shown extended up to the current column. If the net has pins to the right of the current column, the net is shown extended towards the next column with an arrowhead.

GREEDY ROUTING ALGORITHM

Route Channel From Left to Right: **For** each column $i = 1, 2,...,$ **until** $i \geq n$ and there are no split nets **do:**

(a) Make Feasible Top and Bottom Connections: if $TOP[i] \neq 0$ or $BOT[i] \neq 0$, "bring in" that net if possible to the nearest possible track which is either empty or already assigned to this net; add that track to $Y(TOP[i])$ or $Y(BOT[i])$. (Pic. A). Note that a net E is not routed to the nearest track in $Y(E)$ if there is a nearer empty track (Pics. B, C). In the case of all tracks being occupied nothing is done at this time (Pic. D). If

$TOP[i] \neq 0$ and $BOT[i] \neq 0$, try to "bring in" both nets, but if vertical segments will overlap then just bring in the net with least wire (Pic. E). As special case, if there are no empty tracks, and $TOP[i] = BOT[i] \neq 0$, then run a vertical wire from to bottom of this column (Pic. F).

(b) *Free Up As Many Tracks As Possible By Collapsing Split Nets:* Add doglegs in this column to collapse split nets in a pattern that will create the most empty tracks for use in the next column. Find an admissible system of doglegs which creates the largest number of empty tracks by an exhaustive search (it is affordable because of the limited number of admissible patterns for doglegs). Chosen dogleg pattern will free up one track per dogleg it contains, plus one additional track for every net it "finishes". Any ties between patterns that free up most tracks are resolved by choosing the pattern which leaves the outermost uncollapsed split net as far as possible from the channel edge (Pic. I). Any remaining ties are resolved by choosing the pattern with largest sum of dogleg lengths (Pic. J). For every dogleg in the chosen pattern connecting track $y1$ with track $y2$ ($y1 < y2$) for some net E , delete $y1$ from $Y(E)$. (This is an arbitrary choice that might get modified later). Note that this step will typically collapse a net that was temporarily brought in to an empty track in step (a) when that net had a previously assigned but more distant track.

(c) *Add Doglegs To Reduce The Range of Split Nets:* For each uncollapsed split net E, the range of tracks assigned to the net is reduced by adding doglegs that have the effect of moving the net: (i) from the maximum track in $Y(E)$ to the lowest empty track and (ii) from the minimum t track in $Y(E)$ to the highest possible empty track (Pic. K). Since this is performed after step (b), no collapsing will occur, but the difficulty of collapsing the remaining split nets may be reduced. If a dogleg for net E is made from track $y1$ to track $y2$ replace $y1$ by $y2$ in $Y(E)$.

(d) *Add Doglegs to Raise Rising Nets and Lower Falling Nets:* Consider all the unsplit rising and falling nets being routed in order of decreasing distance from their track $y \in Y(E)$ to their "target edge". A dogleg will be introduced (if possible) moving the net to an empty track closes to it's target edge. If a dogleg for net E is made from track $y1$ to track $y2$ replace $y1$ by $y2$ in $Y(E)$.

(e) *Widen Channel If Needed To Make Previously Infeasible Top Or Bottom Connections:* If a net $TOP[i]$ or $BOT[i]$ could not be brought in to a track in step (a), create a new track for this net and bring the net in to this track. Place this track as near the center of the channel as possible between existing tracks, but so that desired connection to the edge of the channel can be made (Pic. M). Re-label the tracks. Add the new track to $Y(TOP[i])$ or $Y(BOT[i])$ as appropriate.

(f) *Extend To Next Column:* For each net E such that $|Y(E)| = 1$ and E has no pins after column i , set $Y(E) = \emptyset$. Then for each track $y \in Y(E)$ for some E, extend the "dangling end" for net E along track y into column $i + 1$ with appropriate horizontal wiring (Pic. N).

This completes the description of Greedy Router. The Router usually starts with the number of tracks equal to channel density, however the ***initial-channel-width*** can be used as a controlling parameter of the algorithm.

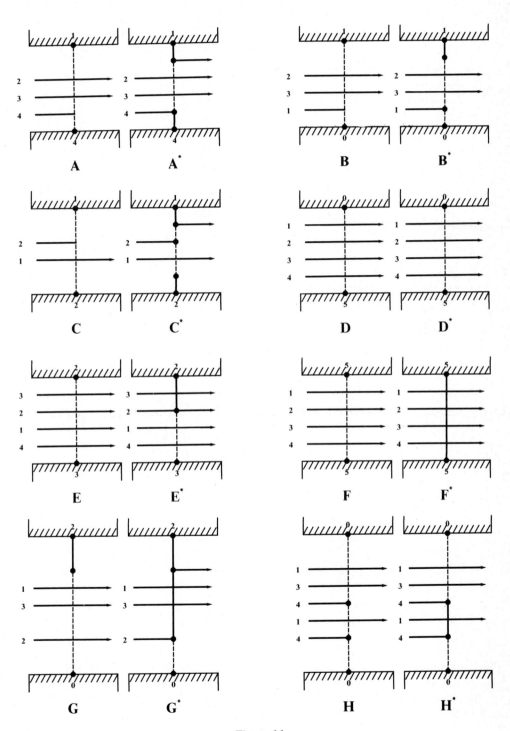

Figure 11.

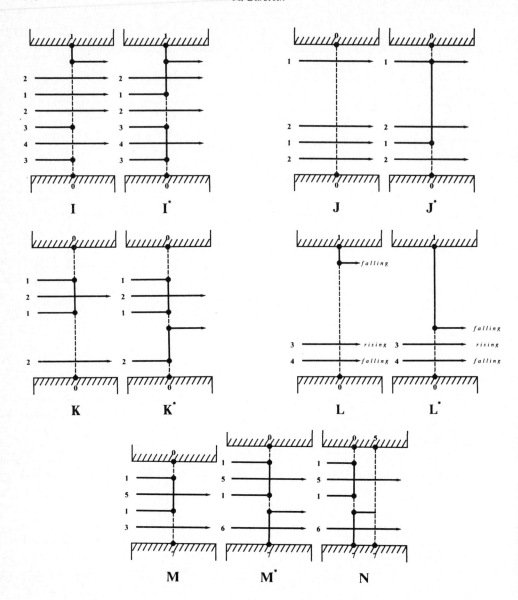

Figure 11 (continued).

Another useful controlling parameter enabling (along with **steady-net-constant** stc) to reduce number of vias is **minimum-jog-length** - lower bound for the dogleg length alllowed at phases (b), (c), (d). The Greedy Router may add columns beyond the channel length in order to collapse dangling ends.

"This algorithm is the result of long series of experimentation and evaluation of variations on the basic idea of scanning down the channel from left to right and routing everything as you go"([14]).

One nice feature of the Greedy Router is that its control structure is very flexible and robust: it is easy to make variations in the heuristics employed to achieve special effects or to rearrange priorities.

The Greedy Router was tested on numerous channels from practical chips as well as program-generated test examples. It performed satisfactorily achieving channel width equal to density in most cases. When applied to Deutsch's "Difficult" Channel the Greedy Router produced 20 track wiring (although it used more vias than the router of [9]). Note, that the data on "Difficult" Channel presented in [14] is not complete, it presents only about 'three quarters' of the "Difficult" channel. Nevertheless, the router completed the whole channel in 20 tracks as well.

5. HIERARCHICAL ROUTER

In this section we present another approach to channel routing problem applying 'divide-and-conquer' technique [16, 17, 18]. We call the resulting router 'Hierarchical'. It was the first router that automatically completed the "Difficult" Channel in 19 tracks. In order to describe the router we have to introduce the

GENERALIZED CHANNEL ROUTING PROBLEM.

The main reason behind this generalization is that there exists a mechanism for reduction of the generalized problem as well as the original channel routing problem, to a generalized problem of smaller size. An essential part of the generalized problem is that wiring is viewed in terms of grid cells rather then grid points and lines.

Suppose we are given a channel, i.e. a pair of vectors

$$TOP = t(1), t(2), \ldots, t(n)$$
$$BOT = b(1), b(2), \ldots, b(n)$$

Consider the rectilinear $(m \times n)$ grid $G = G(i, j)$, $i = 1, 2,\ldots, m$, $j = 1, 2, \ldots, n$. Here $G(i, j)$ denotes a cell of the grid at the intersection of the $i -$ th and $j -$ th strips. Grid lines are comprised of cell boundaries. We distinguish between horizontal and vertical boundaries. The horizontal boundary $H(i, j)$ $(i = 1, 2,\ldots, (m - 1)$, $j = 1, 2,\ldots, n)$ is merely a segment separating $G(i, j)$ and $G(i + 1, j)$. Similarly, the vertical boundary $V(i, j)$ $(i = 1, 2,\ldots, m, \cdot j = 1, 2,\ldots, (n - 1))$ is a segment separating $G(i, j)$ and $G(i, j + 1)$.

We assume that all boundaries of the grid are weighted by nonnegative integers; the weight of a boundary B will be denoted $W(B)$. We call these weights **boundary capacities** and impose on them the following restrictions:

(a) for every horizontal boundary H:

$$0 \leq W(H) \leq 1$$

(b) capacities of the top and bottom vertical boundaries are zeros:

$$W(V(k,j)) = 0, \text{ for } k = 1, m \quad (i = 1, 2,..., n).$$

The **graph** of the grid has cells as its vertices, and two cells are adjacent if they share a common boundary (vertical or horizontal). Any subtree R of this graph is called a **route.** We say that the route R crosses a boundary B if a pair of cells attached to this boundary constitutes an edge in R. We say that the route R turns within a cell G if it crosses two non-opposite boundaries of G. Figure 12 illustrates these notions. We assume that the numbers $TOP = t(1), t(2), \dots , t(n)$ are assigned to the top strip of cells of the grid - $G(1,1), G(1,2), \dots , G(1,n)$ and similarly that $BOT = b(1), b(2), \dots , b(n)$ are assigned to the bottom strip $G(m, 1), \dots , G(m, n)$. In other words, these cells are labeled by the integers from TOP and BOT.

A **routing** is a set of routes $R(1), R(2), \dots , R(N)$, such that $R(i)$ contains all cells carrying the label $i, i = 1, 2, \dots , N$. Suppose a certain routing is specified and consider an arbitrary vertical boundary $V(i, j)$. Let LL (left load) denote the number of routes not crossing $V(i, j)$ but turning within the cell $G(i, j)$, and similarly RL (right load) denote the number of nets turning within $G(i, j + 1)$ and not crossing $V(i, j)$. The load of the vertical boundary $V(i, j)$ is defined as

$$L(V) = \max \{LL, RL\}.$$

For any boundary B let $C(B)$ denote the number of routes crossing B.

Now we are ready for the final   **definition.**

Routing $R(1), R(2), ..., R(N)$ is called   **legal** if and only if

(i)   for any boundary B:

$$C(B) \le W(B);$$

(ii)   for any vertical boundary V:

$$C(V) + L(V) \le W(V).$$

Figure 12.   Figure 13.

Generating a legal routing for a given grid with appropriate labels from TOP and BOT and with capacities assigned to the boundaries satisfying (a) and (b), presents a generalized channel routing problem. If all boundary capacities are equal to 1 (except those ruled out in (b)), the legal (generalized) routing corresponds to traditional channel routing, because condition (ii) rules out "overlapping vias" (Figure 13).

HIERARCHICAL ROUTING

Our approach to channel routing is based on the reduction to a generalized problem for a $(2 \times n)$ grid. The reduction is performed on every level of hierarchy in a way outlined below.

Consider the generalized problem for an $(m \times n)$ grid. Partition the grid into two parts - $([m/2] \times n)$ and $(]m/2[\times n)$ subgrids. Consider vertical strips of these subgrids as single supercells, i.e. factorize the cells of the subgrids making all cells with the same abscissa equivalent (consider them as one).

We end up with two horizontal strips, i.e. a $(2 \times n)$ grid. Vertical boundary capacities of this grid will be the sums of corresponding boundary capacities of the original grid. If the original capacities were all equal to 1, then the new capacities of the $(2 \times n)$ problem will represent the numbers of horizontal tracks crossing the boundaries. The process is illustrated on Figure 14. The $(2 \times n)$ wiring procedure is described in the next section.

Assuming that the routing within this $(2 \times n)$ grid is obtained, we now partition each of the horizontal strips of the grid into two parts, thus generating two $(2 \times n)$ subproblems (second level of hierarchy).

The global routes from the previous level define terminal positions for wiring of new $(2 \times n)$ subproblems. Figure 15 illustrates this step. We proceed with the resulting $(2 \times n)$ routing in the same manner until single cell resolution is reached. At this level the routing will be completed.

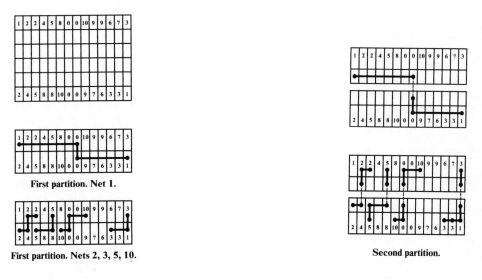

First partition. Net 1.

First partition. Nets 2, 3, 5, 10.

Second partition.

Figure 14. **Figure 15.**

It is worth mentioning that at every level of hierarchy we obtain routes of all nets, so a net is never positioned exactly before all other nets are routed globally at least on previous levels of hierarchy.

If the $(2 \times n)$ router fails to obtain a legal routing, i.e. the routing satisfying (i) and (ii) of the previous section, then condition (ii) is relaxed. In other words the value $W(V) - C(V) - L(V)$ may become equal to (-1). In this case all vertical capacities of boundaries in the same horizontal strip as V are increased by 1. The increase is equivalent to the addition of a new horizontal track to the original problem.

WIRING WITHIN $(2 \times n)$ GRID

Denote $G(i, j)$ the cells of our $(2 \times n)$ grid, $i = 1, 2$, $j = 1, 2, ..., n$. Wiring within the grid of thickness 2 is performed by the algorithm TBN (two-by-n) which allocates routes for all nets, one at a time.

With each interconnection route of a net we associate a cost, which is comprised of the costs of boundary crossings; it is simply the sum of those costs.

The TBN wiring algorithm is a linear algorithm that finds the minimal cost tree interconnecting a set of elements located on a $(2 \times n)$ grid.

The input for the algorithm consists of

1) The terminal locations on the $(2 \times n)$ grid: EL is a $(2 \times n)$ Boolean matrix; if $EL(i,j) = TRUE$ then the corresponding cell is a terminal cell.

2) Costs of boundary crossings :
   - HC is an n element vector which stores the costs of crossing the horizontal boundaries; $HC(j)$ indicates the cost that must be added if a wire crosses the boundary between $G(1, j)$ and $G(2, j)$;
   - VC is a $(2 \times (n - 1))$ matrix that stores the costs of vertical boundary crossings; $VC(i, j)$ indicates the cost which must be added if a wire crosses the boundary between $G(i, j)$ and $G(i, j + 1)$, $i = 1, 2$.

Note : if the barrier cost is greater than LRG, which is set to a large number, the corresponding boundary is blocked and a wire can not pass through this cell boundary. In other words, cost of crossing of the blocked boundary in isfinitely high.

Matrices HC and VC represent our cost-functions. They should reflect the boundary capacities and via conditions. Clearly, $VC(i, j)$ and $HC(j)$ must be a decreasing functions of the number of wiring tracks crossing the corresponding boundaries. Selection of the proper cost functions is extremely important because of its influence on the quality of the final routing. We discuss this point in the next section.

The output of TBN consists of the minimal interconnection tree.

The TBN algorithm is a modification of an algorithm developed in [19], which finds a Steiner tree that interconnects a set of elements located on a $(2 \times n)$ grid. But our algorithm solves the problem for an arbitrary cost-function associated with crossing a boundary (matrices HC and VC), whereas the algorithm of [19] assumes the rectilinear Steiner tree, which is the case when all these costs are equal to 1.

We need the following definitions.

1) Let $T^1(k)$ denote the minimal cost tree which interconnects the following set of cells:
$$\{G(i,j) : (j \le n) \& (EL(i,j) = TRUE)\} \cup \{G(1, k)\}.$$

2) Let $T^2(k)$ denote the minimal cost tree which interconnects the following set of cells:

$$\{G(i,j) : (j \le n)\&(EL(i,j) = TRUE)\} \cup \{G(2, k)\}.$$

3) Let $T^3(k)$ denote the minimal cost tree which interconnects the following set of cells:

$$\{G(i,j) : (j \le n)\&(EL(i,j) = TRUE)\} \cup \{G(1, k) ; G(2, k)\}.$$

4) Let $T^4(k)$ denote the minimal cost forest, consisting of two different trees T^* and T^{**} : T^* uses cell $G(1, k)$, T^{**} uses cell $G(2,k)$ and the set

$$\{G(i,j) : (j \le n)\&(EL(i,j) = TRUE)\}$$

is interconnected by either one of them (this means that the trees have to be joined later).

We compute the trees $T^i(k + 1)$, $i = 1, 2, 3, 4$ recursively from $T^i(k)$ (Figure 16).

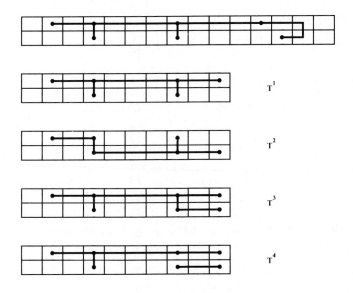

Figure 16.

This procedure is sometimes referred to as dynamic programming. First we need to construct initial trees to start the recursion.

Denote by *FIRST* and *LAST* the abscissas of the leftmost and rightmost terminal cells, i.e.

$$FIRST = \min \{ k : EL(1,k) \lor EL(2,k) = TRUE\},$$

$$LAST = \max \{ k : EL(1,k) \lor EL(2,k) = TRUE\}.$$

Trees $T^i(k)$ for $k \le FIRST$ are computed trivially and serve as a basis for recursion. In fact, for $k \le FIRST$ $T^l(k)$ $(l = 1, 2)$ consists of a single vertex : $G(l, k)$, $T^4(k)$ consists of the disjoint pair of vertices : $G(1, k)$ and $G(2, k)$. However, $T^3(k)$ is obviously a path $G(1, k), \dots , G(1, s), G(2, s), \dots , G(2, k)$ where $1 \leq s \leq k$ and

$$H(s) + \sum_{i=s}^{i=k-1} V(1,i) + V(2,i)$$

is **minimal**.

In rectilinear case, when all costs are the same, $T^3(k)$ is the adjacent pair of vertices $-G(1, k)$ and $G(2, k)$. But in general, the cost $H(k)$ might be too high and detouring might result in cheaper route (see Figure 17).

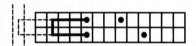   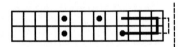

Figure 17.   **Figure 18.**

Suppose that $FIRST \leq k \leq LAST$ and the trees $T^i(k)$, $i = 1, 2, 3, 4$ are constructed. To construct $T^j(k + 1)$ we enumerate all possible extensions from $T^i(k)$, $i = 1, 2, 3, 4$, and select the cheapest one. For example, to construct $T^1(k + 1)$ we select the cheapest way of adding $G(1, k + 1)$ to each of the $T^i(k)$, $i = 1, 2, 3, 4$, and select the cheapest among them. Note, that $T^4(k + 1)$ always becomes either trivial extension of $T^4(k)$ or one of the vertices $G(1, k + 1)$ or $G(2, k + 1)$ is isolated whereas the other component is identical to $T^1(k + 1)$ or $T^2(k + 1)$.

When the value $k = LAST$ is reached our trees are covering all terminal cells. We temporarily select the cheapest among $T^1(LAST)$, $T^2(LAST)$, $T^3(LAST)$ and denote it by T. In rectilinear case it is obviously the interconnection we need. But it is possible that the costs $H(s)$ for $s \leq LAST$ are so high that it is cheaper to take a right side detour. In a way symmetric to the construction of $T^3(FIRST)$ we select a path $-G(1, LAST), \dots , G(1, s), G(2, s), \dots , G(2, LAST)$, where $LAST < s \leq n$ and

$$H(s) + \sum_{i=LAST}^{i=s-1} V(1,i) + V(2,i)$$

is **minimal**.

Then we join the components of $T^4(LAST)$ by this path, (see Figure 18), compare the cost of resulting interconnection with the cost of T and finally select the cheapest.

Note that at each stage of recursion we make a constant number of comparisons, and the total computation time is $O(n)$ because the initiation stage - computing $T^3(FIRST)$ - takes $O(FIRST)$ time, construction of T is performed in time $O(LAST - FIRST)$ and final right side detour is computed in at most $O(n - LAST)$ time.

In case when the interval [*FIRST, LAST*] is small in comparison with [1, *n*] we can limit the detours from either side (Figures 17, 18) in order to eliminate unnecessary computations. We usually limit the detour outside of the [*FIRST,LAST*] segment by 2 or 3 , so in most cases the time spent for 2×*n* routing will be $O(LAST - FIRST)$. On the other hand, we may be forced at some point to cross a boundary having an infinite cost LRG. This will result in non resolvable conflict for routing because it means utilization of the same piece of the vertical track for two different nets. In order to avoid this conflict we permit to wire outside of the channel, in other words we add additional vertical tracks at the left or right ends of the channel. This is achieved by introduction of additional boundaries $V(1, i)$, $V(2, i)$, $H(i)$ for $i \leq 0$ and for $i \geq n$ and assigment of high but not infinite cost to their crossing.

REFINEMENTS

Costs of crossings of horizontal boundaries - *HC* - are chosen very simply. Because, by definition, the capacity *W* of a horizontal boundary $B(j)$ is either 0 or 1, $HC(j)$ is set to LRG if $W(B(j)) = 0$, i.e. if this boundary is already crossed by one of already routed nets, and $HC(j) = C1$ otherwise, where $C1$ is a positive constant.

Costs of vertical crossings - *VC* - are more complex. If *B* is a vertical boundary $B(i, j)$ ($i = 1, 2$, $j = 1, 2 ,..., n$), $W(B)$ its initial capacity, $C(B)$ the number of already routed nets crossing *B*, and $L(B)$ the load, then set

$$X = W(B) - C(B) - L(B)$$

and

$$VC(i,j) = (C2)^{(-X)} + F,$$

where $C2$ is another positive constant and *F* is an additive corrector, which is determined by the vertical constraints graph *VG* and is not dependent on previously routed wires. The corrector *F* reflects relative positions of the nodes of *VG* corresponding to nets whose intervals in *HG* cross the *j* -th vertical line. It makes an influence on wiring only if there are several minimal interconnections of equal cost. In order to avoid random picking of the wire route at the starting stage of TBN the values of *F* can be set to force nets to be either above or under the cut line. If the currently routed wire belongs to a directed cycle in a subgraph of *VG* spanned by nets crossing the j-th vertical line, *F* is set to 0. If it does not belong to a directed cycle then we can estimate how many nets should be above or below the current net in this subgraph, and set *F* accordingly (*F* can be both negative and positive). Setting *F* to 0 all the time also does not hurt, if the net will be rerouted later, as is mentioned in the next section. But proper setting of correctors may result in good wire packing the first time.

The constants $C1$ and $C2$ are defined experimentally during the "fine-tuning" process of algorithm implementation.

In case when the TBN is unable to generate an interconnection of the total cost less than infinite (LRG) we allow detouring outside the channel. Costs of crossing the boundaries $V(1, i)$, $V(2, i)$, $H(i)$ for $i \leq 0$ and for $i \geq n$. are all equal to some big number, but significantly less than LRG; they are chosen in such a way, that a finite cost interconnection within the channel limits is always cheaper than an interconnection crossing these outer boundaries, but crossing of those is still cheaper than LRG (Figures 17, 18).

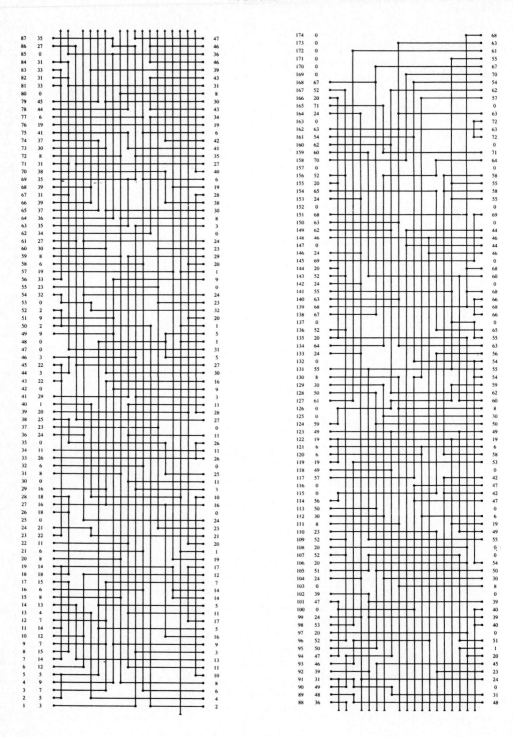

Figure 19.

Because the $(2 \times n)$ wiring algorithm performs wiring of one net at a time, it is dependent on the order in which the nets are routed. But the rerouting procedure decreases this order dependency. Once all the nets are imbedded into the $(2 \times n)$ grid we remove one of them, update the boundary capacities, compute new boundary costs (all other nets remain imbedded) and again run the TBN procedure (against the background of all the other wires). We perform this rerouting for every net in random order. Very often the nets choose exactly the same route they had before. But several of them in fact get rerouted, improving the distribution of the remaining boundary capacities. Sometimes the rerouting allows us to get rid of negative boundary capacities, thus eliminating the necessity of introduction of new horizontal track. We experimentally observed that in repeated rerouting (third and subsequent applications of TBN) nets tend not to choose different routes (which is not the case when maze-type rerouting is applied for a general grid). So, more than one cycle of rerouting appears impractical for the case of the $(2 \times n)$ grid. The small thickness of the grid - 2 strips - is probably the reason why the rerouting process converges so fast.

Another trick, which sometimes helps a lot, is rerouting within a sliding $2 \times n$ window; at any level of hierarchy we may select a pair of consecutive rows, usually with largest number of negative boundary capacities, and try to reroute the net segments passing through them.

R E S U L T S

The complexity of the presented algorithm can be estimated as follows. The time spent for routing of a single net at any level of hierarchy is proportional to the total length of horizontal segments of this net (assuming that a constant time is spent for "detouring"). It is observed that the total length of a net at any level of hierarchy is bounded by $O(n)$, where n the length of the channel (number of columns). Since there are N nets, and $\log_2 m$ levels of hierarchy, where m the width of the channel (number of tracks), the upper bound for algorithm complexity can be expressed as $N \cdot n \cdot \log_2 m$.

This figure should be considered as a worst case behavior, when almost all nets expand through the whole channel.

The algorithm was tested on several artificially created random examples with approximately 500 nets, channel length of 1000. The results were close to optimal, in 7 out of 10 examples the lower bounds for channel thickness were achieved, in 2 of them the achieved thickness was 1 track above the lower bound, and only 1 required two additional tracks.

The Hierarchical Router is the only router (presently) that completed the Deutsch's "Difficult" Channel in 19 tracks. The 19 track wiring [17] is reproduced on the Figure 19. It uses 336 vias, 31 doglegs, the total wire length equal to 5023. Only one net, namely #51, is detouring, i.e. using columns outside of it's interval span, however several other nets are detouring "internally". The solution was obtained after significant amount of experimentation (more than 20 program runs, one run taking about 22 CPU seconds of 3081) with different cost settings, tuning the corrector (using the results of analysis of Vertical constraints Graph structure) and rerouting within the sliding $2 \times n$ window.

6. NEW LOWER BOUNDS

The importance of the channel density as the lower bound of the channel width is now apparent. So, it is natural to ask ourselves a question: is it possible to derive an upper bound for channel width in terms of density alone ? The answer to this question is negative, because, as it will be demonstrated below, channel density is not the only factor determining the width; we must also consider how many nets must "switch columns" in order to be routed. These considerations lead to another lower bounds for the width. We are presenting here the results of D. J. Brown and R. L. Rivest [20].

Suppose we are given a channel routing problem $TOP = p_1, p_2, \ldots, p_n,$ and $BOT = q_1, q_2, \ldots, q_n,$ such that every net is 2-terminal and connects TOP to the BOT, i.e. every net can be represented by a pair of integers (p_i, q_j). We further refer to channels satisfying the above as to **2-point net** channels. A net (p_i, q_j) is called trivial if $p_i = q_j$. Let m denote the number of nontrivial nets, i.e. nets that have to change the columns (to be "moved"). Denote t the thickness of the channel, the minimal possible number of tracks for which a routing exists.

Lemma. $t \geq -(n - m) + \left] \sqrt{(n - m)^2 + 2m} \right[$.

Proof. We going to scan track by track from TOP to BOT and keep count of nets that can be moved into their target columns on each track. We consider a sequence $l_0, l_1, \ldots, l_s$ of numbers, l_i denoting the number of nontrivial nets that get to their target column positions at the track i. Denote also $m_i = m - l_i$ - number of nets that are to be moved after the track i. Naturally, $m_0 = m$ ($l_0 = 0$), $m_t = 0$ ($l_t = m$).

Note, that we are allowed to use columns outside of the channel span; we refer to columns $1, 2, , \ldots, n$ as to **regular** columns. However, irregular columns (i, with $i \leq 0$ or $i > n$) may be used for wiring. Let e_i denote the number of regular empty columns between tracks i and $i + 1$. When a net makes a horizontal jog within a track to target column c then this column must have been empty between the current track and the previous one. This means that

$$m_i - m_{i+1} \leq e_i$$

or

$$m \leq \sum_{i=0}^{t-1} e_i.$$

Note that when a wire jogs horizontally within a track i then one empty column ("source") between tracks i and $i + 1$ is matched exactly with one empty column ("target") between tracks $i - 1$ and i. The only way to change e_i from one track to the next is to route wires from a regular column to irregular (outside of channel span) column (which increases e_i by one) or vice versa (which decreases e_i by one).

Note also, that since at most two nets can be routed the from regular to irregular columns within one track - one to the left, another to the right - we have

$$e_i - 2 \leq e_{i+1} \leq e_i + 2.$$

This means that $e_i \leq e_0 + 2i$ and $e_i \leq e_t + 2(t - i)$ for any i. This implies that, for $t \geq 3$:

$$m \leq \sum_{i=0}^{[t/2]} (e_0 + 2i) + \sum_{i=[t/2]+1}^{t-1} (e_t + 2(t - i))$$

But it is easy to see, that under our assumptions about the channel, $e_0 = e_t = n-m$, so

$$m \leq t(n - m) + \sum_{i=0}^{[t/2]} i + \sum_{i=1}^{[(t-1)/2]} i$$

or

$$m \leq t(n - m) + \frac{t^2}{2}$$

and this implies the stated lower bound.

The bound of the Lemma is tight in that for any n and m values, there is an example which is routable in the specified number of tracks [20, 23].

If in addition to the above conditions the utilization of irregular columns is not allowed, then $e_0 = e_1 = \ldots = e_t$ and the lower bound is transformed:

$$t \geq \frac{m}{n - m}.$$

1 2 3 4 5 6 7 8 9 10 11 12 13

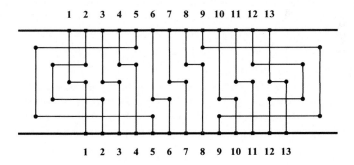

1 2 3 4 5 6 7 8 9 10 11 12 13

Figure 20.

Let us consider a specific channel, which is called "shift-right-one" channel:

$$TOP = 1, 2, 3, \ldots, n, 0$$

$$BOT = 0, 1, 2, \ldots, n - 1, n$$

The density of this example equals 2, but according to the lemma at least $-1 + \sqrt{2n+1}$ tracks are required for routing; in case when it is not permissible to route outside of channel span, n tracks are required. Figure 20 presents optimal routing of shift-right-one example for $n = 13$.

7. CHANNEL FLUX

Motivated by the argument of the previous section, and following [21, 22], we introduce here the concept of **channel flux** which provides another fundamental bound for the channel width. Flux and density together characterize the difficulty of channel routing problem. Baker, Bhatt and Leighton [21] presented a routing algorithm with provable upper bound in terms of density and flux.

While channel density provides a global limitation on channel width, it fails to capture the local congestion inside a channel. Consider the segment of consecutive pins on one side of the channel that all have to be permuted or at least change columns; we need certain number of horizontal tracks in order to realize the permutation, i.e. "move" the nets.

Suppose that instead of making vertical cuts in the channel, we make a horizontal cut which isolates a set of contiguous columns from one side of the channel. We can vary the size of the cut (number of columns within the cut) as well as it's position. We say that a net is split by the a horizontal cut if it contains terminals both within the cut and outside. For any given position of a cut we can measure the number of distinct nets split by the cut. The greater the number of distinct nets split by a cut, the greater the congestion is within the cut. The larger the size of a congested cut, the larger the channel width, because if the region of local congestion is very large, then so is the overall global congestion of the channel.

We further assume that the channels considered in this section do not have trivial nets. If a channel contains a trivial net, then a straight top to bottom connection is performed, after which the net and the column can be removed, so that we concentrate on nontrivial nets.

Definition. *The flux of a channel is the largest integer f for which there exists a horizontal cut of size $2f^2$ which splits at least $2f^2 - f$ nontrivial nets.*

Using similar arguments as for the lemma of the previous section, we prove that flux is a lower bound for channel width.

Theorem 1. *Every channel with density d and flux f requires channel width of at least max (f,d) tracks.*

Proof. Let C be a horizontal cut of the channel which spans $2f^2$ columns and splits $\geq 2f^2 - f$ nontrivial nets. For each nontrivial net split by the cut, choose any two terminals from different columns that lie on opposite sides of the cut. Consider the channel formed by the set of chosen terminals, i.e. assume that all columns which do not contain a chosen terminal are blank. This new channel consists of at least $2f^2 - f$ nontrivial two point nets. Moreover, at most f of the $2f^2$ columns spanned by the original cut may be empty. It is clear that no more than $f + 2$ nets can be routed into the correct column on the first track: f into empty columns and one out of each side of the cut. After the first track there are at most $f + 2$ empty columns, the extra two having possibly been created by wires exiting across the side of the cut in the first track. So, at most $f + 4$ nets can be routed into correct column on the second track. In general, at most $f + 2i$ nets can be routed into the correct column on the track number i.

Suppose there exists a wiring of this channel in t tracks. The total number of nets that can change columns anywhere in the channel is bounded by

$$\sum_{i=1}^{t} (f + 2i) \; = \; tf + t(t + 1).$$

But since at least $2f^2 - f$ nets must eventually be routed, we have: $tf + t(t + 1) \geq 2f^2 - f$ or $t \geq f$, which proves the theorem.

Flux is usually small for practical problems. There are three reasons for this. First, practical problems tend to have columns that contain less then two terminals. Second, practical problems often have nets containing terminals that are close together and on the same side of the channel. Third, flux is a local phenomena and thus is less likely to grow with the length of the channel. Flux value for Deutsch's "Difficult" Channel is merely 3. That is why good heuristic algorithms achieve routing solutions in number of tracks close to density. However the notion of flux enabled Baker, Bhatt and Leighton to develop an upper bound for channel width along with an algorithm that guarantees the number of tracks not exceeding the upper bound.

Theorem 2. *The "Manhattan" Routing Algorithm of* [21] *routes every problem with density d and flux f in a channel of width* $2d + O(f)$. *The running time of the Algorithm is linear in the area of routing.*

Theorem 3. *The "Manhattan" Routing Algorithm of* [21] *routes every 2-point net problem with density d and flux f in a channel of width* $d + O(f)$. *The running time of the Algorithm is linear in the area of routing.*

The importance of "Manhattan" Routing Algorithm of [21] is of fundamental theoretical nature. For practical problems it is easily outperformed by heuristics.

8. SWITCHBOX ROUTING

Switchbox Routing problem can be viewed as more restrictive version of Channel Routing. Switchbox sometimes referred to as "Four Sided Channel". Switchbox is a rectangular routing area with no obstructions inside and with nets entering from all four sides of the rectangle (Figure 21).

The problem can be simply described by labeling the terminal pins by net numbers; pins having the same positive label have to be interconnected. The switchbox routing in practical cases turns out to be more difficult than general channel routing. Mainly because it is not clear what to do in case of failures; unlike in channel routing addition of tracks is not allowed. Theoretically it obviously falls into the category of NP-hard problems.

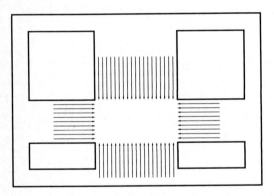

Figure 21.

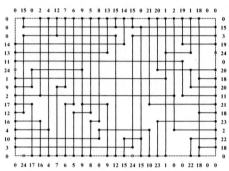

Figure 22.

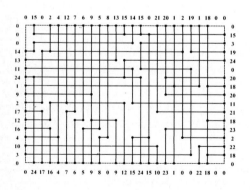

Figure 23.

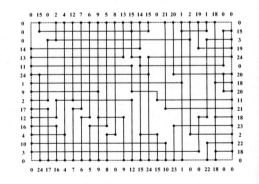

Figure 24.

Several heuristics are developed for the problem. J. Soukup [5] briefly described an unpublished algorithm due to S. Akers. Minimum Impact Router [24] can be applied to switchbox problem. One of the best practical switchbox routers belongs to Hamachi and Ousterhout [25]. The Two-Dimensional Router [26, 27] can be converted to a switchbox router. The ideas of Greedy channel router can be naturally extended to do the switchbox routing [28] (similar algorithm was also reported by J. Koschella in his B.S. Thesis at MIT [29]). Hierarchical channel router can also be modified for switchbox problem and in a fairly straightforward way: instead of partitioning along the length of the channel, alternate partition in both vertical and horizontal directions should be introduced [16, 17].

In [17] we published a small example of the switchbox problem to be used as a benchmark for comparisons of different heuristics. It has 24 nets; grid size - 16 by 24. Our switchbox router completed all but one net (Figure 22). The example seemed to be difficult, so we [17] conjectured that it may not be wireable. We were wrong. T. Stanion [29] was the first who produced the solution manually applying greedy routing technique. The switchbox router of [25] also left out one net, but completed the example with limited interactive guidance (one net manually prewired). D. Deutsch [30] reports two manual solutions, one of which does not use the leftmost column (Figure 23). K. Gardiner [31] reports that his router, which is similar to the router by C.-P. Hsu [26], produced the solution completely automatically (Figure 24). This router uses a heuristic maze search and a complex series of cost functions to optimize wireability (however is very slow).

This concludes our survey of routing algorithms for the classical ("Manhattan") wiring model.

9. OTHER WIRING MODELS

This section is merely a bibliographical survey of results on channel routing for different wiring models.

YACR

YACR stands for "Yet Another Channel Router"; the router proposed by Sangiovanni-Vincentelli and Santomauro [32] operates under the assumption that a user may add vertical tracks when they are needed anywhere within a channel (similar to [15]). Clearly one can achieve the channel width equal to density under these assumptions, since a vertical column may be added every time there is a vertical conflict, after the horizontal segments are packed by Left Edge routine. But YACR attempts to minimize (heuristically) the number of added columns. YACR does not use dogleging. Experimental results demonstrated improvement in overall routing area over routers operating within classical model; YACR also requires fewer vias. It can also be applied to "three sided channels" (switchbox with no pins on one side).

YACR-II

YACR-II is the follow-up version of YACR, developed by the same authors plus Reed [33]. It operates under different assumptions, however. Unlike YACR, YACR-II is not allowed to add columns. But it is allowed to perform horizontal "jogs" on the "vertical" layer (and, optionally, vertical "jogs" on a "horizontal" layer), which may result in wire overlap. The algorithm works in two phases: horizontal track assignment and maze running. The first phase is performed in the manner similar to it's precursor - YACR: horizontal segments are packed in channel of width equal to density in such a way that scores of vertical conflicts are minimized. All admissible (non conflicting) vertical segments are imbedded. The rest of interconnections are realized as a result of sequential invocation of three different maze running routines; wire overlap may occur at this stage.

YACR-II is very fast and practical in most cases, since most of the technologies allow limited wire overlap. From the theoretical point of view the results of YACR and YACR-II are not very surprising. YACR-II achieved 19 track routing of Deutsch's "Difficult" example in 5 CPU seconds on a VAX 11/780. However, 19 tracks is no longer a lower bound for the width in this case, since horizontal overlap is allowed.

FREE WAY - NO OVERLAP ROUTER

The router proposed by Marek-Sadowska and Kuh [34] operates under the assumptions that wires can be routed on both layers in both directions, but no wire overlap is allowed. For the case of 2 terminal nets only the authors introduce net classification as follows.

> **Net type A: both terminals on the top side of a channel;**
> **Net type B: both terminals on the bottom side of a channel;**
> **Net type C: trivial – terminals on the same vertical track;**
> **Net type D: rising – bottom pin to the left of the top;**
> **Net type E: falling – top pin to the left of the bottom.**

The router restricts the possible ways the nets can be routed as it is shown on Figure 25. Solid lines represent connections on the first layer, dashed lines - on the second.

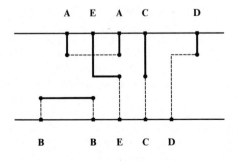

Figure 25.

The algorithm is based on an order-graph $G_o = (V,E)$, which is a digraph whose nodes correspond to all nets and a directed edge $(x,y) \epsilon E$ iff net x and net y horizontally overlap and the horizontal segment of x must be placed above that of y in order to avoid 'shorting' intersection on the same layer. Unlike vertical constraints graph, the graph G_o is always acyclic. The algorithm routes nets top down by following the order-graph. The nets corresponding to nodes with no ancestors are packed first by Line Packing routine "packing" the first wiring track. After the first track is completed, the problem is reduced, order-graph updated and the routine is called again, and so on.

The general case of multi-terminal nets can be reduced to the above [34]. The experimental results indicated mixed performance of the router in terms of the achieved channel width. The main contribution of [34] is the departure from the classical model and the demonstration of the flexibility in the routing problem.

WIRING WITH OVERLAP AND KNOCK-KNEE WIRING

Rivest, Baratz and Miller [35] considered 2 layered wiring model where wires are allowed to run on both layers in both directions, are free to change layers through vias and even run parallel to each other on different layers - overlap. Channel density d is no longer a lower bound for channel width in this case; all we can trivially guarantee is $d/2$. But the authors of [35, 36] presented an algorithm that will never use more than $2d - 1$ tracks for 2-point nets problem. Donna Brown [23] points out that the description of Algorithm 1 in [35] in not precise and claims that $(7d/4 + const)$ suffice for the routing in this case.

Papers [37, 38] deal with multilayer overlap models, also present provably good linear time algorithms ("provably good" usually means "not arbitrarily bad", i.e. an upper bound for channel width can be proved; but in this case upper bounds are really close to known lower bounds).

"Knock-knee" model is multilayered routing model where wires are free to be routed in both direction on each layer, but overlap is forbidden; however allowed is corner overlap - "Knock-Knee". Tom Leighton [39] established some lower bounds for channel width for this model. Franco Preparata and others [40−43] made major contributions to this field in terms of upper bounds, bounds on number of layers and algorithms.

Lack of time and space prevents us from going into details of the results mentioned in this section. Most of these results may have very important practical applications.

RIVER ROUTING

River routing is an instance of routing within a channel (or switchbox) when there is only one layer available for interconnections. In case of 2-point problem pins must appear exactly in the same order on both sides of the channel in order to assure planarity. The planarity issue is extremely restrictive and it makes the problem significantly simpler. Optimal solutions however involve more geometrical rather than combinatorial considerations. Main contributors to this area are Tompa [44], Dolev et al. [45, 46] , Leiserson and Pinter [47, 48] and Hsu [49].

MISCELLANEOUS

W. Chan [50] suggests another heuristic algorithm operating within classical model. It extends several ideas of [9]; cyclical vertical constraints are not handled. Deutsch's "Difficult" channel takes 20 tracks to wire by this algorithm.

Very interesting extension of the channel routing problem was proposed by Leong and Lui [51]. The set of terminals of every net is partitioned into disjoint subsets. The problem is to connect one representative from each subset together. The idea of proposing the problem like that is inspired by possibility of iteration between placement phase of the design with the routing stage. [51] also suggests a method of reduction of the generalized problem to the classical one, which is subsequently solved by the "greedy" router [14].

R. Pinter [52, 53] studies interesting and useful properties of wire layouts within the classical and other models.

I conclude with the list (far from complete) of recent papers [54−63] also addressing various aspects of the channel routing problem.

ACKNOWLEDGEMENTS

Mary Youssef, Ravi Nair and Jerry Kurtzberg communicated many suggestions and remarks. Correspondence with Donna Brown, Ted Stanion, Kevin Gardiner, Dave Deutsch, Alberto Sangiovanni-Vincentelli was very helpful. I wish to thank them all.

I am also indebted to Ralph Otten for his numerous reminders that this paper is long overdue; his reference #11 of my last reference [64] (hereby creating perpetual cyclical constraint!) was very inspiring.

R E F E R E N C E S

[1] Hashimoto, A., Stevens, J., Wire Routing by Optimizing Channel Assignment within Large Apertures Proc. 8th Design Automation Workshop, 1971, 155-169

[2] Nan, N., Feuer, M., A method for automatic wiring of LSI chips, Proc. IEEE Int. Symp. Circuits and Systems, New York, 1978, 11-15.

[3] Heller, W. R., Hsi, C. G., Mikhail, W. F., Wirability - Designing Wiring Space for Chips and Chip Packages, Design and Test of Computers, V. 1, 3, 1984, 43-51.

[4] Persky, G., Deutsch, D. N., Schweikert, D. G., LTX - A minicomputer-based system for automatic LSI Layout, J. Des. Automation and Fault-Tolerant Computing, V. 1, 3, 1977, 217-256.

[5] Soukup, J., Circuit Layout, Proc. of the IEEE, V. 69, 10, 1981, 1281-1304.

[6] Berge, C., Graphs and Hypergraphs, Dunod, Paris, 1970.

[7] Kernighan, B. W., Schweikert, D. G., Persky, G., An Optimum Channel-Routing Algorithm for Polycell Layouts of Integrated Circuits, Proc. 10th Design Automation Workshop, 1973, 50-59.

[8] LaPaugh, A. S., Algorithms for Integrated Circuit Layout: An Analytic Approach, Ph.D. Theses, Laboratory for Computer Science, MIT, (November 1980).

[9] Yoshimura, T., Kuh, E. S., Efficient Algorithms for Channel Routing, IEEE Trans. on CAD of Integrated Circuits and Systems, V. CAD-1, 1, 1982, 25-35.

[10] Yoshimura, T., An efficient channel router, Proc. of the 21st Design Automation Conference, 1984, 38-44.

[11] Deutsch, D. N., A dogleg channel router, Proc. of the 13th Design Automation Conference, 1976, 425-433.

[12] Szymanski, T. G., Dogleg channel routing is NP-complete, IEEE Trans. on CAD, CAD-4, 1, January 1985, 31-40

[13] Johnson, D. S., The NP-Completeness Column: An Ongoing Guide, Journal of Algorithms, 3, 1982, 381-395.

[14] Rivest, R. L., Fiduccia C. M., A "Greedy" Channel Router, Proc. 19th Design Automation Conf., 1982, 418-424.

[15] Kawamoto, T., Kajitani Y., The Minimum Width Routing of a 2-Row 2-Layer Polycell Layout, Proc. 16th Design Automation Conf., 1979, 290-296.

[16] Burstein, M., Pelavin, R., Hierarchical Channel Router, INTEGRATION, the VLSI Journal, 1, 1983, 21-38, also in the Proc. of 20-th Design Automation Conference, 1983, 591-597.

[17] Burstein, M., Pelavin, R., Hierarchical Wire Routing, IEEE Trans. on CAD., Special Issue on Automatic Wire Routing, Ed. E. S. Kuh, CAD-2, 4, October 1983, 223-234.

[18] Burstein, M., Hong, S. J., Pelavin, R., Hierarchical VLSI Layout: Simultaneous Placement and Wiring of Gate Arrays, Proc. International Conf. VLSI-83, Trondheim, Norway, 1983, 45-60.

[19] Aho, A., Garey, M. R., Hwang, F. K., Rectilinear Steiner Trees: Efficient Special Case Algorithms, Networks, 7, 1977, 37-58

[20] Brown, D. J., Rivest, R. L., New Lower Bounds for Channel Width, Proc. 1981 CMU Conf. VLSI Systems and Computations, 1981, 178-185.

[21] Baker, B. S., Bhatt, S. N., Leighton, F. T., An Approximation Algorithm for Manhattan Routing, 15-th Annual Symp. Theory of Computing, 1983.

[22] Bhatt, S. N., The Complexity of Graph Layout and Channel Routing for VLSI, Ph.D. Theses, Dept. Electrical Engineering and Computer Science, MIT, (February 1984).

[23] Brown, D. J., Private Communication, January 1985.

[24] Supowit, K. J., A Minimum-Impact Routing Algorithm, Proc. 19-th Design Automation Conf., 1982, 104-111.

[25] Hamachi, G., Ousterhout, A., A Switch Box Router with Obstacle Avoidance, Proc. 21st Design Automation Conf., 1984, 173-179.

[26] Hsu, C.-P., A New Two-Dimensional Routing Algorithm, Proc. 19-th Design Automation Conf., 1982, 46-50.

[27] Hsu, C.-P., Theory and Algorithms for Signal Routing in Integrated Circuit Layout, Ph.D. Theses, University of California, Berkeley, May 1983.

[28] Luk, W. K., A Greedy Switchbox Router, CMU-CS-84-148 Technical Report.

[29] Stanion, T., Private Communication, December 1983

[30] Deutsch, D. N., Solutions to a Switchbox Routing Problem, IEEE Trans. on CAD, (to appear), 1985.

[31] Gardiner, K., Private Communication, May, 1984.

[32] Sangiovanni-Vincentelli, A., Santomauro, M., YACR: Yet Another Channel Router, Proc. Custom Integr. Circuits Conf., Rochester, NY, 1982, 460-466.

[33] Sangiovanni-Vincentelli, A., Santomauro, M., Reed, J., A New Gridless Channel Router: Yet Another Channel Router the Second (YACR-II), Proc. ICCAD, 1984.

[34] Marek-Sadowska, M., Kuh, E. S., A New Approach to Channel Routing, Proc. ISCAS-82, Rome, 1982, 764-767.

[35] Rivest, R. L., Baratz, A. E., Miller, G., Provably Good Channel Routing Algorithms. Proc. 1981 CMU Conf. VLSI Systems and Computations, 1981, 153-159.

[36] Bolognesi, T., Brown, D. J., A Channel Routing Algorithm with bounded Wire Length, Unpublished Manuscript, Coordinated Science Lab., Univ. of Illinois at Urbana-Champaign, 1982.

[37] Brady, M. L., Brown, D. J., Optimal Multilayer Channel Routing with Overlap, Technical Report, Dept. Electrical and Computer Engineering, University of Colorado, Boulder, 1984

[38] Hambrusch, S. E., Channel Routing Algorithms for Overlap Models, IEEE Trans. on CAD, CAD-4, 1, 1985, 23-30.

[39] Leighton, F. T., New Lower Bounds for Channel Routing, Unpublished Manuscript, Lab. for Computer Science, MIT, 1981.

[40] Preparata, F. P., Sarrafzadeh, M., Channel Routing of Nets of Bounded Degree, Amalfi VLSI Conf., January 1984.

[41] Sarrafzadeh, M., Preparata, F. P., Compact Channel Routing of Multiterminal Nets", Technical Report ACT-44, Coordinated Science Lab., University of Illinois, October 1983 (also to appear in Annals of Discrete Mathematics).

[42] Mehlhorn, K., Channel Routing in Knock-Knee Mode: Simplified Algorithms and Proofs, Technical Report, SFB 124, B2, Univ. Saarlandes, West Germany, November 1984.

[43] Preparata, F. P., Lipski Jr., W., Optimal Three Layer Channel Routing, IEEE Trans. on Computers, C-33, 5, (May 1984), 427-437 (also in Proc. 23rd Foundations of Computer Science Conf., pp. 350-357).

[44] Tompa, M., An Optimal Solution to a Wire-Routing Problem, J. Computer and System Sciences, 23, 1981, 127-150,

[45] Dolev, D., Karplus, A., Siegel, A., Strong, A., Ullman, J. D., Optimal wiring between rectangles, Proc. 13th ACM Symp. Theory of Computing, May 1981, 312-317

[46] Siegel, A., Dolev, D., The Separation for General Single Layer Wiring Barriers, Proc. 1981 CMU Conf. VLSI Systems and Computations, 1981, 143-151.

[47] Leiserson, C. E., Pinter, R. Y., Optimal Placement for River Routing, Proc. 1981 CMU Conf. VLSI Systems and Computations, 1981, 126-142 (also SIAM Journal on Computing, V.12, 3, 1983).

[48] Pinter, R. Y., River Routing: Methodology and Analysis, 3rd CALTECH Conference on VLSI, March 1983, 141-163

[49] Hsu, C.-P., General River Routing Algorithm Proc. 20th Design Automation Conf., 1983, 578-583

[50] Chan, W. S., A New Channel Routing Algorithm, 3rd CALTECH Conference on VLSI, March 1983, 117-139

[51] Leong, H. W., Liu, C. L., A New Channel Routing Problem, Proc. 20th Design Automation Conf., 1983, 584-590

[52] Pinter, R. Y., On Routing Two-Point Nets Across a Channel, Proc. 19th Design Automation Conf., 1982, 894-902

[53] Pinter, R. Y., Optimal Routing in Rectilinear Channels, Proc. 1981 CMU Conf. VLSI Systems and Computations, 1981, 160-177.

[54] Hightower, D., Boyd, R., A Generalized Channel Router, Proc. 17th Design Automation Conf., 1980, 12-21

[55] Wiesel, M., Mlynski, D. A., Two-Dimensional Channel Routing and Channel Intersection Problems, Proc. 19th Design Automation Conf., 1982, 733-739.

[56] Dupenloup, G., A Wire Routing Scheme For Double-Layer Cell Arrays, Proc. 21st Design Automation Conf., 1984, 32-35.

[57] Chadrasekhar, M. S., Breuer, M. A., Bounds on Channel Width and a Routing Algorithm for Classical Channel Configuration, Proc. IEEE ICCC-82, September 1982, 250-255.

[58] Ishii, M., Harada, N., Ido, S., Koyama, M., Inoue, T., A Channel Router Having Layer Assignment Capability for Arbitrarily-Sized Rectangular Building Blocks, Proc. IEEE ICCC-82, September 1982, 260-264.

[59] Heyns, W., The 1-2-3 Routing Algorithm or the Single Channel 2-Step Router on 3 Interconnection Layers, Proc. 19th Design Automation Conf., 1982, 113-120.

[60] Sato, K., Shimoyama, H., Nagai, T., Ozaki, M., A "Grid-Free" Channel Router, Proc. 17th Design Automation Conf., 1980, 22-31.

[61] Terai, M., Kanada, H., Sato, K., Yahara, T., A Consideration of the Number of Horizontal Grids Used in the Routing of Masterslice Layout, Proc. 19th Design Automation Conf., 1982, 121-128.

[62] Young, M. H., Cooke, L., A Preprocessor for Channel Routing, Proc. 18th Design Automation Conf., 1981, 756-761.

[63] Wada, M. M. A Dogleg "Optimal" Channel Router with Completion Enhancements, Proc. 18th Design Automation Conf., 1981, 762-768.

[64] VanGinneken, L. P. P. P., Otten, R. H. J. M., Stepwise Layout Refinement, Proc. IEEE ICCD-84, October 1984, 30-36.

Chapter 5

GLOBAL ROUTING

E.S. KUH, M. MAREK-SADOWSKA

*Department of Electrical Engineering and Computer Sciences
and the Electronics Research Laboratory
University of California, Berkeley, CA 94720, U.S.A.*

1. INTRODUCTION

Circuit layout is a crucial part of VLSI design. It deals with the physical design of chips to fulfill the device, circuit and interconnection specifications, and in the meantime, to satisfy design rules for a given technology. Soon, chips with a million transistors will be a reality. The task of putting together a circuit of this size on a tiny chip is obviously complex. So, the layout problem is usually divided into subproblems of chip planning and wirability analysis, partitioning, plcement, global routing and detailed routing. This paper deals with global routing. It assumes the placement of devices or a collection of devices and circuits in the form of modules which are rectangular or rectilinear in shape. It is to be followed by detailed routing, i.e., the final interconnection specifications among the terminals (pins) of modules in terms of layer, via and track assignments.

Global routing (sometimes called losse routing) is the preliminary step of the complete routing process. It calls for a routing plan in which each net is assigned to particular regions on the chip reserved for routing. The aim is not only to make 100% assignments of nets to regions for routing, but also to minimize, for example, the total wire length leading to a minimal chip area.[1] The regions allocated for routing depend on the layout style. There are three approaches to current layout design: the gate array, the standard cell and the building block. The global routing problem for each is somewhat different.

In gate array and standard cell design, routing space is provided principally by horizontal routing channels. Channels are open-ended regions with pins on the top and bottom boundaries. So, the global routing problem amounts to assigning nets to channels. The interchannel connection is by means of vertical avenues on the sides or center of a chip and by feedthroughs between gates or cells on each row. The main difference between the two approaches is that the gate array chip is prefabricated and has channels of fixed height whereas the standard cell design has channels whose heights are unspecified and the aim is to minimize the sum of channel heights, thus the total chip area. We will discuss global routing for gate array and standard cell in Sections 3 and 4, respectively.

In the building block design, the modules are of irregular shape and size. Routing regions are only defined after placement is completed. Defining the routing regions is a crucial first step in global routing. Regions can be channels, switchboxes and/or L-shaped, etc. In general, channels are considered the most desirable because channel routing (two-layer detailed routing) is well understood. Not only can we route channels efficiently, i.e., with high routing density, but also we can estimate accurately the required channel width in advance. Unfortunately, it is generally insufficient to employ only channels in building block layout. At present there is active research on the problem of routing region definition and ordering. Global routing in building block layout will be discussed in Section 5.

One common problem for global routing in all three approaches is to find the short-est path in connecting a net. Standard methods are available to connect a two-pin net, for example, the Lee-Moore algorithm [1]. For a multi-pin net, the shortest path problem amounts to the well-known Steiner tree problem [2]. In the following section we will discuss briefly the Steiner-tree-on-graph problem along with its application to global routing.

2.   SEQUENTIAL ROUTING AND THE STEINER PROBLEM ON THE GLOBAL ROUTING GRAPH

The most common approach to global routing is sequential routing, i.e., to route one net at a time and to choose the shortest path whenever possible. The Lee-Moore grid expansion algorithm and many of its modified versions have been widely used, for example see [3]. The problem of finding the shortest connection for an n-pin net is called the Steiner problem. It is related to the spanning tree problem but it allows, in addition to the n-pins, connecting points called the Steiner points. Figure 1 illustrates the shortest spanning tree, the shortest Steiner tree and the shortest rectilinear Steiner tree for a 3-pin net. For layout we use rectilinear geometry, thus the shortest rectilinear Steiner tree is of principal interest. Hanan [4] showed a simple algorithm for determining the shortest rectilinear tree for $n \leq 5$. The solution for $n > 5$ is still unknown; however, good heuristic algo-rithms are available [5].

In global routing, the problem calls for a 100% routing completion given the spec-ified routing regions. It is, in general, not possible to route all the nets with shortest connection paths because they compete with each other for paths to be used. There are two approaches to solve this problem. One is to route all nets

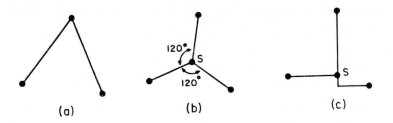

Figure 1

3-pin nets: (a) shortest spanning tree, (b) shortest Steiner tree in Euclidean geometry, and (c) shortest rectilinear Steiner tree.

independently, and if some routing regions become too crowded, reroute the nets which cause congestion [8]. The second approach is to route nets sequentially and update the information about routing regions after completing each net. Obviously, the routing order becomes a crucial factor. There are arguments for routing the shortest nets first. There are also arguments for routing nets with the most pins first. It is almost impossible to devise a general procedure which works the best in all situations.

The global routing problem can be represented by a global routing graph to depict relationships between routing regions and pins to be connected. There are different ways of defining the global routing graph. For example, edges of a graph could represent routing regions and be given weights to exhibit routing conditions in the sequential routing process. The weight can include the length factor of a channel and the current information on congestion due to nets previously routed. Let ω be the edge weight, L be the length of a routing channel, and T be the current number of available tracks in the channel. Then the weight can be specified by

$$\omega = aL + \frac{b}{c^{\frac{T+1}{}}} \tag{1}$$

where a, b and c are constants to be selected. Clearly, T changes after each net is routed, and the weight needs to be updated. The problem of assigning a path on a graph with the least total edge weight is thus of interest. With the weight chosen in Eq. (1), the channels already congested will be less likely picked.

Formally, we define $G = (V, E, \omega)$ be a graph where V is the set of N vertices which include, among others, the n-pins of a net being routed, E, the set of edges

representing routing regions, and ω the set of positive real numbers representing
the edge weights. The precise ways of defining the global routing graph will be
discussed later when we consider the different layout approaches. The Steiner
problem on graph can be stated as follows:

Given $G = (V,E,\omega)$ and a subset $V_1 \subseteq V$, find a connected subgraph G_1 which spans V_1
and has a minimum total edge weight. The problem is illustrated in Fig. 2 where
$V = \{a,b,c,d,e\}$ and $V_1 = \{a,b,d\}$. The edge weight is marked and the solution is
given in Fig. 2(b).

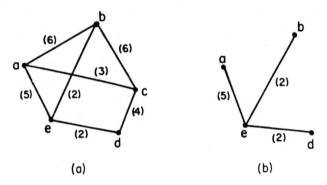

(a)   (b)

Figure 2

With $V_1 = \{a,b,d\}$, the minimum Steiner tree is shown where the total edge weight is
equal to 9. The vertex e is the Steiner point.

It is of interest to note that if $|V_1| = 2$, the problem is reduced to the 2-point
shortest path problem, for which there exist many good algorithms [7]. If $|V_1|$
$= |V|$, the problem is reduced to the familiar minimum spanning tree problem, for
which there are the Prim and the Kruskal algorithms [8,9]. The general Steiner
tree problem on graph is NP complete. Heuristic algorithms for solving the
problem of Steiner tree on graph often make use of the exact solutions for the
above extreme cases.

One approach uses an "extended" shortest-path-on-graph algorithm. This basic
algorithm is executed for each vertex being added to the gradually constructed
Steiner tree. The algorithm to find an approximation of Steiner tree by repeated
application of the Dijkstra's shortest path algorithm is as follows:

Single Component Growth Algorithm

IF $|V_1| = 0$ or $|V_1| = 1$, THEN RETURN;
Initialize component $G_1 \leftarrow (\{s\}, \phi, \omega)$ where $s \in V_1$;
Initialize $\underset{\sim}{V}_1 \leftarrow \{s\}$;
WHILE $\underset{\sim}{V}_1 \neq V_1$ DO
BEGIN
do expansion on G_1 until it meets a vertex $t \in V_1 - \underset{\sim}{V}_1$; $G_1 \leftarrow G_1 \cup S(G_1, t)$,
$\underset{\sim}{V}_1 \leftarrow \underset{\sim}{V}_1 \cup \{t\}$
END

Here $S(G_1, t)$ denotes a shortest path from component G_1 to the vertex t. When the algorithm terminates, G_1 is the obtained approximation of the Steiner tree.

A better approximation of Steiner tree can be found if, in using the above method, both the initial component and its growth are selected more carefully. In [10] one of the proposed algorithms is a variation of the single component growth method with the following modifications. First, the initial component is chosen to be the shortest Seiner tree for 3 vertices in V_1, out of all possibilities. Then at each pass three carefully selected components are connected by applying the same algorithm as in the initial step.

A second approach applies a philosophy similar to that of Kruskal's spanning tree construction process. Like the single component growth method, it uses the expansion as the basic operation. However, it allows multiple components to grow at the same time and to combine with each other. The algorithm terminates when all components are combined into a single one.

Multiple Component Growth Algorithm

Let $V_1 = \{v_1, v_2, \ldots v_n\}$,
IF $|V_1| = 0$ or $|V_1| = 1$, THEN RETURN;
Initialize component $G_i \leftarrow (\{v_i\}, \phi, \omega)$, $1 \leq i \leq n$;
$r \leftarrow n$;
WHILE $r > 1$ DO
BEGIN
do expansions on existing $\{G_i\}$ until two components meet;
let these two components be G_m and G_n, then
$G_m \leftarrow G_m \cup G_n \cup S(G_m, G_n)$, $G_n \leftarrow \phi$, $r \leftarrow r-1$;
END

The modification of the multiple component growth algorithm reported in [10] calls for connecting three components with an optimum solution at each pass. The more components combined with an optimum solution at each step, the better solution obtained among those components, and the better final solution.

3. Global Routing in Gate-Array

Consider an artificial example of a 3-row gate array configuration as shown in Fig. 3(a). Each row consists of four gates. There are four horizontal routing channels and three vertical avenues each specified in terms of the maximum number of tracks allowed. Each gate has four pins on top and four pins on the bottom. Given the net-list, that is, the interconnection specification for the pins and nets, the routing problem is to make a 100% connection subject to the track requirements. For the purpose of global routing, we can divide the chip into 4x4 global cells as shown in Fig. 3(b). The track requirements on the channels are translated into a boundary capacity requirement called cross supply. We assume that the cross supply for each boundary of the 4x4 global cells in equal to 2.

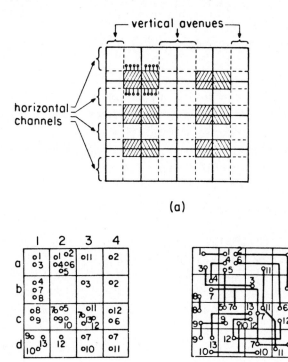

(a)

(b)

(c)

Figure 3
(a) Gate array configuration, (b) global cells and pin location, and
(c) completed routing.

The pins in each global cell, designated by the net numbers, are also shown. Pins
specified by the same number are to be connected; their precise locations within
each cell are of no consequence. The global routing problem is to determine the
path for each net in terms of the global cell it goes through without violating the
cross supply constraint at the boundaries. A solution to the example is illustrated in Fig. 3(c).

It should be mentioned that global routing in gate array is to be followed by
vertical assignment; after which, the specification for channel routing is
precisely determined. Detailed routing then follows. This is beyond the scope of
the present chapter.

Global routing for gate array can be accomplished by various methods. The sequential routing based on the Steiner tree is a popular approach. The graph model
used is obtained directly from the global cell configuration and the cross capacity
information. Thus, for each global cell, we have a vertex, and for each boundary
between global cells, we have an edge. The edge weight needs to be updated after
each net is routed to give the most current data on the number of available tracks.
The method depends very much on net ordering and, in general, leaves many nets
unconnected. Thus a rerouter is needed to complete 100% routing. One rerouting
method is discussed in Sec. 3.2.4.

There are other methods developed to handle global routing for gate arrays. In the
following we will discuss some of these approaches, which can be classified into
three groups: those based on optimization, using hierarchical decomposition, and
those using a constructive approach. A rerouting algorithm can supplement any of
these methods.

We should also mention the existence of hardware routers, e.g., the wire-routing
machines of IBM. They are based on the familiar Lee-Moore grid expansion algorithms and use parallel computation to speed up the process. The machines can be
constructed using existing VLSI parts. The acceptability of hardware routers rests
on the cost nad accessibility of such machines as well as the quality of future
software algorithms for global routing.

3.1. Optimization

The global routing problem can be formulated as an optimization problem. However,
in order to make it solvable, some simplifying assumptions are necessary.
Typically, they are:

A1. The net list is restricted to contain 2-pin nets only.

A2. The shape of the routes is limited to few feasible patterns. Usually only 1 or 2 bends per wire are allowed.

In most practical cases the above assumptions are not too restrictive. Any global routing problem for gate array with net list L_1 having multi-pin nets can be transformed into a problem with net list L_2 having 2-pin nets only; and whenever the second problem is solved so is the first. The restriction on the number of bends is also desirable because usually a bend introduces a via which is not only more costly, but is also less reliable. Intuitively, it is obvious that the global routing problem, when transformed into an optimization problem, will call for integer solutions. Thus, we may expect an integer programming approach, and that indeed is the case [11]. The other approach is to observe an analogy of the problem with statistical physics of random systems and use heuristic optimization techniques [12]. In the following we will give a brief sketch of the two approaches.

3.1.1. Integer programming

With assumptions A1 and A2, the global routing problem can be formulated as follows:

Given a net list $L = \{(s_1,t_1),\ (s_2,t_2),\ \cdots\ (s_n,t_n)\}$ containing n terminal pairs that are to be joined by paths, let $G = (V,E,\omega)$, and each edge $e_i \in E$ be assigned an integer weight ω_i, $i = 1,2,\cdots m$, where $m = |E|$ is the total number of edges in the graph G. To complete global routing it is required that no more than ω_i of the n paths use edge e_i. Associated with each terminal pair (s_j,t_j) is a set P_j of paths, one of which must be used to join s_j to t_j. Let

$$a_{jk}^i = \begin{cases} 1, & \text{if the k-th path in } P_j \text{ is selected, which uses edge } e_i \\ 0, & \text{otherwise} \end{cases}$$

A solution to the problem can be specified by the indicator variable x_{jk}, interpreted as follows:

$$x_{jk} = \begin{cases} 1, & \text{if the k-th path in } P_j \text{ is selected} \\ 0, & \text{otherwise} \end{cases}$$

Then

The linear forms which are to be optimized are:

$$(i) \quad \sum_{i=1}^{11} X_i = min$$

$$(ii) \quad \sum_{i=1}^{4} y_i = max \text{ in descending order of priorities.}$$

The integer programming problem can consider some more restrictions, for example, number of bends of wires [15]. Also in [15] it is reported that the integer programming problem considered here can be solved by rounding off solutions to a corresponding linear programming problem.

It should be mentioned that a solution to the linear programming problem indicates only for a given net type, how many nets of this type are routed in what way. The decision on how to route particular nets can be made in an arbitrary way, usually by some kind of heuristics.

3.2.2. The top-down scheme

In the top-down approach we start from the top with 2x2 supercells representing the whole chip, i.e., we divide the whole chip into 2x2 structure. We start by routing the 2x2 structure first. We then consider the next hierarchy and determine the implication of the previous routing in terms of its current level of routing. The process then continues until we reach the bottom of the hierarchy, i.e., the routing at the level of the primitive cells. The process is illustrated by means of a simple example with 4x4 cells and a single net as shown in Fig. 6. In Fig. 6(a) we show the 4 terminals on the 4x4 cells to be connected. In Fig. 6(b) we show the 2x2 supercells and a possible connection. Note that each supercell contains a maximum of one terminal pertaining to the single net. To further simplify the discussion, we consider the next hierarchy in the horizontal direction first. This leads to the successive connections shown in Figs. 6(c), 6(d), and 6(e). Note in Fig. 6(d), there exist two possible connections in each of the 2x2 primitive cells. In Fig. 6(e) we show one of the possible connections. We next proceed to the vertical hierarchy and make the necessary connections across the boundary shown as (a,b) and (c,b) in Fig. 6(f). The final routing is shown in Fig. 6(g). The details of the top-down scheme is given in [15].

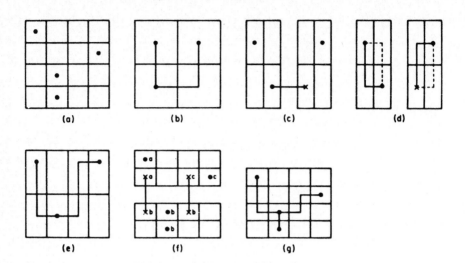

Figure 6
Example illustrating the top-down hierarchical routing approach.

3.2.3. The bottom-up scheme

The bottom-up hierarchical router works as follows. On each level of hierarchy it
considers a sequence of routing problems for arrays of 2x2 supercells to obtain a
preliminary sketch. It uses a maze-type router to determine the actual shape of
the connections. Let us consider level 1, i.e., the bottom level. Here supercells
are cells. We complete connections only inside consecutive 2x2 cell structures.
See Fig. 7(a), as an example. These 2x2 cells are formed by every other vertical
and horizontal lines bounding the cells. In Fig. 7(a) thick lines correspond to
closed lines, i.e., lines which nets are forbidden to cross. Note, connections can
only be made in 2x2 cell arrays bounded by thick lines. Next, we move to level 2
as shown in Fig. 7(b). We do it by opening every other previously closed line and
connections are completed inside the regions bounded by closed (thick) lines.
Each supercell is divided into cells by thin broken lines representing lines which
bound the primitive cells. When we move to the next level, we open every other
closed divided line as shown in Fig. 7(c). This process is continued until all
lines become opened. On each level, each 2x2 supercell problem is routed
independently. Connections are made between supercells only, i.e., if there are
2 pins or partial connections of the same net inside one supercell, they are left
untouched. First, we look globally at each 2x2 supercell problem, i.e., we dis-
regard the particular cells which form the supercells. If there is a partial
connection of net i in some supercell, it appears as a pin in this cell in the
global view as illustrated in Fig. 8.

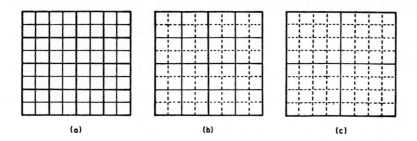

Figure 7
Supercell and cell structure on different levels of hierarchy; (a) Level 1,
(b) Level 2, (c)Level 3.

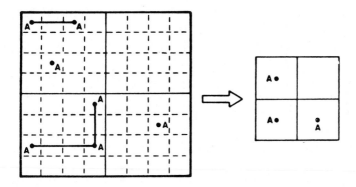

Figure 8
Net A on the 3rd level of the hierarchy before routing has taken place. On the
right is its global view.

As it was mentioned in Sec. 3.2.1, there are 11 different possible pin configurations which may occur for 2x2 cell array. If we take a closer look at possible
routing patterns as depicted in Fig. 4, we note that some routes are marked with
heavy lines while others with light lines. Preferred routings are the shortest
connectons drawn with the heavy lines; light lines are chosen when the preferred
pattern cannot be realized because of lack of cross-capacities. For each 2x2 array
of supercells we decide the rough shapes of nets depending upon their configuration.
First, for nets which are seen in the supercells as configurations $K(1)$, $K(2)$,
$K(3)$ or $K(4)$ we assign the preferred routes. If the preferred routes cannot be
assigned for some nets, we chose the other possible patterns. Next, we assign
nets for the configurations in $K(7)$-$K(10)$ and follow with configurations $K(5)$, $K(6)$
and $K(11)$. Note that this way of assignment leads to a wiring with minimum total
number of connections between supercells.

Afer the rough patterns of connections have been decided, a maze router similar
to the one described in [3] is used to find a fixed pattern of connections inside
the considered region of 2x2 supercells. If a connection cannot be completed
according to the chosen rough sketch, it is left aside. After an attempt is made
to connect all the nets inside the currently routed 2x2 supercells following the
chosen patterns, the failed connections are routed by a conventional maze-router
operating on the entire array of cells contained in the 2x2 supercells. Now, if
some net i fails to be connected on some level of the hierarchy, it will remain
disconnected even though such a connection may be possible to make. Such a situation may occur when a wire cannot be routed with a minimum number of connections
between supercells.

In [16] the bottom-up approach was discussed in detail along with a rerouting
scheme, which will be briefly mentioned in the next subsection.

3.2.4. Rerouting

In the global routing problem our goal is to find a solution, i.e., a set of connections realizing all the nets for which cross capacities are not exceeded.
However, what we usually get from a global router is a partial routing, i.e., a
set of connections which realize some nets and parts of other nets, but at least
one net is unconnected. In general, no unconnected net can be completed before at
least one existing connection is rerouted or removed. Now, for a given partial
routing, a partial solution is its maximal subset which is contained in some solution of the global routing problem. Thus, the rerouting problem can be formulated
as follows. Given a partial routing, remove a minimal subset of interconnections
such that the remaining routing is contained in some solution S of the global
routing problem, and find S.

The rerouting problem is extremely difficult, because there is no way to determine
a priori which connections in the partial routing belong to a partial solution.
The only rerouter which can either find a solution or detect that there is no
solution to the given global routing problem is a rerouter with the ability to
search through all partial routings. It is not practical to scan through all
possible partial routings hoping that a solution will appear somewhere at the
beginning of the sequence. Instead, a partial routing can be modified according to
certain rules that in most cases leads to a solution. One set of rules can be
derived from the following simple observations.

Let us consider a partial routing. It contains at least one net N_x which cannot
be completed due to congestion in some regions. Thus the net N_x consists of at

least two disconnected parts: $N_x^{(1)}$ and $N_x^{(2)}$. Let $R_x^{(1)}$ be the set of cells which can be reached from $N_x^{(1)}$ and $R_x^{(2)}$ a set of cells which can be reached from $N_x^{(2)}$. $R_x^{(1)} \cap R_x^{(2)} = \phi$, because N_x cannot be connected in the existing situation. Let $B_x^{(1)}$ and $B_x^{(2)}$ be the boundaries of $R_x^{(1)}$ and $R_x^{(2)}$, respectively. The cross capacities along $B_x^{(1)}$ and $B_x^{(2)}$ have been exhausted and some connections crossing $B_x^{(i)}$, $i = 1,2$ must be removed in order to complete the net N_x. Let us consider any one on $B_x^{(i)}$, $i = 1,2$; let it be $B_x^{(1)}$. Suppose that all the nets which cross $B_x^{(1)}$ have pins in $R_x^{(1)}$ and in $A - R_x^{(1)}$, where A denotes all cells on the boards, and that each net crossing $B_x^{(1)}$ crosses it only once. In such a case, a solution does not exist, because none of the nets crossing $B_x^{(1)}$ can be rerouted in a different way. Now, suppose that there is exactly one net crossing $B_x^{(1)}$ more than once. In such a case, we know that this net has taken a wrong pattern and it can be ripped up. However, if there is more than one net crossing $B_x^{(1)}$ twice or more times, then we know that at least one of them has taken a wrong pattern. On the other hand, we do not know which one. We could remove all such nets, but it may lead to too long a rerouting process. Thus, the following reasoning can be applied in a rerouting algorithm.

Rule 1 calls for ripping such nets from the blocking boundaries, which can be detected as those having taken wrong routes.

Rule 2 deals with rerouting of some nets away from the congested areas and removing some of the redundant portions that cause an opening of the blocked boundary.

Let us consider a situation shown in Fig. 9(a). Inside the boundary B where cross capacities are exhausted, there is a pin A of the net N_a. There is another pin of net N_a on the other side of B. There is also some net N_c which crosses B more than once and which can be rerouted as marked by a dotted line. The dotted line added to N_c causes closing of some cycle, thus one edge from it can be removed. Figure 9(b) shows a situation after the rerouting of net N_c has taken place. The closed boundary B has disappeared and in Fig. 9(b) and its previous position is now marked by a dotted line.

In [16] the details of a rerouting algorithm implementing the above rules along with experimental results obtained from the computer program are reported.

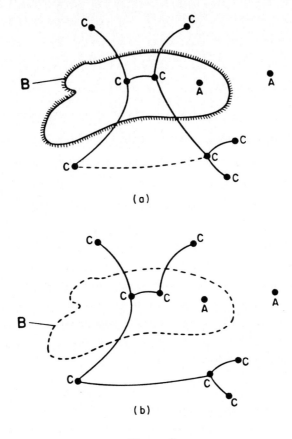

(a)

(b)

Figure 9
Example of application of rule 2.

3.3. Constructive Method

Another approach, which eliminates the net ordering problem and alleviates the
need for rerouting, is to construct a solution continuously, based on pin distribution and congestion analysis. As it can be expected, in most situations, an
algorithm based on such an approach is much slower than the other methods.
However, it is better than other apporoaches for solving difficult problems where
large portions of the solution are uniquely specified. Such unique patterns are
not detected by other methods and may also cause major problems for the rerouter.
An algorithm reported in [17] is based on some general properties of solutions to
the global routing problem. Here we give a brief discussion of that approach.
Let P be a global routing problem. P has a solution if there exists such routing
patterns for nets in P that all the nets are connected and all the cross supply
constraints are satisfied. Suppose P has m feasible solutions in total,

S_1, S_2, $\cdots$ S_m. If $U = \overset{i=m}{\underset{i=1}{\cap}} S_i \neq \phi$, then U contains the unique portions of

solutions, i.e., the unique patterns of some nets in P. A routing R_j^i of net N_i is
said to be a feasible pattern, if there is a solution S_j in which net N_i takes the
pattern R_j^i. Thus, each net N_i, i = 1,2, $\cdots$ n, can take one of the feasible

patterns $\{R_j^i\}$, j = 1,2, $\cdots$ m. In case N_i has a unique routing path,

$R_1^i = R_2^i = \cdots = R_m^i$. If only a part of N_i is unique, $\overset{j=m}{\underset{j=1}{\cap}} R_j^i \neq \phi$. If we are

unable to identify the unique portions of solutions, finding a solution for P may
become an exhaustive search. We will give the essence of the approach by introducing a set of definitions, lemmas and theorems.

Definition 1. A set of cells F_i forms a <u>feasible routing area</u> for net N_i, if for
each pair of cells C_a and C_b in F_i with a common boundary of non-zero cross
supply, there is a feasible routing R_k^i passing through the common boundary.

It is obvious that a feasible routing area for a net is determined by the current
routing situation on the chip. In other words, once a net is connected, even
partially, the feasible routing areas of other nets may be changed.

Definition 2. A cell is said to be a <u>non-pass-through</u> (NPT) cell, if the number
of unconnected pins inside the cell is equal to or less by one than the total
remaining cross supplies on its boundaries. A non-pass-through region can be also
defined as a closed region within which the number of different nets needed to be
connected outward is equal to or less by one than the total cross supplies along
its boundary.

Lemma 1. Suppose P has a solution. If there is a 2-pin net N_i with pins in two
adjacent cells with non-zero cross supply on the common boundary, then there
exists a solution S_j for P in which N_i is routed using the shortest connecting
path.

Lemma 2. If P has a solution, and the cross supply on a closed region boundary is
μ and there are ν pins inside, $\nu > \mu$, then at least $\nu-\mu$ of them must be connected
within this region.

Lemma 3. If P has a solution, and there are two adjacent NPT cells, C_a and C_b,
with a non-zero cross supply μ on the common boundary, and there are μ or less

multi-pin nets which have pins in both C_a and C_b; then there is a solution in which these pins are connected by the shortest path between C_a and C_b.

Theorem 1. The routing R_j^i has a unique pattern if and only if net N_i is forced by the current situation on the chip (i.e., cross supply constraints, previous routed patterns, etc.) to take the unique shape.

Definition 3. A barrier is a boundary of a non-pass-through region.

Definition 4. A boundary of a closed region is said to be a barrier for net N_i if net N_i cannot cross this region.

Theorem 2. Suppose P has a solution with a net N_i to be connected. The feasible routing area F_i of net N_i is a region B_i bounded by the barriers for net N_i.

Therefore, in order to find a solution for P, we need to find the barriers first to guide the net connection, i.e., check all possible dissections of the routing area. An algorithm, applying the barrier checking principle directly, would lead to an exhaustive search. However, the above stated properties of solutions and barriers are useful to design a practical heuristic algorithm. We need one more definition.

Definition 5. In a particular planar drawing of the global routing raph G, we define two types of meshes.

(1) Outermost mesh which bounds the drawing of G from the outside, i.e., the circuit separates the exterior face from G.

(2) Inner meshes which bound the areas from the inside where edges are missing because the cross capacity which they represent has been already exhausted.

Since finding barriers is time consuming, the algorithm only locates the following barriers in G: (i) a cutset whose end vertices are NPT cells which can be grouped like a wavefront in G (Fig. 10(a)-(b)). (ii) G is disconnected (Fig. 10(c)).

Since the routing area is fixed and cross supplies are limited in a gate array chip, we try to use up first the cross supplies in the outer area, then the inner area, to avoid the congestion in the center. The algorithm first completes the 2-pin-adjacent-cell connections (Lemma 1) and the unique routing patterns for some nets which are forced by NPT cells. Then it makes connections along the outermost

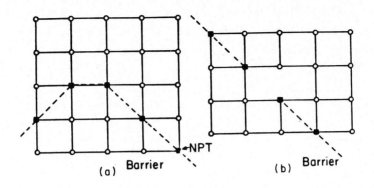

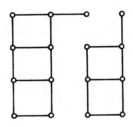

Figure 10
Barriers detected by the algorithm: (a) (b) the NPT cells group like a wavefront
barriers, (c)the cell graph is disconnected.

and inner meshes. It searches for 2 closest pins which belong to an uncompleted
net on these meshes, and chooses a path to connect them along the meshes. Each
net grows from the outer area toward the inside gradually in the style of a
spanning tree. Therefore nets need not compete for the internal routing area with
each other, and wiring density distribution can be evened out.

4. Global Routing for Standard Cell Layout

In standard cell design, cells of equal heights are arranged in a row. The
configuration is similar to the gate array; however, the cells can be of different
width and, in addition, routing channels are not fixed. After placement is done
in the form of rows of cells, global routing is to be carried out to minimize the
total routing area in terms of feed-through assignments and judicious use of
equivalent pins which belong to a given cell. The problem is of great interest in
practice, yet not much has been written on the topic. Global routing for standard
cell layout can be considered a multi-channel optimization problem. A net list

which specifies the terminal connection is the given input together with information on equivalent pins and allowed positions for feed-throughs. Each net which
spans more than a single channel is to be decomposed into subnets defined in
individual channels.

A common figure of merit for the result of a global routing is its total density,
defined as follows. The local density of a channel at a given x-coordinate is
the number of nets that intersect a vertical line segment which passes through that
channel at coordinate x. The channel density of a channel is its maximum local
density over all x. The (total) density of a standard cell layout is the sum of
channel densities over all channels. Clearly, the total density is a lower bound
on the number of horizontal tracks required for routing. The ultimate objective
is to decompose all the nets in an optimal way so that the total density of layout
is minimized. Thus, for a net N, we define the output as a set of subnets,
$\{s_1^N, s_2^N, \cdots, s_k^N\}$ such that

$$(i) \quad \bigcup_{i=1}^{k} s_i^N = N, \text{ and}$$

$$(ii) \quad \text{All terminals of } s_i^N \text{ are on either side of a channel.}$$

We call $\{s_1^N, s_2^N, \cdots, s_k^N\} := d(N)$, a decomposition of net N. Let D_i be the
maximum density for channel i, thus the objective is to find such decomposition
for all the nets that

$$\sum_i D_i = \min \qquad\qquad\qquad\qquad (6)$$

An approach which deals with gate array global routing based on multi-channel
optimization has been proposed [18]. It can be transformed to optimization on
graph using spanning and Steiner tree algorithms. The approach can be extended to
standard cell layout by using (6) as an objective function.

Another approach developed to handle a special case of the general problem, where
only equivalent pin assignment is being optimized, has been reported in [19]. In
this method it is assumed that placement is obtained by folding one row of
linearly placed cells into a set of parallel rows. In such a case a majority of
nets have to connect terminals in one row (linear nets) or between 2 neighboring
rows (bi-linear nets). It is also assumed that each cell is a double-entry, that
is, each cell has each of its signals appearing on both its top and bottom side.

In [19] it is discussed how to connect linear and bilinear nets by making good choices between equivalent pins.

Obviously, more research is needed in global routing for standard cell layout.

5. Building Block Layout

Global routing for building block layout is considerably different from that of gate array. First, the configuration does not have a regular geometry. Second, there exist various ways of defining the routing regions. Third, placement of modules is not fixed; routing space for regions can be expanded or contracted. We will address the above issues and will present some approaches to the problem. In all of the approaches developed for global routing for building block layout, a global routing graph reflecting routing conditions is generated and nets are routed sequentially by some Steiner-tree-on-graph algorithm. Thus the major differences among the various methods are in the ways of constructing a global routing graph. Also, some methods allow for small dynamical updating of the graph structure reflecting placement modifications occurring in order to accommodate routing. There are two basic approaches to defining such graphs. In both cases, the routing area is divided into regions. In the first approach routing regons are represented by vertices and adjacency of routing regions corresponds to edge in the graph. In the second approach certain routing regions are represented by vertices while others are represented by edges. Additionally, the graph structure is modified before a particular net is routed in order to add vertices corresponding to terminals of a pertinent net being routed. In the following we will show examples of methods applying the two approaches.

5.1. Regions Adjacency Graphs [21]

Let us consider a layout depicted in Fig. 11(a). We may divide the routing area into a set of routing regions by extending each horizontal and vertical line bounding a block until it intersects another block or the external boundary. Now, each routing region into which the original routing area was divided is represented by a vertex. If two routing regions are adjacent, the corresponding vertices are adjacent in the graph. The graph for the placement in Fig. 11(a) is shown in Fig. 11(b). For the same placement the same routing area can be divided into different routing regions, which obviously lead to a different graph, see Figs. 11(c)-(d). Graph edges may be assigned weights which describe congestion in corresponding regions explained earlier. There are many approaches in specifying routing regions. Although the choice may not be important to the quality of global routing results, the same routing regions are usually used for detailed

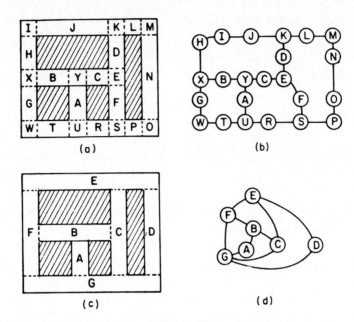

Figure 11
Global routing graph reflecting routing regions adjacencies. Different definitions of routing regions for the same building block layout structure lead to
different graphs. Compare (a) (b) and (c), (d).

routing for which a good division into routing regions becomes crucial. Thus, in
an overall strategy of generating layout for building block design it is important
how the entire routing area is separated into routing regions. A brief discussion
of the subject is given in Section 5.3.

5.2. Bottleneck Graph [22,23]

In this approach, routing area is divided into two types of regions: bottlenecks
and switchboxes. A normal type of bottleneck exists between opposite edges of two
different blocks if the following conditions are met:

   (i)   The opposite edges are vertical (horizontal) and there exists a
         horizontal (vertical) line intersecting both edges.

   (ii) In the case of vertical (horizontal) edges, every horizontal (vertical)
         line intersecting the two edges does not intersect any other vertical
         (horizontal) edge which lies between these two edges.

The concept of normal bottleneck, together with the terms active edges and active
region of a bottleneck, is illustrated in Fig. 12.

Now, routing area of a whole chip may be divided into active regions of normal bottlenecks and the rest called switchboxes. Note that the active region of a normal bottleneck is always a regular channel. An example of such a division is shown in Fig. 13(a). The switchbox regions are cross-hatched. The bottleneck graph is defined as follows:

(i) Each switchbox corresponds to a vertex in a graph, and

(ii) if two switchboxes are adjacent to a common normal bottleneck region then there is an edge joining the corresponding vertices.

Figure 13(b) shows the bottleneck graph corresponding to the layout configuration of Fig. 13(a). The details of an algorithm for generating a bottleneck graph as well as how to update such a graph when some block is to be moved during the routing process are discussed in [23].

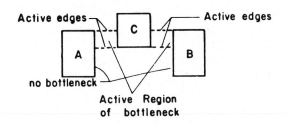

Figure 12
Example of bottlenecks.

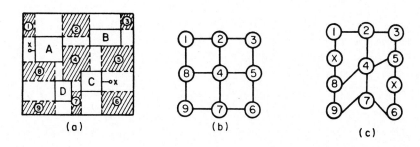

Figure 13
(a) Bottlenecks and switchboxes for a simple layout configuration, (b) the corresponding bottleneck graph, and (c) the expanded bottleneck graph with added nodes corresponding to terminals of a net marked in (a).

Next, if a net is to be routed, the bottleneck graph is modified to represent its
terminal configuration. A terminal of a net which is not covered by any bottleneck is located in a switchbox. It will be represented on the global routing
graph by the node corresponding to that switchbox. For a covered terminal, i.e.,
it lies within the active region of a bottleneck, a new node will be inserted in
the graph. This new node will split the edge corresponding to its covering
bottleneck into two. Each split edge represents one part of the active region of
the bottleneck divided at the terminal position. A global routing graph for
routing a net whose terminals are marked as crosses in Fig. 13(a) on blocks A and
C is shown in the Fig. 13(c). The vertices corresponding to the location of the
pins are marked with X.

The bottleneck graph is used in BBL, the Berkeley Building-Block Layout System for
custom chip design [22]. An efficient Steiner-tree-on-graph algorithm is used for
global routing [10].

5.3. Routing Region Definition and Ordering (RRDO) [24,25]

The role of global routing for building block layout is to make assignments of nets
into specified routing regions based on a global consideration. Pertinent to the
overall layout process is the way that routing regions for detailed routing are
defined. It is well-known that detailed routing based on ordinary channel routers
usually yields excellent results in terms of area usage. Furthermore, channel
routers have the unique property that routing space can always be provided after
the completion of the routing by simply widening the channel when necessary. The
subject of channel routing is discussed in detail in the succeeding chapter. In
fig. 14(a), we illustrate an ordinary channel formed by two Blocks A and B. The
three nets are routed as shown and three tracks are used. Had we allocated only
two tracks for the channel, either Block A can be moved up or Block B can be moved
down to provide the extra track. No rerouting is necessary. This is a desirable
property, crucial in building-block layout.

In Fig. 14(b) we show a "T" junction which can be divided into two ordinary
channels. Channel I is the "base" channel which must be routed before channel II
the "bar" channel. There are two simple reasons. First, in routing channel II,
we need the net information at the boundary line between the channels. That, of
course, is not available before channel I is routed. Second, if Block A or Block
B had been moved horizontally in routing channel I, it would change the pin position of the upper boundary of channel II.

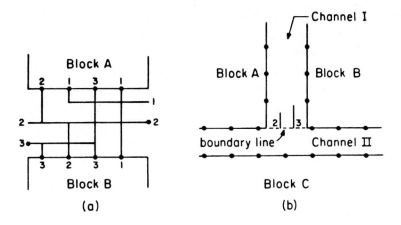

Figure 14
(a) An ordinary channel, and (b) an up-side-down "T" junction which is decomposed into two ordinary channels.

Unfortunately, in building block layout, it is not possible, in general, to rely only on ordinary channels. That is why in BBL we divide routing regions into ordinary channels and switchboxes. In the following we present a different approach.

Depending on the placement of blocks we can divide the floorplan into two types. One has a slicing structure [26] as shown in the example in Fig. 15(a). By slicing, we mean that we can use horizontal or vertical lines to divide the chip into two parts. Thus, line 1 separates Block A from the rest. We apply slicing again on the remaining structure. Thus line 2 separates Block B from the rest. Finally, line 3 is used to separate Block C from Block D. A slicing structure refers to those floor plans for which slicing can be continued until every block is separated. It is interesting to note that such slicing scheme gives the reverse order of channels to be routed. In this example, channel 3 must be routed before channel 2; and channel 2 must be routed before 1. Thus, for slicing structure, we need only ordinary channels for detailed routing, and a feasible routing order can be easily determined.

The other type of floor plan is a nonslicing structure as illustrated by the example in Fig. 15(b). The four "T" junctions create a cyclic precedence relation [20,24]. If we call the four channels in the figure a, b, c, and d as shown; then channel a must be routed before b, channel b must be routed before channel c, channel c must be routed before channel d, and channel d must be routed

before channel a. This is clearly impossible. Therefore, we propose the use of
an L-shaped channel as shown in Fig. 16(a). Although the routing for an L-shaped
channel has not been throughly investigated, it has a similar feature as an
ordinary channel in that the channel width can be adjusted. In Fig. 16(a), if we
move Block A in the direction shown, we can create more tracks. We now illustrate
in Fig. 16(b) the use of an L-shaped channel. The algorithm to determine the
order of routing region is similar to the previous example. Here, we first use a
process called "corner removing" to separate Block A from the rest. The remaining
structure becomes a slicing type, and the order is given by 2, 3, and 4 shown in
the figure. The routing order is again the reverse, i.e., channel 4, channel 3,
channel 2, and, finally, the L-shaped channel 1.

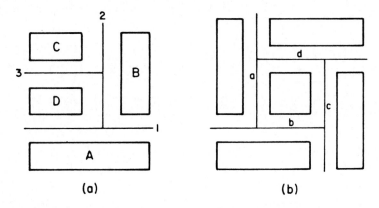

Figure 15
Floor plans: (a) a slicing structure, and (b) a general structure which contains
a cycle.

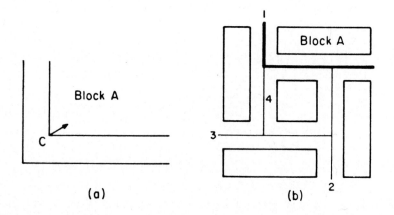

Figure 16
(a) An L-shaped channel, and (b) an illustration of "corner removing" which leads
to the use of an L-shaped channel.

In [25], an algorithm is proposed which can be used for any given floor plan. The algorithm minimizes the use of L-shaped channels and, in the meantime, determines the routing region ordering.

6. Conclusion

Global routing is a crucial part of layout design. It is intimately tied to both its predecessor, the placement of blocks or cells, and its follower, detailed routing. The main objective for global routing is to develop a good routing plan based on placement in such a way that detailed routing can be completed efficiently. The details of the problem, however, vary according to what layout style is used. Fortunately, it can be formulated from a mathematical point of view, for example, based on Steiner tree and a global routing graph. In the case of gate array design, because of its regular layout structure, we can translate the problem into mathematical formulation rather precisely. Consequently, there exist many approaches and good results. In the standard cell and building block layout, not much has been written on global routing, and we believe that more research is definitely in need on both the formulation of the problem and creative approaches to solve it.

Acknowledgement

The authors wish to acknowledge research support of the National Science Foundation through Grant ECS-8201580, the Semiconductor Research Corporation, the MICRO program and its participating companies.

References

[1]   Lee, C., An algorithm for path connection and its applications, IRE Trans. Electronic Computers, EC-10 (1961) 346-365.

[2]   Hakimi, S. L., Steiner's problem in graphs and its implications, Networks, 1 (1971) 113-133.

[3]   Soukup, J., Fast maze router, Proc. Design Automation Conf., (1978) 100-101.

[4]   Hanan, M., On Steiner's problem with rectilinear distance, SIAM Journal of Applied Mathematics, 14 (March 1966) 255-265.

[5]   Hwang, F. K., An O(nlogn) algorithm for suboptimal rectilinear Steiner trees, IEEE Trans. on Circuits and Systems , CAS-26, (1978) 75-77.

[6]   Ting, B. S. and Tien, B. N., Routing techniques for gate array, IEEE Trans. on Computer-Aided Design, CAD-2 (1983) 301-312.

[7]   Dijkstra, E. N., A note on two problems in connection with graphs, Numar. Math., 1 (1959) 269-271.

[8]   Prim, R. C., Shortest connecting networks and some generalizations, Bell System Tech. J., 36 (1957) 1389-1401.

[9]   Kruskal, J. B., On the shortest spanning subtree of a graph, Proc. Amer. Math. Soc., 7 (1956) 48-50.

[10] Chen, N. P., New algorithms for Steiner tree on graphs, Proc. IEEE ISCAS, (1983) 1217-1219.

[11] Karp, R. M., Leighton, R. L., Rivest, R., Thompson, C. D., Vazirani, U., and Vazirani, V., Global wire routing in two dimensional arrays, Ann. Symp. on Foundations of Comp. Sci., 24 (1983) 453-459.

[12] Kirkpatrick, S., Gelatt, Jr., C. D., and Vecchi, M. P., Optimization by simulated annealing, Science, 220 (1983) 671-680.

[13] Binder, K., editor, The Monte Carlo method in statistical physics, (New York: Springer Verlag, 1978).

[14] Vecchi, M. P. and Kirkpatrick, S., Global wiring by simulated annealing, IEEE Trans. on Computer-Aided Design, CAD-2, (1983) 215-222.

[15] Burstein, M. and Pelavin, R., Hierarchical wire routing, IEEE Trans. on Computer-Aided Design, CAD-2 (1983) 223-234.

[16] Marek-Sadowska, M., Global router for gate array, Proc. IEEE Int. Conf. on Computer Design, (1984) 332-337.

[17] Li, J-T. and Marek-Sadowska, M., Global routing for gate array, IEEE Trans. on Computer-Aided Design, CAD-3, (1984) 298-307.

[18] Aoshima, K. and Kuh, E. S., Multi-channel optimization in gate array layout, Proc. 1983 IEEE ISCAS, (1983) 1005-1008.

[19] Supowit, K. J., Reducing channel density in standard cell layout, Proc. 20th Design Automation Conf., (1983) 263-269.

[20] Kani, K., Kawanishi, H. and Kishimato, A., ROBIN: A building block LSI routing program, Proc. IEEE ISCAS, (1976) 658-661.

[21] Preas, B. T. and Gwyn, C. W., Methods for hierarchical IC layout, Proc. IEEE ISCAS, (1979) 482-485.

[22] Chen, N. P., Hsu, C. P., Kuh, E. S., Chen, C. C. and Takahashi, M., BBL: A building block layout system for custom chip design, Proc. IEEE Int. Conf. on Computer-Aided Design, (1983) 40-41.

[23] Chen, N. P., Routing System for Building Block Layout, Ph.D. Thesis, Dept. EECS, University of California, Berkeley (1983).

[24] Kajitani, U., Order of channels for safe routing and optimal compaction of routing area, IEEE Trans. on Computer-Aided Design, CAD-2, (1983) 293-300.

[25] Dai, Weiming and Asano, T., A new routing region definition and ordering scheme using L-shaped channels, to appear Proc. IEEE ISCAS (1985).

[26] Otten, R. H., Automatic floorplan design, Research Report RC 9656, IBM T. J. Watson Research Center (1982).

Footnote

It should be mentioned that 100% assignments in global routing do not always guarantee 100% interconnection of all specified nets. The final routing completion needs to be addressed again at the detailed routing step.

LAYOUT DESIGN AND VERIFICATION
T. Ohtsuki (Editor)
© Elsevier Science Publishers B.V. (North-Holland), 1986

Chapter 6

LAYOUT COMPACTION

Dieter A. MLYNSKI*, Chen-Han SUNG**

University Karlsruhe, Karlsruhe, F.R.G.
** San Diego State University and
University of California at San Diego, San Diego, U.S.A.*

Layout compaction is one of the most challenging problems
of VLSI layout. It is the final step in the flow of layout
design after initial layout has been done by placement and
routing techniques. Its aim is a final mask layout with
minimum chip area and consistent with design rules. Compaction prerequisites are proper representations of initial
layout and design rules. Today's compaction methodologies
and different compactor implementations are discussed and
future outlooks in this area are given.

1. INTRODUCTION

1.1 A Global View on Layout Compaction

Since the early times of integrated circuits the layout design is done in two
successive steps: placement and routing, and followed by a third step: the layout
verification. One of the main concerns of layout design is minimum chip area or
maximum chip density. Although it is not the only objective it well reflects many
others, for example minimum signal delay time or maximum yield. When those three
steps are properly done either by a skilled designer or by appropriate CAD tools,
the resulting layout is both valid and compact.

With the increasing complexity of LSI and especially VLSI computer aided layout
design has become indispensable. On the other hand available CAD tools for placement and routing tend to use more chip area than actually needed. In other words,
they unnessarily waste chip area. Automatic layout currently still cannot compete
with handlayout as far as the chip area is concerned.

Therefore, another step has come up and settled at the end of the flow of layout
design, i.e. layout compaction. Strictly speaking, compaction is not new. A skilled designer is always using compaction as a inseparable part of placement and
routing. However, as a CAD tool layout compaction is the ultimate step in the flow
of layout design after placement and routing has already been done. This step has

gained much interest since the time of LSI. Today compaction is one of the most
challenging problems of VLSI layout design.

The aim of layout compaction can be stated as follows: Starting from an initial
layout and without changing its topology, a final mask layout has to be achieved
with a minimum chip area and consistent with design rules. The restriction to
invariance of topology is necessary in order not to render the previous steps of
placement and routing obsolete. It is achieved by maintaining relative neighbours,
i.e. adjacency of layout elements. Elements are either on the bottom-most mask
level, e.g. diffussion windows, or on higher design levels, e.g. transistors and
cells. These elements are not allowed to jump across each other during compaction.

This final compaction step is essential in the flow of layout design in case that
(1) the initial layout is not a physical mask layout but only a symbolic layout
     demanding for a translation into final mask geometry (<u>mask layout compilation
     problem</u>);
(2) the initial layout is a mask layout but not efficiently packed, i.e. has too
     much wasted area demanding for a denser packing (<u>chip area minimization pro-
     blem</u>);
(3) the initial layout is densely packed but found by verification not to be
     valid for the given technology, i.e. violates certain technology design rules
     demanding for a corrected redesign (<u>redesign problem</u>);
(4) the initial layout is valid for the technology which its design was based on,
     but needs to be rescaled to fit either a set of current technology defined
     design rules, or other current design requirements like user defined design
     rules (fan-in, fan-out, power bus or shape of functional blocks) when re-
     using the layout in a different environment (<u>rescaling problem</u>).

To summarize, a compactor is a compiler translating an initial layout into a final
layout consistent with the given initial topology and given design rules. It
therefore consists of three parts: the <u>data base</u> with the input data of the
initial layout, the <u>knowledge base</u> including technology defined and user defined
design rules and the <u>compaction strategy</u>. Discussed in section 2, data base and
knowledge base are prerequisites for compaction. The compaction methodology shall
be discussed in section 3, implementation and performance of available compactors
in section 4. A final section is devoted to future outlooks.

1.2 Short Survey on Compactor Development

The earliest known compactor if not the earliest at all originated already in 1970 due to Akers, Geyer and Roberts [01]. Their general approach to split the two-dimensional problem into two one-dimensional problems is still valid today and used by most of the available compactors. Therefore, the largest section within this chapter on compaction will cover the one-dimensional approach. The basic idea in [01] is to search for various horizontal and vertical "cuts" through the chip containing wasted area. Therefore, throughout this chapter this compactor shall be referred to as CUT.

There was no progress on compaction until 1977. But during this time design rule check has been further developed [02] and symbolic layout has become a major tool for a preliminary layout design stage [22,13]. Both of them are used as prerequisites for compaction.

In 1977 FLOSS [05] has started an unending series of contributions to layout compaction and of compactor implementations (see section 4). STICKS [40] in 1978 includes already a knowledge base with design rules and layout heuristics, an exhaustive automatic compaction strategy and various interactive functions, such as local optimization and jogs.

CABBAGE [15,16,17] has introduced in 1978 another basic idea for compaction, the constraint graph and the longest path search technique. This has been further developed, modified and implemented in several other compactors since then. Today it is the dominating compaction technique at least for the one-dimensional approach but also widely used in the two-dimensional approach.

In 1983 Schlag, Liao and Wong [32] have published a two-dimensional compactor of VLSI layouts based on boolean variables and functions. Branch and bound search is applied to a family of boolean decision trees. Boolean decision variables have been also used by Kedem and Watanabe [18,19]. Their compactor allows only local decision making between horizontal and vertical constraints. It is not truly two-dimensional but may be classified as semi-2-dimensional and shall be referred to as ROCHESTER as has been done earlier in the literature [04].

Two-dimensional compaction is still a problem for future research (see section 5). It may be aided by the technique of simulated annealing and by the future development of expert systems.

2. COMPACTION PREREQUISITES

As already mentioned in the introduction a compactor consists of three parts: the data base, the knowledge base and the compaction strategy. The latter one is the central part of the compactor and to be discussed in section 3.

In this section we shall discuss the data and the knowledge base which are the necessary prerequisites for compaction. The initial layout is represented by a set of elements and adjacency information on the elements. It is stored in the data base. The design rules stored in the knowledge base are either defined by techno- logy or by the user. They are consulted by and ruling over the compaction strategy.

2.1 Initial Layout Representation

The initial layout may be represented either in the bottom-most mask level or in an upper hierarchical cell level, either as a physical mask layout or as a sym- bolic layout, either on grid locations or on relative locations. Different repre- sentations shall be outlined in this subsection as far as compaction rules and compaction strategy are concerned with. A summary of different representations of the initial layout for a number of compactors is given in Table 1.

For the rescaling problem, the redesign problem and the chip minimization problem the initial layout is a physical mask layout either on the bottom-most mask level or on an upper hierarchical cell level. To begin with we shall assume bottom- most mask level (denoted by B in Table 1) and shall compare it with the cell level later on. Then the initial mask layout (denoted by M in Table 1), which may be generated either manually or automatically, is represented by the full set of mask layers necessary for chip production. Each mask layer consists of physical mask structures, i.e. of structures with given geometric dimensions and locations. Generally these mask structures are simple rectangles, e.g. diffusion windows, contact holes and horizontal or vertical wire segments, which are oriented paral- lel to the chip edges. Each rectangle is represented by a pair of horizontal line segments and a second pair of vertical line segments. Since compaction may rescale the rectangles resulting in larger or smaller rectangle dimensions the individual line segments are the basic elements for compaction. The positions of line seg- ments and spacing between line segments is given in integer numbers based on a mask grid (denoted by MG in Table 1). The grid unit is the smallest distance, depending on technology, between any pair of line segments on a single mask layer

Compactor	Ref.	Hierarch. Level		Layout Style		Initial Positions			
		B	C	M	S	MG	CG	VG	R
CUT	01	B			S		CG		
FLOSS	05		C		S				R
STICKS	40		C		S				R
SLIP	09	B		M		MG			
CABBAGE	15	B			S				R
SLIM	11	B		M		MG			R
MULGA	38	B			S			VG	
Liao/Wong	23	B			S				R
Schlag et al.	32		C		S				R
SQUASH	27	B			S			VG	
ROCHESTER	18	B			S				R
VPACK	04	B			S			VG	
Wolf et al.	41	B			S				R
PLOWING	34	B		M		MG			
Kingsley	20		C		S				R
Schiele	30	B		M		MG			

Table 1
Initial layout representation for different compactors
B = Botttom–Most Mask Level   MG = Mask Grid Positions
C = Cell Level   CG = Coarse Grid Positions
M = Mask Layout   VG = Virtual Grid Positions
S = Symbolic Layout   R = Relative Positions

or any pair of mask layers.

Symbolic layout has been introduced in the early seventies $[22,13]$ and since then has become a major tool in order to reduce the immense number of layout data (line segments) handled in a preliminary layout design stage $[13,40,09]$. An initial symbolic layout (denoted by S in Table 1) is represented by a subset of layers instead of the full set of mask layers. Usually two layers are sufficient: one layer for devices (polysilicon, diffusion, implantation) and the other layer for wires (metal). Interaction between the two layers occurs at contacts. Moreover a set of simple symbols is used to represent complex structures of the different layers. For example, a transistor may be represented by "X", a metal run by "I" and an interlayer contact by "O" $[09]$. Using these symbols the initial layout can be represented on a single reference layer. This symbolic representation is used as a high-level design language for the initial layout $[40]$. Especially it is used in the early design stage when a completely new layout is designed either manually or interactively. The resulting symbolic layout still demands for a compilation into a final mask layout (mask layout compilation problem). The symbols representing devices or wires are the basic elements for compaction. Thus compaction complexity is considerably reduced. It should be noted that the data reduction by symbolic layout has been mainly achieved by avoiding geometric dimensions. For example a physical wire is represented by its center line. Its width is not used during the initial symbolic layout but shall be called during compaction either from the data base, in case of a fixed wire width, or from the knowledge base, in case of a technology depending wire width.

Another advantage of symbolic layout is, it does not necessarily use geometric positions. It may be done without any fixed grid thus allowing relative positions of devices and interconnections independent of their physical shape and size $[40]$. Hence, during the early design stage the designer always has enough area available in order to squeeze in larger devices or wider interconnects without having disastrous effects rippling over the whole layout out to its edge. Using symbolic layout with relative positions shall be denoted by R in Table 1.

Symbolic layout can be also based on a coarse grid spread on the reference layer $[01,13]$. The grid unit usually is about 10 times larger than for the mask grid. This is possible since complex physical structures with several overlapping mask layers are represented as single symbols. Each grid unit is then either occupied by a symbol or unoccupied. The coarse grid contains less adjacency information on the initial layout than the mask grid but more than relative positions. Therefore, in case of initial layout representation using relative positions, the compaction strategy shall need special tools to extract and model adjacency information. Using symbolic layout with a coarse grid shall be denoted by CG in Table 1.

The virtual grid as introduced by Weste [38] is denoted by VG in Table 1. It is a grid the layout elements are snapping to but which does not initially have any physical mask related spacing. Hence, it has the same advantages for the early layout stage as relative positioning combined with the advantage of treating the adjacency information on a grid basis. Therefore, virtual grid symbolic layout may be classified as initial layout using relative positions but with an integrated ability to extract and model adjacency information.

Now we shall consider initial layout on an upper hierarchical level, the so-called cell level, denoted by C in Table 1. With the increasing complexity of circuits and systems, hierarchical design has become a generally accepted design style for VLSI. Besides the bottom-most mask level a hierarchy of several upper cell and macrocell levels is available. Usually a cell is a standardized component stored in and to be recalled from library. But it also may be any fixed part of a mask layout. Using the cells for initial layout on cell level they are considered and handled like black boxes. Only their contours including pin positions (contact holes) for interconnections between cells have to be handled. Therefore, an initial mask layout on cell level is represented by a subset of layers: one layer with contact holes of cells which outline the invisible contour of cells and a second layer with cell interconnects. Since the cells are standardized, i.e. have fixed dimensions, and usually also the interconnections have fixed width, the basic elements of compaction in this case are center points of cells and center lines of wires. If nothing else is mentioned we shall use this initial mask layout representation on cell level throughout the compaction methodology in section 3. Of course positions of cell and wire centers as well as spacing are given in integer numbers based on the mask grid similar as in the case of mask layout on the bottom-most mask level.

Finally, symbolic representation can be used for an initial layout on cell level. Then symbols are used to represent cells and wire segments either in relative positions or on a coarse or virtual grid. Again the symbols are the basic elements for compaction.

Table 1 shows for a number of compactors a classification according to the initial layout representation they use, i.e. classifying their input data. An exhaustive listing of all known compactors has not been aimed at. SLIM [11] is classified as R and MG in Table 1 as in a first step the relative positions (R) of an initial

symbolic layout are translated into mask grid positions (MG) of a mask layout;
both are available to and used during the compaction procedure.

2.2 Design Rules

Due to practical reasons, the final mask layout coming out of the compaction
procedure must stay in the limitation of the fabrication technology and the specification of the potential users. Thus a set of design rules is needed for the
layout compaction. For simplicity, these design rules usually are enforced at the
input phase instead of being scattered around in the compaction procedure.

On the other hand, when the process technology is changed or improved, the design
rules will accordingly be modified and/or updated simply by adding, deleting, or
modifying the appropiate rules. This can be done easily without any adverse consequence to the compaction procedure.

The design rules are either defined by the technology or by the user. Both technology and user defined design rules are represented in a format of lower-bound,
equality, or upper-bound constraints. E.g. for two elements with center coordinates (x_1, y_1) and (x_2, y_2) and a horizontal constraint value d the formats of the
three types of constraints are:

$$x_1 - x_2 \geqq d, \tag{1}$$

$$\text{or} \quad x_1 - x_2 = d, \tag{2}$$

$$\text{or} \quad x_1 - x_2 \leqq d. \tag{3}$$

Vertical constraints are equally formulated with y-coordinates instead of x-
coordinates.

2.2.1 Technology Design Rules

In general, technology design rules come in four categories (see Figure 1): the
minimum size of an element in a single mask layer, the minimum spacing between two
elements either in a single mask layer or in different mask layers, and the
overlap among elements on different mask layers.

When we deal with predefined devices or cells, such as transistors or gates,
overlap rules are automatically satisfied. Line-width for each connection is

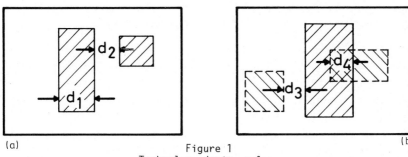

Figure 1
Technology design rules
(a) on a single mask layer and (b) on two mask layers
(elements with solid lines are on layer 1, with dotted lines on layer 2)

assumed to be fixed at its minimum size and enters into the calculation as constant. Thus technology design rules for cell level compaction mostly indicate a lower bound type relationship within or among elements, in the sense that typically the size of elements or the spacing between two elements on given mask layers must be greater than or equal to certain prescribed values. These are implemented as technology defined parameters for the individual elements, and are stored in the knowledge base. It should be mentioned that in a more complex situation the design rule can be expressed by two or more lower bound type constraints. Figure 2 shows an example where cell B is adjacent to cell A either to the left or to the bottom with a minimum spacing d. This constraint can be expressed as [19]:

$$x_A - x_B \geq d + (w_A + w_B)/2 \quad \underline{OR} \quad y_A - y_B \geq d + (h_A + h_B)/2. \qquad (4)$$

If the adjacency between A and B is possible on either of the four sides of A the constraint is expressed by four inequalities [32].

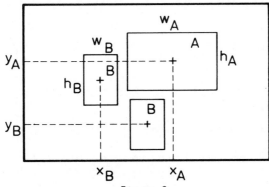

Figure 2
Two-dimensional adjacency constraint

2.2.2. User Specified Design Rules

Besides the basic technology design rules, it is useful to include the customer's requirements, or designer's "tricks of trade" developed through years of experience [40]. In general, these <u>user-specified design rules</u> offer the designer additional control over layout parameters. The designer may explicitly specify the absolute sizes of critical circuit elements or absolute spacing between two elements. For a selected pair of elements, not necessarily being adjacent, the designer can set an upper-bound, equality, or lower-bound constraint between them.

Example 2.1

In order to abut two cells directly without generating wires for inter-cell routing, the designer usually set some fixed equality constraints on the relative distances of the ports so as to align the ports to be connected together.

Example 2.2

When the length of an interconnection between two elements is critical to the signal speed, the designer can impose an upper-bound type constraint on the length of the connection so that the wire capacitance can be kept within a reasonable limit.

Example 2.3

In Figure 3 wire A is connected to element B. The designer may allow wire A to slide along the edge side of B as long as it doesn't violate the line-width and

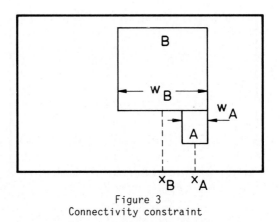

Figure 3
Connectivity constraint

the minimum spacing rules. This type of constraint comming from connectivity requirement can be expressed by two upper-bound type constraints [19]:

$$x_B - x_A \leq d \qquad \underline{AND} \qquad x_A - x_B \leq d \qquad\qquad (5)$$

with $d = (w_B - w_A)/2$.

In such a way, the designer can interact with the compactor smoothly, i.e., additional constraints imposed by the user or other programs could easily be incorporated into the compactor.

3. COMPACTION METHODOLOGY

3.1 General Classification

The problem of IC layout compaction is described by Sastry and Parker [28] as a partial order problem. It is presented in graph theoretic terms and its complexity is analyzed. They found out that the compaction of two-dimensional IC layouts is an NP-complete problem. Later on, Schlag, Liao and Wong [32] also proved the above result, which is independent of the compaction approach.

A common restriction to all different classes of compactors is, that topology of the initial layout has to remain unchanged. Strictly speaking, layout elements like cells and wire segments may move in any direction but must not jump over each other. Since connectivity as given by the initial circuit diagram also has to be preserved, the wires are considered to be strechable.

Further general restrictions are, that the initial layout contains orthogonal structures only, i.e. devices and cells are composed from rectangles, routing is done horizontally and vertically only (HV-routing) and layout grids are always rectangular. According to these restrictions we shall classify the compaction strategies as one- and two-dimensional strategies, respectively.

Most of the available compactors are based on some one-dimensional approach which solves the two-dimensional compaction problem by two one-dimensional procedures, i.e. a horizontal compaction and a vertical compaction. These two procedures are applied successively. Thus, the layout elements are moved in one direction at a time changing either their x-coordinates only during horizontal compaction or their y-coordinates only during vertical compaction. This one-dimensional approach is discussed in the following section 3.2.

A simultaneous compaction in both the horizontal and the vertical direction is preferable, since it avoids the tradeoff of moving an element in either direction. Although a general two-dimensional compaction strategy is not yet available, all attempts to treat x- and y- positions simultaneously at least during some part of the compaction procedure shall be classified as two-dimensional approach and discussed in section 3.3.

A classification of the compactors given in Table 1 with respect to the one- or two-dimensional approach is shown in Table 2 (see section 4).

3.2 One-Dimensional Approach

Throughout this section we shall consider a horizontal compaction only. Cells and vertical wire segments are moved horizontally, changing their x-positions only. Horizontal interconnections must be stretched accordingly in order to maintain connectivity. We also assume the left edge of the chip to be fixed at the position $x=0$. Then all elements are compacted from right to left. Vertical compaction is working identically on the y-positions with the bottom of the chip fixed at $y=0$ and compacting the elements from top to bottom.

All compactors basically consist of two successive steps: a search for movable elements and a moving strategy. These are discussed in the following subsections.

3.2.1 Search for Movable Elements

The problem of search for movable elements can be stated as follows: Given an initial layout, find elements which are movable horizontally to the left and for each element the possible maximum amount of move m consistent with design rules. In case a geometric initial position x_o is not given and therefore the move m is not defined, lower- and upper-bounds h and h' for the range of positions consistent with design rules have to be calculated. When we allow m=0 as "trivial move", then all elements are "movable". Hence, we only have to find the m-values (or h-values) of all elements.

In the following we shall distinguish two different classes of search strategies:
- A search strategy resulting in the m-values (or h-values) for all elements is called a global search. The subset of elements with $m \neq 0$ contains all really movable elements.

- A search strategy resulting in a non-empty subset of elements with m ≠ 0 is
  called a <u>local search.</u>

The amount m of a tolerable move of a certain element j depends on its neigh-
bour(s) i to the left and on design rules concerning the pair(s) of elements
(i,j). These pairs of adjacent elements can be extracted from the initial layout.
Williams [40] applies an exhaustive "row call" for this. Starting with the leftmost vertical column of elements and going on column by column from left to right,
each element j is called, then its neighbour i to the left is determined, the
design rule concerning the pair (i,j) is looked up and the maximum move for
element j consistent with the design rule calculated.

To ease the search procedure, two different representations of adjancency information have been developed: a matrix representation is used for an initial layout on
a grid basis and a search for "cuts" is executed in the matrix; a graph representation is used for an initial layout with relative positions and a search for
"longest paths" is executed in the graph. This is discussed in the following
subsections.

3.2.1.1 Basic Cut Search

The cut search technique has been first introduced by Akers e.a. [01] for symbolic
initial layout based on a coarse grid. The grid size is determined by the design
rules, i.e. by minimum wire width plus minimum wire spacing. Moves of elements are
restricted to 1 grid unit, i.e. $m = \left\{0, 1\right\}$.

A chip with n_x x n_y grid units is one-to-one mapped into a n_x x n_y matrix as
follows. Each grid unit (i,j) which is occupied either by a circuit element or by
a vertical wire segment is not available for horizontal moves of other elements
and therefore the corresponding matrix element $m_{ij} = 0$. For all grid units which
are either unoccupied or occupied by a horizontal wire segment, i.e. a strechable
wire segment, the corresponding matrix element is $m_{ij} = 1$. Figures 4a and b give
an example of an initial layout and its matrix representation. Obviously, since we
assume compaction to the left, the matrix element m_{ij} gives the m-value for the
layout element in grid position (i+1,j). The generation of the matrix results in
the m-values of all elements except the left chip edge on grid position i = 0
which is fixed. Therefore, the cut search strategy is a global search.

The matrix now contains all the necessary adjacency and spacing information. A
necessary condition for moving an element to the left is that it is bordering on a

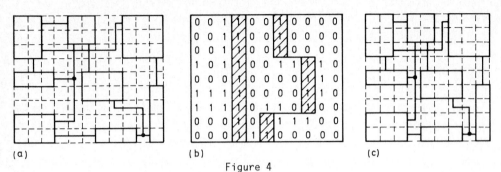

Figure 4
Cut search technique
(a) initial layout, (b) cut search matrix, (c) compacted layout

series of 1-elements in the matrix to its left. However, elements are not movable
independent of each other but only in groups of elements connected by vertical
wire segments. (A vertically connected group can be split into subgroups by introduction of jogs as already mentioned by Williams [40]. This shall be discussed
later.)

The basic idea of cut search technique is to search for a "cut" separating the
chip, i.e. the $n_x \times n_y$ matrix, respectively, into two parts. A cut is a series of
n_y 1-elements. They either form a column of the matrix (so-called <u>simple cut</u>) or a
series of subcolumns, the bottom of each being joined to the top of the next by a
horizontal rift line (<u>rift line cut</u>). In Figure 4b a simple cut and a rift line
cut is shaded. The rift line must not intersect 1-elements belonging to the same
group of vertically connected elements. Each removal of a cut results in a
horizontal compaction of 1 grid unit.

For complexity calculation we shall denote the number of grid units by n^2 and the
number of layout elements by N. This is proportional to the number of grid units
n^2. Matrix generation takes $O(n^2) = O(N)$ steps. In the worst case the number of
cuts is n and for each cut search n^2 matrix elements have to be checked. Therefore, the complexity of the search procedure is $O(n^3) = O(N^{1.5})$.

3.2.1.2 Modifications of Cut Search

The main advantage of the cut search procedure based on a coarse grid is the
simple and effective matrix representation of adjacency information resulting in a
global search strategy. However, this is at the expense of mask space since the
coarse grid does not allow individual spacing according to individual design
rules.

Therefore, Dunlop [09] has modified the cut search technique for application to an initial mask layout based on a mask grid. Of course, this does not allow a matrix representation of adjacency information as the number of mask grid units is much too high. Instead, a "working window" is used during the cut search procedure for testing design rules, i.e. calculating m-values, between an element i and other elements. The window has same height as element i and horizontally the full width of the chip in order to get hold of all other elements which might conflict with element i. The cut search starts at any element at the bottom of the layout with m≠0 and continues by moving the working window vertically. If the cut search reaches a dead end (m=0), a rift line is searched and the cut search continued. If a rift line is not found, the cut search starts backtracking. Obviously, this is not a global but a local search. Its complexity is estimated by Dunlop to be $O(N^2)$.

Weste [38] has introduced the "virtual grid" in order to combine both the advantages of treating adjacency information on a coarse grid basis and design rule calculation on a mask grid basis. Design rule calculation is restricted to adjacent columns or vertical grid lines which is done in n steps per column. As n columns have to be checked in the worst case, the complexity obviously is $O(n^2) = O(N)$. However, this is achieved by using simple cuts only.

3.2.1.3 Basic Longest Path Search

The longest path search technique has been first introduced by Hsueh and Pederson [15] for symbolic initial layout with relative positions. In this case the elements initially do not have geometric positions x_o and hence maximum move-values m are not defined. Therefore, minimum x-positions h have to be calculated for the elements instead of m-values. The basic idea of longest path search technique is to represent both the adjacency information from the initial layout and lower bound constraints on pairs of elements by a weighted graph. Longest path search in this graph is a global search strategy, resulting in the calculation of h-values for all elements.

The first step is a one-to-one mapping of the initial layout and its horizontal constraints into a directed <u>constraint graph</u>. This is accomplished as follows. Each group i of vertically connected elements, as it is to be moved as a whole, is mapped into a vertex v_i. Design rules are applied to calculate the minimum spacing between center lines of group i and each group j bordering it on the rigth. A lower bound type constraint between groups i and j

$$h_j - h_i \geq d_{ij} \qquad (6)$$

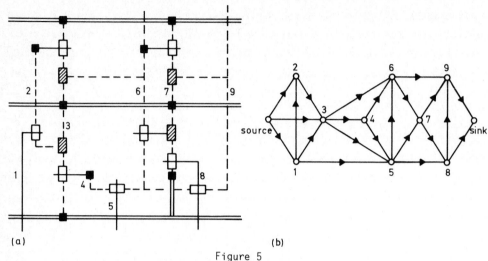

Figure 5
Longest path search technique
(a) symbolic layout and (b) constraint graph for a T-flip flop

is mapped into a directed edge (v_i, v_j) weighted by the constraint value d_{ij}. If more than one constraint are applying to a pair of groups (i, j), the maximum constraint value is considered as a weight to the corresponding edge (v_i, v_j). Figures 5a and b give an example of an initial layout and its graph representation. (Edge weights are not shown in the example.) Obviously, the left chip edge is the source and the right chip edge the sink of the directed graph. The graph now contains all the necessary adjacency and spacing information.

In the next step of the search procedure a <u>longest path</u> is calculated from the source vertex to each other vertex. The length of a path P_k from the source to vertex k is defined as the sum of weights d_{ij} for all edges $(v_i, v_j) \in P_k$. Usually several paths P_k will exist between source and vertex k. The h-value of vertex k, i.e. of a vertically connected group of elements, is determined by the length h_k of the longest path:

$$h_k = \text{Max}_{P_k} \left\{ \sum_{(v_i, v_j) \in P_k} d_{ij} \right\} . \tag{7}$$

Figure 6 shows again the example graph from Figure 5b with arbitrarily chosen edge weights and resulting h-values of vertices. Edge weights and vertex h-values are integer multiples of the mask grid unit which depends on technology. Longest paths are denoted in Figure 6 by heavy lines. The longest path from source to sink is called <u>critical path</u>. Its length gives the minimum chip width achievable by compaction. When the right chip edge is fixed to the position determined by the critical path, then the x-positions of all groups on the critical path are fixed to their h-values, too. For all groups not belonging to the critical path the h-values only give a lower-bound:

$$x_k \geqq h_k . \tag{8}$$

Their x-positions are to be determined by the moving strategy. The calculation of an upper-bound shall be discussed later.

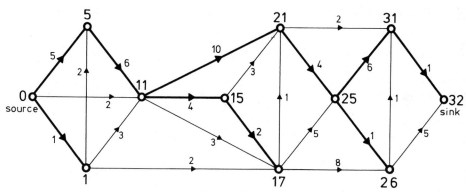

Figure 6
Constraint graph from Figure 5b
with edge weights and h-values for vertices
(heavy lines denote longest paths)

3.2.1.4 Longest Path Algorithms

The main part of the longest path search technique is a proper algorithm for determination of longest paths in weighted directed graphs. Numerous algorithms have been developed in graph theory and in operations research for solving this longest path problem or the equivalent shortest path problem [08].

In the following three types of graphs have to be distinguished: acyclic graphs, graphs with only negative cycles and graphs which also possess positive cycles. Cycles are caused by upper bound constraints (see section 2.2) because they can be rewritten into opposite directed lower bound constraints with a negative constraint value:

$$h_j - h_i \leqq d \iff h_i - h_j \geqq -d \quad \text{with} \quad d > 0. \tag{9}$$

Therefore, if an upper bound constraint is applied to a pair (i,j) of adjacent groups with j being to the right of i, a back directed edge (v_j, v_i) is added to the graph weighted by the negative constraint value $-d$. Each negative weighted edge causes at least one cycle in the graph. When the lower and upper bound constraints do not conflict each other, then the cycles have negative length and are simply called <u>negative cycles</u>. Otherwise they have positive length and are called <u>positive cycles</u>.

The simplest case is an acyclic graph which is achieved by the restriction to lower bound type spacing constraints [16]. Then the following well-known longest path algorithm can be applied to calculate the h-values of vertices.

LONGEST PATH ALGORITHM

```
Initialize h_j = 0 for all vertices 0 ≦ j ≦ N ;
            put source v_o into the empty queue Q ;
WHILE Q not empty DO
BEGIN
    take v_i from Q ;
    for each of its successors v_j DO
    BEGIN
        h_j = Max { h_j, h_i + d_ij } ;
        v_j = v_j - 1 ;
        IF v_j = 0   THEN   add  v_j to the end of Q ;
    END
END
```

The queue Q is used in the longest path algorithm to list vertices the h-values of which are already calculated. At the beginning Q contains the source vertex v_o only. Vertices are added to one end of Q and taken again from the other end of Q. When a vertex v_i is taken from Q, the preliminary h-values of all its successors v_j are updated. At the beginning $h_j = 0$ for all vertices of the graph. When the number v_j of those predecessors of vertex v_j, which have not yet been considered, equals 0, then updating of h_j can be stopped and therefore vertex v_j is added to the end of the queue.

For complexity calculation it should be noted, that the number of updating steps for each vertex equals the number of its predecessors. As for constraint graphs the average vertex degree is independent of the number N of vertices or groups, the complexity of the longest path algorithm is theoretically as well as experimentally O(N). However, this simple longest path algorithm cannot be directly applied to cyclic graphs. This is because in a cycle the predecessor of a certain vertex is also its successor and hence the updating loop cannot be finished.

The next case to be discussed is a cyclic graph with negative cycles only. Liao and Wong [24] have proposed a search strategy for this case as follows. First, the longest path algorithm is applied to the acyclic subgraph obtained by removing all edges with negative weight. After re-imbedding the n negative edges, (n+1) iterations are necessary in the worst case in order to adjust the calculated h-values according to the upperbound constraints. Therefore, the worst case complexity of this search procedure is O(nN). Liao and Wong estimate it as a linear raising with N for n ≪ N. However, in general n will raise linearly with N resulting in a worst case complexity $O(N^2)$. Schiele [29] has adapted a longest path algorithm due to Bellman and Ford [03,12] to this case. The procedure is iterative. The calculated h-values cannot be considered as final before the iteration comes to its end. For the average complexity Schiele obtains theoretically $O(N^{1.5})$ and experimentally $O(N^{1.3})$. It must be expected that the search strategy by Liao and Wong has about the same average complexity.

The third and most difficult case is a graph with positve cycles. Liao and Wong [24] and Starzyk [35] have worked on this case which can be treated by exhaustive search techniques. Schiele [30,31] has adapted another longest path algorithm to this case which is due to Dantzig, Blattner and Rao [07]. This DBR-algorithm is able to detect and locate positive cycles. The further treatment of positive cycles is discussed with the moving strategy in a later subsection. According to Schiele the DBR-algorithm has theoretically a complexity $O(N^2)$ and experimentally $O(N^{1.1})$.

3.2.1.5 Modifications of Longest Path Search

Modifications of the longest path search procedure are mainly concerned with different mapping of the inital layout into a constraint graph.

Within the basic procedure as discussed so far groups of vertically connected elements have been mapped into vertices of the constraint graph. A vertically connected group can be split into two or more subgroups by introduction of jogs. Figure 7 gives an example of a jog splitting group 6 from the initial layout of Figure 5a into two subgroups and the corresponding vertex 6 from Figure 5b into two vertices 6a and 6b. Strictly speaking, a jog breaks a vertical interconnection between elements by introducing a horizontal stretchable wire segment thus allowing the two subgroups to jog in either horizontal direction. Each subgroup is to

be mapped into a different vertex. Interactive jog generation as a tool for
achieving higher mobility of elements during compaction has been first introduced
by Williams [40] and later on also used by Hsueh [16] and others. Although jog
insertion is difficult to control and therefore has been sometimes criticized [38], it is used today in many compactors [27,30], either interactively or
automatically. Application of jogs for breaking positive cycles shall be discussed
later.

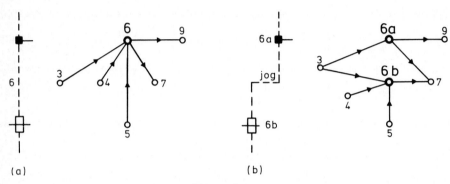

Figure 7
Jog insertion for group 6 from Figure 5a
(a) before jog insertion, (b) after jog insertion

Kedem and Watanabe [19] model a vertically connected group of m elements into m
vertices and (m–1) pairs of opposite directed edges, each pair representing a
connectivity constraint (see example 2.3 in subsection 2.2.2).

Other modifications are concerned with the mapping of adjacency information and
spacing constraints from the initial layout into edges of the constraint graph.
Dunlop [11] has used a mapping of horizontal interconnections between two vertically connected groups into edges between the corresponding vertices. In order to
take into account not only horizontally spacing constraints, e.g. between elements
A and B in Figure 8, but also diagonal or oblique constraints, e.g. between
elements A and C in Figure 8, Weste [38] uses arcs swept out by the lower and
upper corner of element A. This results in the dotted protection frame and the
shaded protection zone. The concept of visibility or a picket is used by several
authors [16,32,33,30] in order to simplify the adjacency check and limit the
number of edges in the constraint graph to those pairs of elements actually
constraining each other. A set of pickets, one for each mask layer, has been also
used by Boyer and Weste [04] in their so-called "most recent layer algorithm" in
order to keep track of the spacing constraints between different mask layers.

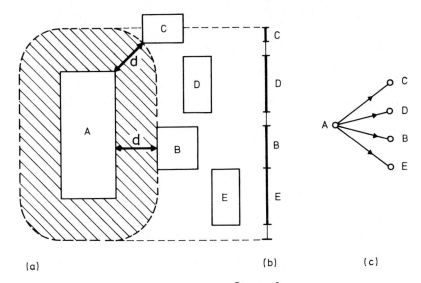

Figure 8
Group of elements constrained by orthogonal and oblique spacing rules
(a) protection frame, (b) picket representation, (c) graph representation

Finally, a modification of the basic longest path search shall be outlined which has not been published before. As has been mentioned in subsection 3.2.1.3, for all groups not belonging to the critical path the h-values of vertices give lower-bounds for x-positions only. An upper-bound h' consistent with design rules can be calculated by a longest path search starting at the sink and searching in opposite edge directions. Let be h_s the h-value of the sink and P_k' a path from sink to vertex k. Then the upper-bound h_k' of vertex k is given by:

$$h_k' = h_s - \text{Max} \left\{ \sum_{(v_i, v_j) \in P_k'} d_{ij} \right\}. \qquad (10)$$

Figure 9 gives the constraint graph for the T-flip-flop of Figure 5 with both the h- and h'-values as weights for vertices outside the critical path. For all vertices k belonging to the critical path the horizontal lower- and upper-bound values are equal. The corresponding groups are not movable but fixed to the positions

$$x_k = h_k = h_k' \qquad \text{for} \quad k \in \text{critical path.} \qquad (11)$$

All other groups are movable within a horizontal <u>range of tolerance</u> consistent with design rules:

$$h_k \leqq x_k \leqq h_k' \qquad \text{for} \qquad k \not\in \text{critical path.} \qquad (12)$$

Therefore, this extended longest path search results in the full information of movability consistent with design rules. Determination of position x_k for movable elements within the horizontal bounds has to be done by the moving strategy.

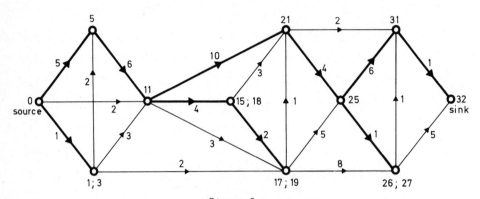

Figure 9
Constraint graph for the T-flip flop from Figure 5
with vertex weights h and h;h' respectively
(heavy lines denote longest paths)

3.2.2 Moving Strategy

The main problem of compaction is to decide in which direction and how far to move each element in order to obtain an "optimal" final layout.

During the search procedure (subsection 3.2.1), all movable elements are found either by a global strategy or by a local strategy. The design rules stored in the knowledge base are consulted, and the adjacency information derived from topology or geometry of the initial layout is used. Here, the moving strategy will be used to determine how to move each element.

In the one-dimensional approach, each movable element can only be moved from right to left during the horizontal compaction, and from top to bottom during the vertical compaction. Thus the question about in which direction to move the elements is not existing in the one-dimensional approach. It is predefined by the method.

3.2.2.1 Moving Strategy After Cut Search

In case of the initial layout being represented in a matrix form with some kind fixed grid, the question of how far to move elements is not existing either. Elements are moved to the left (or bottom) by one grid unit, when a simple or rift line cut is removed during horizontal (or vertical) compaction. Hence, the amount of move is always one grid unit. The removal of all vertical cuts results in a horizontal compaction, and the removal of all horizontal cuts results in a vertical compaction. The horizontal compaction is followed by a vertical compaction. After that, a new cut search will be performed. Then the process of moving will be repeated until no more cut can be found . This simple moving strategy is used in CUT $[01]$, SLIP $[09]$, MULGA $[38]$, VPACK $[04]$ and partially in SLIM $[11]$.

3.2.2.2 Moving Strategy After Longest Path Search

In case of the initial layout being represented by a constraint graph G_x for horizontal constraints among vertically connected groups, only relative positions are given for horizontal compaction as no grid is used. Same is for vertical compaction when the initial layout is represented by another constraint graph G_y for vertical constraints among horizontally connected groups. Thus the search for all longest paths in the constraint graphs G_x and G_y can only find movable elements and the ranges of horizontal and vertical tolerance for each movable element

$$h \leqq x \leqq h' \quad \text{and} \quad v \leqq y \leqq v'. \tag{13}$$

Here, h and h' stand for the horizontal bounds of tolerance, while v and v' stand for the vertical ones, respectively. In this case deeper moving strategy than the last one in the above subsection is needed to determine where each element to be moved to.

As noted before, elements with h = h' or v = v' form the critical paths in G_x and G_y, respectively. They are not movable. The lengths of those two critical paths give the minimum chip width and heigth, respectively, which will be achieved by the compaction.

For the movable elements, not being on the critical path of G_x or G_y, exact locations (x, y) are to be determined within the ranges of tolerance as given by equ. (13). Three different kinds of moving strategies have been developed and applied in order to obtain an "optimal" final layout. They are called the minimum, maximum and optimum moving strategies and will be discussed in later subsections.

Further, a special strategy is necessary to resolve positive cycles in the constraint graphs. This is discussed in the following subsection.

3.2.2.3 Constraint Graphs with Positive Cycles

Here, we consider the problem of positive cycles and existence of longest paths in
the constraint graphs G_x and G_y. As introduced in subsection 3.2.1.4, positive
cycles are caused by those upper-bound constraints conflicting with lower-bound
constraints. Usually this happens because of conflicts between the user specified
design rules and the technology design rules. Clearly, the longest path search
fails if there is a positive cycle in the constraint graph. In fact, it is known
that there exists a valid and "optimal" final layout for the given initial layout
if and only if there is no positive cycle in the graphs representing the given
initial layout. A proof of this has been given by Liao and Wong [24]. Therefore,
positive cycles have to be detected and resolved. As discussed already in section
3.2.1.4, positive cycles can be detected either by some exhaustive search technique or by the well-known DBR-algorithm [07].

In general, positive cycles can be solved completely after being located, either
through the relaxation of the conflicting design rules [24,20], or by adding jogs.
This can be done either interactively or by automatic jog generation [16,30,06]. A
simple example of positive cycle is given in Figure 10. Minimum wire width and
overlapping requirement for wire and contact are represented by lower-bound constraints, i.e. positive weighted edges. Maximum wire width is represented by an
upper-bound constraint, i.e. an edge with negative weight, resulting in a positive
cycle. The original situation and the resolved form after jog insertion are both
presented in Figure 10.

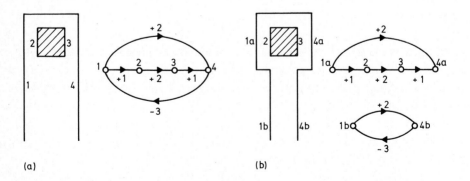

Figure 10
Example for resolution of a positive cycle
(a) before jog insertion, (b) after jog insertion

After proper jog insertion the resulting graph should have no positive cycles any more. Therefore, in the following we shall only consider initial layout which can be represented by graphs G_x and G_y without any positive cycles.

3.2.2.4 Minimum and Maximum Moving Strategy

A moving strategy is called a _minimum moving strategy_ if the amount of move is always the minimum allowance from the design rules. In other words, all movable elements are moved to their _maximum positions_, i.e. for all k:

$$x_k = h_k' \qquad \text{and} \qquad y_k = v_k'. \tag{14}$$

On the other hand, if the moving strategy always choses the maximum allowance from the design rules as the amount of move, then it is called a _maximum moving strategy_. For such a strategy, all movable elements are moved to their _minimum positions_, i.e. for all k:

$$x_k = h_k \qquad \text{and} \qquad y_k = v_k. \tag{15}$$

Both the minimum and the maximum moving strategy are quite simple. They may result in an "optimal" final layout but usually do not. In general, the mimimum moving strategy tends to cause a final layout with high density in the upper right corner but low density in the lower left corner. This is natural as movable elements stay at their right-most and highest possible positions which are tolereted by the design rules. By a similar argument, it is clear that the maximum moving strategy tends to result a final layout with high density in the lower left corner but low density in the upper right corner. These tendencies may result in unnecessary long interconnections with wasted wire length and chip area. Figure 11 illustrates the maximum moving strategy using an example given by Schlag, Liao and Wong [32].

It is Williams who first introduced the maximum moving strategy for compaction in STICKS [40], although it is not a graph theoretic approach as it is presented. Maximum moving strategy has also been used by Liao and Wong [24]. In MAGIC, the compactor PLOWING introduced by Scott and Ousterhout [34] uses the minimum moving strategy.

3.2.2.5 Optimum Moving Strategy

The moving strategy is called an _optimum moving strategy_ if all movable elements are moved into _optimum positions_ within the tolerance interval, i.e. for all k

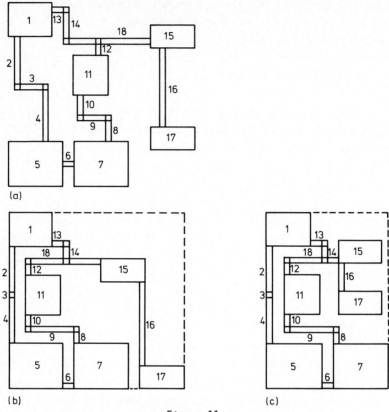

Figure 11
Compaction example by Schlag, Liao and Wong [32]
(a) initial layout, (b) maximum compaction
(c) result of improved compaction from [32]

$$h_k \leqq x_k \leqq h'_k \quad \text{and} \quad v_k \leqq y_k \leqq v'_k , \qquad (16)$$

according to some optimization criteria. The simplest optimum moving strategy is
to move each element to its mean position, i.e.,

$$x_k = (h_k + h'_k)/2 \quad \text{and} \quad y = (v_k + v'_k)/2 , \qquad (17)$$

which does not need any specific optimization criterion. Several optimum moving
strategies have been developed based on different optimization criteria.

It is Hsueh and Pederson who first introduced in CABBAGE [15,16,17], an optimum
moving strategy for compaction. Based on the force-directed placement technique
[14], the optimization criterion is to minimize the force acting on each movable

element. The x-component of force is considered during horizontal compaction, the
y-component of force during vertical compaction only. Depending on different
definitions of force, i.e. different weighting, the optimum positions may be
varied.

Another optimization criterion used by several authors is minimum chip area. An
analytical method for compaction based on such criterion is introduced by Ciesielski and Kinnen [06]. Linear programming (LP) is used as the optimization technique
to minimize chip length and height independent of each other, subject to constraints derived independently for each routing channel. The optimization problem
is formulated as two independent LP problems for the horizontal and vertical
compaction steps. Thus the computational complexity of a single compaction step is
limited by that of LP. The application of this optimization criterion of minimum
chip area to two-dimensional compaction by Kedem and Watanabe [19] shall be discussed in section 3.2.

Finally, it should be mentioned that the criterion of minimum chip area does not
always uniquely determine "optimum" positions for all layout elements. Figure 12
gives an example where chip size is independent of the position of wire segment
A2. Then minimizing rectangle areas in the routing layers can be used as an
additional optimization criterion in order to avoid overlength of routing,
e.g. [20,30]. In fact this reduces both wire width and wire length, but not
necessarily chip area. In Figure 12 the length of wire A to cell B can be reduced
by minimizing the rectangles of wire segments A1 and A3, thus giving maximum
position to wire segment A2.

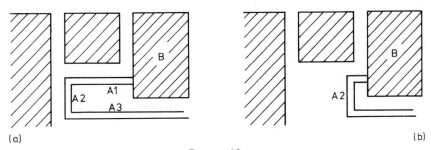

(a)   (b)

Figure 12
Example for avoiding overlength of routing
(a) before and (b) after minimizing wire length

3.3 Two-Dimensional Approach

It has been stated already in section 3.1 that layout compaction belongs to the class of NP-complete problems. Two basic heuristic approaches are available for approximate solution of NP-complete problems within polynomial computation time: divide and conquer strategy and iterative improvement. The one-dimensional approach as discussed in section 3.2 is a divide and conquer heuristic: the two-dimensional problem is devided into two one-dimensional problems.

As the one-dimensional approach give optimal results for one-dimensional compaction problems, attempts have been done to apply it also to the two-dimensional compaction problem more intelligently than by a simple successive iteration. Wolf et al. [41] have proposed a "two-dimensional" compaction strategy with a preferred direction of compaction. It is based on horizontal and vertical critical path analysis. Critical pairs of elements belonging to the critical path in the preferred direction are sheared apart to improve the compaction in this direction. However, there are only minor differences to the technique of breaking a critical path by insertion of jogs as used in CABBAGE.

Two different approaches for the two-dimensional compaction problem are evolving recently and shall be discussed in the following subsections: the boolean decision variable approach and the simulated annealing approach. Both are belonging to the class of iterative improvement heuristics.

3.3.1 Boolean Decision Variable Approach

The strict interrelation between horizontal and vertical constraints inherent to the two-dimensional compaction problem can be formulated using boolean decision variables. A <u>switching variable S</u> controls either a horizontal or vertical spacing constraint on a pair of adjacent cells (i,j) with given constraint value d_{ij}:

$$S_{ij} = \begin{cases} 0 & \text{if constraint on } (i,j) \text{ is relaxed} \\ \\ 1 & \text{if constraint on } (i,j) \text{ is not relaxed.} \end{cases} \qquad (18)$$

Two switching variables S_{ij}^x and S_{ij}^y correspond to each pair (i,j) of adjacent cells, representing the horizontal and the vertical spacing constraint, respectively. These switching variables have been used earlier for placement compaction based on force-directed placement [26], which however is not a layout compaction since routing is not done.

For layout compaction a simpler switching variable has been introduced by Kedem and Watanabe [19]. For certain pairs of cells (A,B), adjacent in the initial layout and specified by the user, two different relative positions shall be allowed as shown in Figure 2. This is controlled by the following two-dimensional spacing constraint:

$$x_A - x_B \geq d_x \quad \underline{OR} \quad y_A - y_B \geq d_y, \tag{19}$$

which itself can be controlled by a single boolean decision variable:

$$S_{AB} = \begin{cases} 0 & \text{if} \quad x_A - x_B \geq d_x \\ \\ 1 & \text{if} \quad y_A - y_B \geq d_y . \end{cases} \tag{20}$$

According to its value the switching variable influences the horizontal and vertical constraint graphs. The edge (A,B) is contained in the horizontal graph G_x and not in the vertical graph G_y if $S_{AB} = 0$ and the other way around if $S_{AB} = 1$. Values of switching variables are determined by solution of a chip area minimization problem. This can be either done by a global but time consuming search tree technique using branch and bound method or by the following simple "initial guess" technique. First, longest path search is done in G_x and G_y. Both graphs contain edges (A,B) for all switching variables. If (A,B) belongs to the critical path of G_x (or G_y) only, then $S_{AB} = 1$ (or 0). By iteration most of the switching variables can be determined in their value. If a switching variable still belongs to both critical paths at the end of the iteration, its value can be chosen according to critical path weights. As this is only a "local" calculation it still can be optimized by the search tree technique.

Because of the restriction to only two out of four possible relative positions for the adjacent pair of cells A and B (see Figure 2) this is not a truly two-dimensional compaction technique but shall be classified as semi-2-dimensional.

A most severe presentation and discussion of the boolean decision variable approach has been given by Schlag, Liao and Wong [32]. They allow the full range of four possible relative positions for any pair of adjacent cells (i,j): cell j may be either to the left or right or bottom or top of cell i. This is expressed by the following set of lower-bound constraints:

$$\begin{aligned} & x_i - x_j \geq d_1 \\ \underline{OR} \quad & x_j - x_i \geq d_2 \\ \underline{OR} \quad & y_i - y_j \geq d_3 \\ \underline{OR} \quad & y_j - y_i \geq d_4 . \end{aligned} \tag{21}$$

Clearly, a necessary and sufficient condition for a legal placement of the pair (i,j), i.e. a placement consistent with design rules, is the validity of one of the four constraints. Four switching variables S_{ij}^k ($k=1,2,3,4$) corresponding to the four constraints are introduced according to equation (18) and a <u>boolean function F</u> over these variables with

$$F = \begin{cases} 1 & \text{for a legal placement} \\ \\ 0 & \text{for an illegal placement .} \end{cases} \qquad (22)$$

Obviously, the conjunctive normal form

$$F = \prod_{(i,j)} (S_{ij}^1 + S_{ij}^2 + S_{ij}^3 + S_{ij}^4) \qquad (23)$$

has this property. A family of decision trees is used to determine the S-values. The vertices of the search tree correspond to adjacent pairs of cells (i,j), the edges correspond to constraints on these pairs. Hence, vertices have at most degree four, corresponding to the four relative positions of pair (i,j). The tree depends on how the search proceeds. It starts with a collapsed layout, violating all minimum spacing constraints. Violations are selected arbitrarily one at a time and removed by proper choice of S-values. Different selection patterns of violations will result in different trees. The exhaustive search for minimum chip size is pruned by branch and bound technique. Obviously, chip size is growing during the search as more and more minimum spacing violations are resolved. Chip size calculated by horizontal and vertical critical path analysis is used as an upper bound on the size during the search. The upper bound is updated as soon as a legal placement with smaller size has been found. Other pruning techniques for the search trees are also discussed in [32].

Because of the exhaustive search the time complexity is exponential in the worst case. A complexity analysis of the average case has not been attempted, since the complexity highly depends on the order of selecting violations and on how efficient pruning for the search tree has been done. Hence, we can only argue by probability. With increasing size of the problem, i.e. number of elements to be compacted, the worst case gets increasingly small probability and the average case will dominate. This arguement of large numbers leads to the stochastic approach in the following subsection.

3.3.2 Simulated Annealing Approach

Optimum compaction is a problem of combinatorial optimization, i.e. finding optimal properties of a very large system. This can be stated as finding the minimum or maximum of a cost function or objective function depending on many variables. The analogy of such combinatorial optimization problems with annealing in solids, i.e. with thermodynamics of systems with many degrees of freedom, has been studied recently. A new technique has emerged: optimization by <u>simulated annealing</u> [21]. Monte Carlo technique is used to simulate a large number of iterative improvement steps randomly chosen in order not to get stuck in a local optimum of the objective function. If the atoms or elements to be handled are not all alike, randomness becomes limited and the optimization is subject to frustrating constraints. With the exemption of gate arrays, layout compaction belongs to this type of optimization for systems with <u>frustration.</u>

The basic technique of simulated annealing is as follows: Small random displacements of the initial layout are generated by Monte Carlo techniques. The resulting change ΔF in the minimization function F of the compaction problem is calculated. The displacement is accepted and the resulting layout used as a starting point for the next iteration, if $\Delta F \leqq 0$. If $\Delta F > 0$ the displacement is accepted with a probability

$$P(\Delta F) = \exp(-\Delta F/kT) \; . \tag{24}$$

It is this stochastic acceptance of displacements leading off the minimum, that guarantees not to get stuck in a local minimum. Parameter T is a control variable for the minimizing process and k is a constant, with kT having the same unit as the objective function F. In statistical thermodynamics F is the energy, T the temperature and k the Boltzmann constant. It should be mentioned that spacing constraints and other design rules can be included in the objective function F. For example the constraint $\Delta x_{ij} = d$ for all pairs (i,j) is added to F by the term

$$\sum_{(i,j)} (\Delta x_{ij} - d)^2 .$$

This has the advantage, that constraints may be violated in early stages of the compaction procedure (at high temperature T) giving more freedom for improvement. When temperature is reduced, the constraints get more and more weight.

The application of simulated annealing techniques to the field of layout in general and layout compaction especially still is at its beginning. There is an
increasing interest in this approach and research is going on at different
places.

4. IMPLEMENTATION AND PERFORMANCE

For a number of compactors which have been already listed in Table 1, implementation and performance data are given in Table 2 and Table 3, respectively. Again,
an exhaustive listing of all known compactors has not been aimed at.

The numbers in the reference columns of the tables are the same as in the listing
of references in section 6. For each compactor the first line refers to the
original publication of that compactor, i.e. to the author(s) of the compactor. If
other sources are available for implementation and/or performance data, they are
added in following lines and the number in the reference column refers to this
second source. If entries are missing in the tables, no data are given in the
literature. The power of N is given in Table 3 for the theoretical and experimental complexity.

5. CONTEMPORARY PROBLEMS

To find the final layout with absolute minimum chip area is an NP complete problem
(see section 3.1), but in reality that is not the end of computer aided layout
design. For layout compaction, a good solution for today's technology means that
no other method currently can produce a smaller size layout. It is from this point
of view, we are considering possible future study on layout compaction.

Today, minimum chip area or maximum chip density is still one of the major layout
problems. It has been improved in a great deal during the last five years. The
one-dimensional compaction approach is almost completely developed to its mature
stage. Stick diagram and the constraint graph are the basic representations of the
topology of layout and the constraints from design rules. Longest path analysis,
as introduced by Hsueh and Pederson [15] for compaction, has become the main
stream of both one-dimensional and two-dimensional compaction since then. Various
improvements and modifications have been done. The complexity of many compactors
is close to linear order now.

Compactor	Ref.	Year	Appr.	Computer	Lang.	Technol.
CUT	01	1970	1D			
FLOSS	05	1977	1D	IBM 370/168	PL/I	COS/MOS
STICKS	40 16 04	1977	1D	HP 3000 CALMA		
SLIP	09 04	1978	1D	HP 21 MX PDP 11/70	C	NMOS
CABBAGE	15	1979	1D	HP 1000 E	FORTRAN	NMOS
SLIM	11	1980	1D			
MULGA	38 04	1981	1D	PDP 11/23 VAX 11/750	C	C/NMOS
Liao/Wong	23	1983	1D			
Schlag et al.	32	1983	2D			
SQUASH	27	1983	1D	VAX		CMOS
ROCHESTER	18	1983	2D	VAX 11/780	PASCAL	NMOS
VPACK	04	1983	1D	VAX 11/780	C	CMOS
Wolf et al.	41	1983	2D			.
PLOWING	34	1984	1D	VAX 11/780	C	C/NMOS
Kingsley	20	1984	1D			
Schiele	30	1984	1D	HP 100 F	PASCAL	C/NMOS

Table 2
Implementation of different compactors
(Approach : 1D = 1-dimensional, 2D = 2-dimensional)

Compactor	Ref	Examples			Complexity	
		Type	Size	Time (sec)	theor.	exper.
CUT	01				1.5	
FLOSS	05				2	
	16	A	1200	29		
		B	4000	100		1.0
STICKS	40				2	
	16	3 - Input Gate	20	20		
		C	300	180 000		3.3
	04		30	1800		
SLIP	09				2	
	16	Inverter	14	28		
		5-Input NAND	28	44		1~2
		D-Flip Flop	52	215		
CABBAGE	15				1.5	
	16	T - Flip Flop	87	21		
		5 T - Flip Flops	407	191		1.4
		Latch Driver	290	220		
SLIM	11		14	24	1.5	
		5 - Input NAND	28	48		
		D-Flip Flop	52	213		1~2.4
MULGA	38				1	
	04	T- Flip Flop		13.5		
		12 T-Flip Flops		336		1.3
Liao/Wong	23			2		
Schlag et al.	32					
SQUASH	27	Pixel Planes	57			
ROCHESTER	18	T-Flip Flop		11.8		
		Priority Q		131.4		
VPACK	04	T- Flip Flop		3.2	1	
		12T - Flip Flops		22		
		192 T - Flip Flops		352		1.0
Wolf et al.	41	Shift Register				
PLOWING	34	ALU Latch		3.0		
		BUS Driver		5.5		
Kingsley	20		235			
		Deutsche's Example	595			
Schiele	30					

Table 3
Performance of different compactors

However, the chip size and the number of elements per chip are going up very quickly while the technology advances allow us to work on much smaller scale. Thus a truely two-dimensional compactor with low complexity is needed badly. Schlag, Liao and Wong [32] have proposed a method. But this is based on a family of decision trees, which is so huge that it looks like a forest. The stochastic approach might be a good answer to the increasingly large size of constraint graphs. The method of simulated annealing developed recently by Kirkpatrick, Gelatt and Vecchi [21] is a useful tool for solving such a large-scale optimization problem like compaction. In two-dimensional compaction both search strategy and moving strategy need more work.

A serious restriction to all today's compaction methodology is the invariant topology of the initial layout. This limits the freedom of compaction and therefore the possible improvement and the rate of chip size reduction considerably. Progress beyond these limitations can be obtained only by a much deeper and closer interrelation between placement, routing and compaction. Experience is available from skilled designers. However, much work is to be done for proper acquisition, representation and computer aided processing of this knowledge.

An expert system seems to be the ultimate goal for solving all these contemporary problems: the NP completeness of the compaction problem which demands for heuristics, the increasingly large size of the compaction problem which demands for stochastics, its two-dimensionality which demands for object-oriented decisions and the aim for a better understanding and usage of available experience and knowledge which demands for knowledge engineering. Moreover, when the technology evolves, the customer's needs evolve and the compaction knowledge evolves, an expert system could guarantee that the most useful layout compactor will also evolve.

6. REFERENCES

[01] Akers, S.B., J.M. Geyer and D.L. Roberts, "IC Mask Layout with a Single Conduct Layer", Proc. 7th Ann. Design Automation Workshop, San Francisco, pp.7-16, 1970.

[02] Baird, H.S., "Fast Algorithm for LSI Artwork Analysis", Proc. 14th Design Automation Conference, pp.303-311, June 1977.

[03] Bellman, R., "On a Routing Problem", Quaterly of Applied Mathematics, Vol. 16, pp.81-90, 1958.

[04] Boyer, D.G. and N. Weste, "Virtual Grid Compaction Using the Most Recent Layers Algorithm", Proc. IEEE International Conference on Computer Aided Design, pp. 92-93, 1983.

[05] Cho, Y.E., A.J. Korenjak and D.E. Stockton, "FLOSS: An Approach to Automated Layout for High-Volume Designs", Proc, 14th Design Automation Conference, pp. 138-141, June 1977.

[06] Ciesielski, M.J. and E. Kinnen, "An Analytic Method for Compacting Routing Area in Integrated Circuits", Proc. 19th Design Automation Conference, pp. 30-37, 1982.

[07] Dantzig, G.B., W.O. Blattner and M.R. Rao, "All Shortest Routes from a Fixed Origin in a Graph ", Proc. International Symposion Theorie des Graphes, pp. 85-90, 1966.

[08] Dreyfus, S.E.,"An Appraisal of Some Shortest Path Algorithms", Operations Research, pp. 395-412, 1968.

[09] Dunlop, A.E., "SLIP: Symbolic Layout of Integrated Circuits with Compaction", Computer-Aided Design, Vol.10, No.6, pp. 387-391, November 1978.

[10] Dunlop, A.E., "Integrated Circuit Mask Compaction", Ph.D. Thesis, Carnegie-Mellon University, Pittsburgh, October 17, 1979.

[11] Dunlop, A.E., "SLIM: The Translation of Symbolic Layouts into Mask Data", Proc. 17th Design Automation Conference, pp. 603-609, June 1980.

[12] Ford, L.R., "Network Flow Theory", Rand Corporation Report, pp. 1-12, 1956

[13] Gibson, D. and S. Nance, "SLIC – Symbolic Layout of Integrated Circuits", Proc.13th Design Automation Conference, pp. 434-440, June 1976.

[14] Hanan, M., P.K. Wolff and B.J. Agule, "Some Experimental Results on Placement Techniques", Proc. 13th Design Automation Conference, pp. 214-224, June 1976.

[15] Hsueh, M.Y. and D.O. Pederson, "Computer-Aided Layout of LSI Circuit Building Blocks", Proc. IEEE International Symposium on Circuits and Systems, pp.474-477, June 1979.

[16] Hsueh, M.Y., "Symbolic Layout and Compaction of Integrated Circuits", Electronics Research Laboratory Memorandum No. UCB/ERL M79/80, University of California, Berkeley, December 1979.

[17] Hsueh, M.Y., "Symbolic Layout Compaction", Report, NATO Advanced Study Institute on Computer Design Aids for VLSI Circuits, Sogesta-Urbino, Italy, June-August, 1980.

[18] Kedem, G. and H. Watanabe, "Graph Optimization Techniques for IC Layout and Compaction", Proc. 20th Design Automation Conference, pp. 113-120, 1983.

[19] Kedem, G. and H. Watanabe, "Graph-Optimization Techniques for IC Layout and Compaction", IEEE Trans. on Computer Aided Design, Vol. CAD-3, No.1, pp. 12-20, January 1984.

[20] Kingsley, C., "A Hierarchical, Error-Tolerant Compactor", Proc. 21th Design Automation Conference, pp. 126-132, 1984.

[21] Kirkpatrick, S., C.D. Gelatt,Jr. and M.P. Vecchi, "Optimization by Simulated Annealing", Science, Vol. 220, No. 4598, pp. 671-680, May 1983.

[22] Larsen, R.P., "Computer-Aided Preliminary Layout Design of Customized MOS Arrays", IEEE Trans. Computers, Vol. C-20, pp.512-523, 1971.

[23] Liao, Y.Z. and C.K. Wong, "An Algorithm to Compact a VLSI Symbolic Layout with Mixed Constraints", Proc. 20th Design Automation Conference, pp. 107-112, 1983.

[24] Liao, Y.Z. and C.K. Wong, "An Algorithm to Compact a VLSI Symbolic Layout with Mixed Constraints", IEEE Trans. on Computer Aided Design of Integrated Circuits and Systems, Vol. CAD-2, No.2, pp. 62-69, April 1983.

[25] McGarity, R.C. and D.P. Siewiorek, "Experiments with the SLIM Circuit Compactor", Proc. 2oth Design Automation Conference , pp. 740-746, 1983.

[26] Rong, Z., "Ein Kräfteplazierungsverfahren für den Schaltungslayout unter Berücksichtigung der Zellenabmessungen", Fortschritt-Berichte der VDI-Zeitschriften, Reihe 9, Nr.38, 1983.

[27] Rosenberg, J., D. Boyer, J. Dallen, S. Daniel, C. Poirier, J. Poulton, D. Rogers and N. Weste, "A Vertically Integrated VLSI Design Environment", Proc. 20th Design Automation Conference, pp. 31-38, 1983.

[28] Sastry, S. and A. Parker, "The Complexity of Two-Dimensional Compaction of VLSI Layouts", Proc. IEEE International Conference on Circuits and Computers, New York, pp. 402-4o6, 1982.

[29] Schiele, W.L., "On a Longest Path Algorithm and its Complexity if Applied to the Layout Compaction Problem", Proc. European Conference on Circuit Theory and Design, Stuttgart, pp. 263-265, 1983.

[30] Schiele, W.L., "Entwurfsregelanpassung der Masken-Geometrie integrierter Schaltungen", PH.D. Thesis, Technische Universität München, München, November 1984.

[31] Schiele, W.L., "Automatic Design Rule Adaptation of Leaf Cell Layouts", to be published in INTEGRATION, VLSI Journal, 1985.

[32] Schlag, M., Y.Z. Liao and C.K. Wong, "An Algorithm for Optimal Two-Dimensional Compaction of VLSI Layouts", INTEGRATION, VLSI Journal 1, pp.179-209, 1983.

[33] Schlag, M., F. Luccio, P. Maestrini, D.T. Lee and C.K. Wong, "A Visibility Problem in VLSI Layout Compaction", IBM Research Report, 1983.

[34] Scott, W.S. and J.K. Ousterhout, "PLOWING: Interactive Stretching and Compaction in MAGIC", Proc. 21th Design Automation Conference, pp. 166-172, 1984.

[35] Starzyk, J.A., "Decomposition Approach to a VLSI Symbolic Layout with Mixed Constraints", Proc. IEEE International Symposium on Circuits and Systems, pp.457-460, May 1984.

[36] Watanabe, H., "IC Layout Compaction Using Mathematical Optimization", PH.D. Thesis, University of Rochester, N.Y., 1983.

[37] Weste, N., "MULGA - An Interactive Symbolic Layout System for the Design of Integrated Circuits", Bell System Technical Journal, Vol.60, No.6, pp. 823-857, July-August 1981.

[38] Weste, N., "Virtual Grid Symbolic Layout", Proc. 18th Design Automation Conference, pp. 225-233, 1981.

[39] Williams, J.D., "STICKS - A New Approach to LSI Design ", Master Thesis, Massachusetts Institute of Technology, June 1977.

[40] Williams, J.D., "STICKS - A Graphical Compiler for High Level LSI Design", Proc. AFIPS, Vol. 47, pp. 289-295, 1978.

[41] Wolf, W., R. Mathews, J. Newkirk and R. Dutton, "Two-Dimensional Compaction Strategies", Proc. IEEE International Conference on Computer Aided Design, pp. 90-91, 1983.

Chapter 7

LAYOUT VERIFICATION

K. YOSHIDA

Toshiba Corporation
Japan

1. Introduction

The automatic layout design method has become an established design tool for semi-custom or custom LSIs, such as gate array or standard cell LSIs, because of the short design time required. However, the chip density, or the number of elements per silicon area, available in automated layout is generally poor, due to the simple layout models usually applied, such as rigid cell structure, coarse grid wiring and regular channel configuration. On the other hand, handcraft design, based upon polygonal shapes on mask layers, has more flexibility in order to obtain a highly optimized design in terms of chip density and of circuit performance. Such handcraft design is often applied to high volume production LSIs, such as full custom or standard LSIs, in spite of the expensive design process. Important tools for handcraft design are automatic layout verification tools as well as interactive graphic systems. Manual inspection of VLSI layout design is becoming almost impossible with its ever-increasing circuit complexity.

Design errors, which may occur in LSI layout design, can be classified into the following categories.

(1) Geometrical design rule error

Each technology has its own geometrical tolerances, such as minimum spacing between shapes, minimum internal width of a shape, and so on. A set of such rules is called the geometrical design rule. A violation of the geometrical design rule usually decreases the chip manufacturing yield, or results in non-functional LSIs.

(2) Topological error, or logical error

Topological errors include mistakes in electrical connections between circuit elements, and improper structure of circuit elements. These errors usually result in circuit malfunctions, and are often detected as logical errors. In most cases, these errors are fatal.

(3) Electrical performance error

The LSI should be designed to meet electrical characteristics requirements, such as power dissipation, timing specifications and others. This kind of error is caused by inadequate device dimensions and disregarded parasitic effects, and results in LSIs with logically correct function but non-satisfactory performance.

Lots of work has been done on verifying above design errors. Especially, geometrical design rule check programs are extensively used in practical VLSI design, due to its relative simplicity in usage and design method independency.

One of the most important problems in verifying today's VLSI layouts is the computer time needed. To handle these huge amount of data, a very efficient algorithm with almost linear time complexity are required.

In the following section, design environments and layout data nature will be discussed. In Section 3, geometrical design rule checking will be described with basic pattern operations. Detailed discussion on various computational algorithms for basic pattern operations will be made in Section 4. Remaining sections will include the discussion on circuit extraction, topological verification and electrical performance verifications, respectively.

2. Design Environments and Layout Data

2.1 Design methods and verification

The occurrence of errors in the above categories depends on design method or design system used. For example, automatic placement and routing seldom results in errors in categories (1) and (2), and only category (3) is subject to being verified. Handcraft design, on the other hand, is most error-prone in every error category, as well as being most difficult to verify, because of poor design regularity and large amount of poorly structured layout data.

There are two ways to use layout verification tools in handcraft design. In the conventional method, layout verification is applied after the entire chip design is finished. The input data to be verified are purely geometrical data in an appropriate format. Usually, batch processing is used with a large amount of computer resources. Another use of verification tools is in an interactive environment. Most of recent graphic layout design systems have verification capabilities that enable partial or entire chip design verification from time to time during the design phase. Since quick response is important, small and simple data and simple verification functions are generally desirable for such an environment.

The following sections mainly discuss entire chip verification for general handcraft design, for which the capability to handle a large amount of data, design method independency and technology independency are desirable features.

2.2 Layout data to be verified

Layout data for integrated circuits consists of mask pattern layers, each of which is a set of <u>figures</u>, or opaque regions, each bounded by a polygon. In most technologies, polygon edges are straight lines or are approximated by straight lines, and, in many cases, they are only orthogonal or 45 degree diagonal. Some designs only use orthogonal edges for simplicity. The polygon is usually represented in the database by a sequence of points defining polygon edges. Opaqueness of a region is represented by the orientation of surrounding edges. It is assumed that an opaque region always lies on the right hand side of edges. A polygon may include holes, as shown in Figure 1. By the nature of the LSI mask, an opaque region may equivalently be represented by the combination of more than one polygons

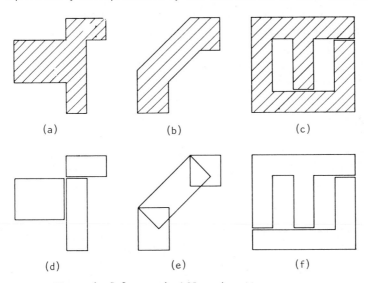

(a)   (b)   (c)

(d)   (e)   (f)

Figure 1 Polygons in LSI mask pattern

which are touching or overlapping each other. Hence, some layout data format may place a limit on the maximum vertex number of a polygon, or may permit only rectangles. Figure 1 also shows examples of such representations.
In actual LSI data, there are around ten layers and the total number of polygons for an LSI is about ten times larger than the number of circuit elements. The number of edges are again about ten times larger than the number of polygons. For a current state-of-the-art VLSI, which has 10^4 to 10^6 elements, the number of edges could fall into the 10^6 to 10^8 ranges. Computer time and memory needed for verification would obviously be a problem.

3. Geometrical Design Rule Check

Integrated circuits are designed as compact as possible within technology limitations, such as minimum line width and spacing, minimum alignment tolerances etc., and regarding device reliability considerations. Such limitations are described in form of a set of geometrical design rules. The geometrical design rule check

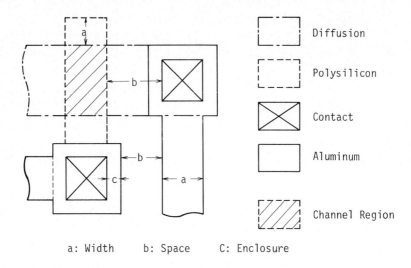

a: Width   b: Space   C: Enclosure

Figure 2   Geometrical design rule (Si-gate MOS technology)

is to verify LSI mask pattern data in terms of this rule[2-10].   Figure 2
illustrates some examples of these rules.

Different process technologies have different design rules.   However, design rule
checking procedure can be described by a series of basic operations.   Typical
basic operations are Boolian operations, checking operations, topological operations, and geometrical operations, as shown in Figure 3.   Some rules are described
in terms of element type, such as "contact holes must be separated from the gate
region on MOS FET by a certain distance".   Such a rule can be successfully processed by combining the above mentioned basic operations.   The MOS FET gate region,
in Si-gate MOS technology, is obtained by AND operation of polysilicon and
diffusion layers.

General purpose design rule check programs (DRC) have been widely used in
industry.   Common problems in using DRC are the computing time involved and
pathology[8].   Although various algorithms with nearly linear time complexity have
been proposed and applied, ever-increasing circuit complexity demands continuous
efforts to improve performance.   The pathology often comes from the fact that the
check is done on elemental figures instead of composite figures.   It occurs as an
unchecked error or a false error.   A false error means an error that is reported
as an error but is actually legal, as shown in Figure 4.   Most false errors,
including those in Figure 4, are solved by carefully combining basic operations.
However, in many cases, there is a resulting penalty of increasing computing time.

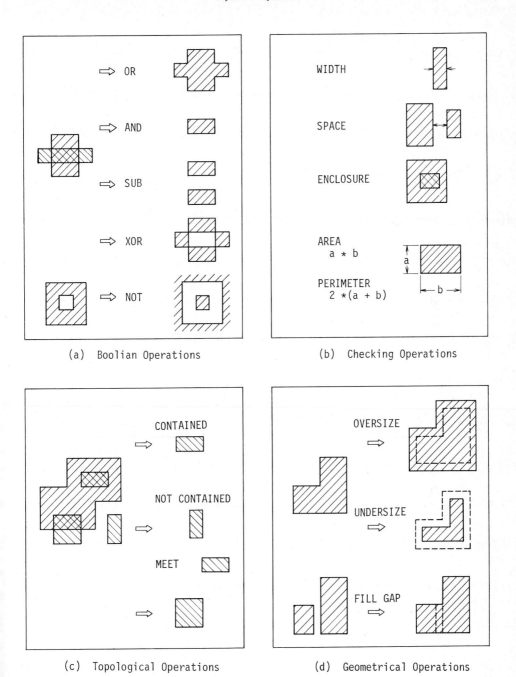

(a) Boolian Operations

(b) Checking Operations

(c) Topological Operations

(d) Geometrical Operations

Figure 3 Basic pattern operations

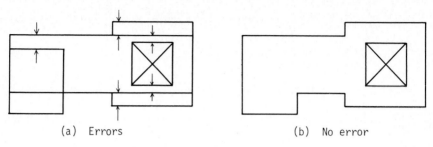

(a) Errors   (b) No error

Figure 4 False errors

4. Basic Algorithms for Pattern Operations

Most basic pattern operations include comparisons of all two figures or two edges, as a most time consuming process. A naive comparison algorithm with $O(n^2)$ time complexity is useless for today's LSI mask patterns, including 10^6 to 10^8 edges, as mentioned before. Thus, many fast algorithms have been proposed[13,19,24]. Those are divided into three groups; (1) Shape based algorithms, (2) Edge based algorithms, and (3) Bit-map algorithms.

The following discussion takes the OR operation of two mask layers as an example. All figures on the layers are polygons consisting of straight edges. The total number of polygons included in those layers is denoted by N and the total number of edges by n. To simplify the complexity consideration, two assumptions will be made which are reasonable for actual LSI masks. First, it is assumed that the polygon density or edge density is almost constant over the square chip area. This leads to the assumption that the average number of polygons or edges inter- secting one vertical (or horizontal) line across the chip area is $O(N^{0.5})$ or $O(n^{0.5})$[13]. In addition, it is assumed that the average number of unique x (or y) coordinates is $O(n^q)$. Considering the LSI design target to pack patterns as densely as design rule permits and its regular and grided layout structure, q is assumed to be much less than 1.0 and to approach 0.5[13,24].

The following discussion assumes a general purpose computer with a relatively limited main memory size and practically unlimited peripheral storage, for which fast access is allowed only sequentially. Also, it is assumed that an external sort routine with $O(n \log n)$ time complexity is available.

4.1 Shape based algorithms

A shape based algorithm usually has two steps; comparison of all figures by their enclosing rectangles, such as shown in Figure 5, and detailed comparison, or pattern operation, between interacting figures[3,6]. The second step is omitted

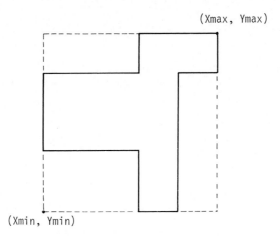

(Xmax, Ymax)

(Xmin, Ymin)

Figure 5   Enclosing rectangle

if two enclosing rectangles have no interaction. It is expected that the number of detailed comparisons is reduced to O(N), since the number of surrounding figures for each figure can be assumed to be constant. However, the number of total comparisons of rectangles is O(N^2) and thus determines total expected time complexity. The target for the following algorithms is to reduce the time complexity for this process.

Detailed comparisons between two figures are performed, for example, as follows. First, all intersection points of edges are calculated and each intersecting edge is split at the intersecting points, thus resulting in a planer graph. Then, by tracing the graph edges, appropriate closed loops of edges are extracted, such that these loops correctly define the intended regions, as shown in Figure 6, for example. If the number of edges per polygon, m, becomes large, the time complexity for the above process is O(m^2). This limits the application of this method.

To avoid the above problem, polygons may be divided into some suitable smaller figures, prior to the operation. Also, a more sophisticated algorithm, similar to those in 4.2 or 4.3, can be applied to two figure comparisons.

4.1.1   Partitioning

Partitioning a problem into smaller problems is a common and effective technique to solve very large problems with super-linear time and space complexity. If a chip is divided into K parts, the expected computing time is reduced to O($\frac{N^2}{K}$), thus resulting in a significant time reduction.

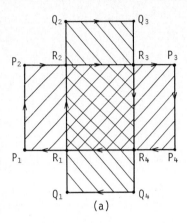 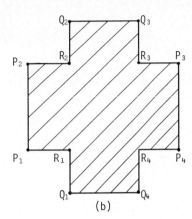

(a)   (b)

Figure 6   OR operation of two polygons

Time complexity seems to be reduced to O(N) by increasing K in proportion to N.
However, this is not true in a practical sense. First, the partitioning process
itself has O(N log K), or O(N log N), complexity. In addition, the border figures
need special treatment with non-negligible computing time. If the average number
of crossing figures to each of K-1 border lines is M, additional computation for
these figures is O (M·N). This can be super-linear complexity, depending on M.

4.1.2 Work-list method

To avoid troublesome border treatment, a simpler algorithm, the work-list method

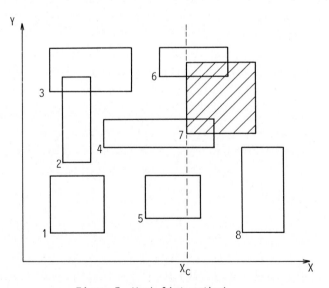

Figure 7   Work list method

was proposed, using a scan line that is swept across the chip[1,3,6,16]. As shown
in Figure 7, only figures whose enclosing rectangles cross the scan line at a
current position Xc are maintained in the work-list as candidates for comparison.

An actual algorithm is described as follows.
1. Sort all figures by their Xmin and Ymin, x and y coordinates of lower left
   corner of the enclosing rectangle, and store them into an external file.
2. Until the file is exhausted, continue the following steps.
3. Extract one figure from the file and let Xc be Xmin of the figure.
4. Delete all work-list entities whose Xmax, x coordinate of its righter corner,
   is smaller than Xc.
5. Compare the remaining work-list entities with the new figure by their enclosing
   rectangles. If they have interaction, carry out a detailed comparison and
   required Boolian operation.
6. Add the new figure to the work list.

As the average number of work-list entities is $O(N^{0.5})$ by assumption, expected
time complexity and space complexity are $O(N^{1.5})$ and $O(N^{0.5})$, respectively. So,
if the computation time for a detailed comparison is constant per figure, overall
time complexity is also $O(N^{1.5})$.

Improvement in computing time can be expected through using a simple primitive
figure partitioned from polygons instead of using original polygons and enclosing
rectangles. A trapezoid, as shown in Figure 8, is an example.[16] While the
average number of work-list entities is almost the same, the detailed comparison
between interacting figures is significantly simplified. Theoretical complexity
in order notation is unchanged. All touching figures must be merged to compose
a polygon at the final stage. This process is also carried out by a similar
algorithm.

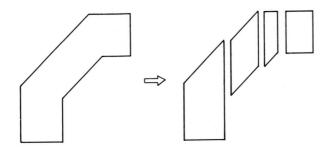

Figure 8 Polygon Partition by trapezoids

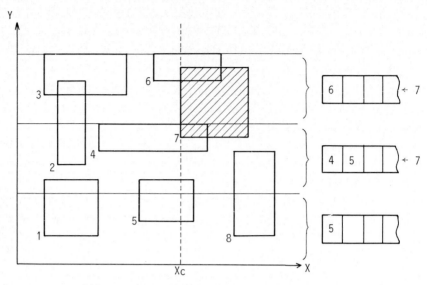

Figure 9 Work list bins

By modifying the data structure for the work-list, time complexity can be improved.
One approach is to partition the work-list into "bins", as shown in Figure 9,
corresponding to partitioned scan lines with equal length of y direction[17].
The bins which the new figure is to be entered in are determined by the minimum
and maximum y coordinates, Ymin and Ymax, for the figure. If the number of bins
is set to be proportional to $N^{0.5}$ and if the average number of entities for each
bin is assumed to be nearly constant, the expected time complexity approaches $O(N)$.
Hence, overall $O(N \log N)$ time complexity can be expected.

4.2 Edge based algorithms

A group of algorithms that handles all the edges of pertinent layers simultaneously
will be discussed. General procedure consists of;

(1) Calculate all intersections between all edges and split them at the intersection points.
(2) Decide the subset of edges that consist of the boundary of opaque and transparent regions (true edges), depending on the operation function.
(3) Reconstruct polygons from the list of true edges.

This approach has a general advantage over shape based algorithms, coming from
data localization.

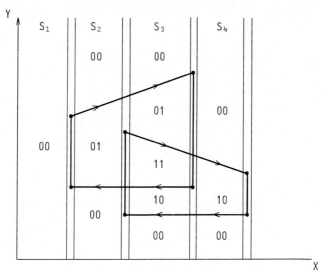

Figure 10   Slit method

4.2.1 Slit method

In this algorithm, the slit method, the entire area is divided into slits by
vertical trasit lines, which include vertical edges, as shown in Figure 10[2,4].
Typical algorithms are as follow.

1. All unique x coordinates for the design are listed and sorted. Vertical transit
   lines are created at a small increment to each side of every one of these x
    occurrences.
2. For every transit line, do steps 3 and 4.
3. Find all intersections in the transit line with non-vertical edges and sort
   their y coordinates.
4. Select true edges for pertinent Boolian operation by scanning edges from the
   bottom to the top along the transit line.
5. Reconstruct polygons from the set of true edges.

True edge selection is performed using two counters corresponding to the two
layers. The non zero value for the counter means opaqueness. During scanning,
if a left (right) bound edge is crossed, one of the counters is increased
(decreased) by one. The opaqueness of the resulted layer at any time is determined
by the Boolian operation of opaqueness for both layers. If the resulting
opaqueness changes when an edge is crossed, the edge is a true edge.

Since the number of transit lines is $O(n^q)$, overall time complexity is expected
to be $O(n^{0.5+q})$, or $O(n^{1.0})$ if q = 0.5, and space complexity is $O(n^{0.5})$.

Reconstructing polygons from the list of true edges involves no particular complexity problem.

In this algorithm, it is assumed that no two non-vertical edges intersect each other. If such intersections are to be checked, a preprocessing may be needed.

4.2.2 Work-list method

The work-list method is also applicable to finding intersections between a set of edges[13]. Similar to the shape based work-list method, all edges are sorted by x and y coordinates of their left-most points. Then, they are sequentially read in for intersection calculation with previous edges maintained in the work-list. Determining true edges for the resulting mask layer is complicated. Baird[13] proposed a vertex scan algorithm, which is based on the consideration of winding order of incident edges to a vertex.

The expected time complexity and space complexity are $O(n^{1.5})$ and $O(n^{0.5})$, respectively.

4.2.3 O(n log n) algorithm

Bentley and Ottmann[18] proposed an algorithm for reporting all intersections between straight lines in the plane, which has O(n log n) time complexity. The algorithm uses a scan line similar to previous algorithms, but employs the fact that the only segments that are adjacent on a scan line can intersect each other and have to be checked for intersection.

Based on this algorithm, Lauther[19] devised a fast algorithm for Boolian operations of mask patterns. In this algorithm, sorted external files and three sorted lists in the main memory are used. The E-file contains all non-vertical input edges of a layer, sorted by Xmin and Ymin, x and y coordinates of their left-most end point, and slope. The Q-list is a buffer for input edges from E-files and new edges split by an intersection, which always maintains the order similar to the E-file. O-list and N-list are linear lists containing all edges crossing the old and new scan lines (Figure 11), respectively. The algorithm is as follows.

1. Sort and store all eges of two layers into E-files.
2. Carry out steps 3 to 6 for subsequent scan line location Xc.
3. Read new edges with Xmin = Xc from the Q-list and insert into the N-list.
4. Calculate y value at x = Xc for all edges in O-list, then merge O-list and N-list into a common O-list preserving vertical order. During the merge, check every new edge for intersection with adjacent edges. If intersection occurs, split them and insert the right parts into the Q-list.
5. Decide true edges using the O-list in a way similar to that described in 4.2.1.

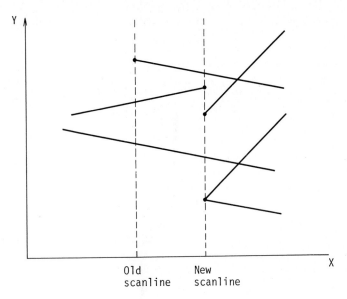

Figure 11   Two scanlines and set of edges

6. Delete edges in the O-list which end at Xc. Whenever, during deletion, two
    edges become adjacent, check edges again for intersection and split them,
    if necessary.
7. Reconstruct polygons from the set of true edges.

In the above algorithm, expected time complexity for intersection checking is
$O(n)$, maintaining O-list and N-list is $O(n^{0.5+q})$ or $O(n)$, maintaining Q-list and
preprocessing for E-file is $O(n \log n)$. Hence, overall expected time complexity
is $O(n \log n)$. Space complexity is $O(n^{0.5})$. Other fast algorithms are also
proposed[9,20].

4.3 Bit-map algorithm

Many LSI patterns include only orthogonal edges. Bit-map representations are
particularly suitable to such patterns.

4.3.1 Simple and essential bit-map representations

In a simple bit-map representation[22], the entire area is divided into grid
squares small enough to represent all figures. Associated with every grid square
is a single bit of computer memory, representing opaqueness of the squares, as
shown in Figure 12 (b). Any Boolian operations of two layers are simply performed
by the Boolian operation of corresponding bits for each layer. The required
computer time and space are proportional to the number of grid squares.

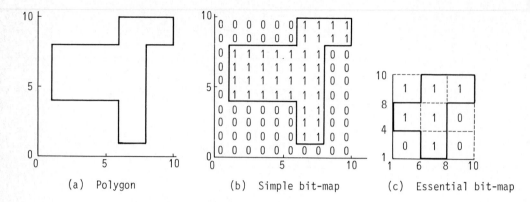

(a) Polygon   (b) Simple bit-map   (c) Essential bit-map

Figure 12 Bit-map representations

Unfortunately, this number is usually prohibitively large. For example, a 1 cm^2 chip, with 1 μ resolution, will need 10^8 bits per layer.

In order to reduce this complexity, the following "essential bit-map" representation is used[11,14]. A list of all unique x and y coordinates for the design is first generated and the entire area is divided into grid rectangles by vertical and horizontal lines having those unique x or y values, as shown in Figure 12 (c). The number of bits is significantly reduced, compared with a simple bit-map. By the previous assumption, expected space and time complexity will be $O(n^{2q})$ or $O(n)$.

4.3.2 Hierarchical bit-map representation

To further reduce the size of bit-map data, a hierarchical bit-map scheme, HIBAWL (Hirarchical, Integrated Circuit, Bit-map Artwork Language) has been proposed by Wilmore[23,24].

For each unique y value, a horizontal scan line (HSL) is defined. To compact the bit-map expression of opaqueness along this HSL, a hierarchy of "region bit-maps" (RBP) is defined. Each RBP is divided into K sectors of equal size with each sector represented by a bit pair, as shown in Figure 13. Any sector which is entirely transparent or entirely opaque has "00" or "11", respectively. A partially opaque sector has "10". Further details of the sector are described by K sectors at the next level in hierarchy. "00" and "11" sectors need no descendant sectors, thus reducing the data size. The above expression is called NOW HSL.

To eliminate repetitive data for vertically unchanged partially opaque regions, only the change information at each HSL is recorded. This is called CHG HSL.

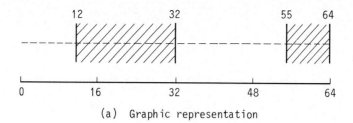

(a)　Graphic representation

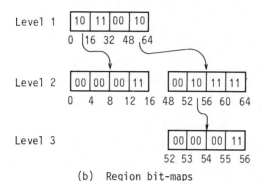

(b)　Region bit-maps

Figure 13　Hierarchical bit-map representation of
scan line (NOW HSL)

In the CHG HSL, "00" represents no change, "10" represents a change from transparent to opaque, while "01" represents a change in the opposite direction, and
"11" represents a partial change which refers to the next hierarchical level.
Figure 14 shows an example.　If CHG HSL at any y position is operationally
combined to NOW HSL just below the y, a new NOW HSL just above the y is obtained,
as shown in Figure 15.　Boolian operation of two mask layers is performed by
sweeping through the two sets of CHG HSL data from bottom to top, calculating
NOW HSLs and performing Boolian operation between them[24].

By the above representation, data size is proportional to the number of horizontal
edges, thus space complexity is O(n).　Computer time required for Boolian operation is also O(n).　Generating HIBAWL data from an usual polygon description
requires O(n log n) time, due to sorting, although this conversion is needed only
once for each LSI mask.

5.　Circuit extraction

General LSI layout data include purely geometric information, but no explicit
circuit information, as assumed in Section 2.　Hence, the equivalent electric
circuit has to be extracted from the set of pattern layers in order to verify the
layout in terms of circuit connectivity or performance[2,16].

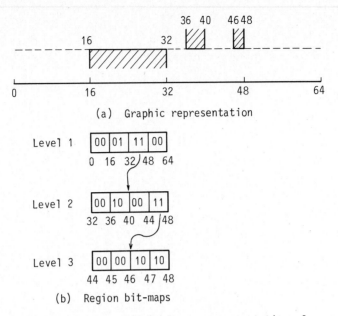

(a) Graphic representation

(b) Region bit-maps

Figure 14 Hierarchical bit-map representation of change information (CHG HSL)

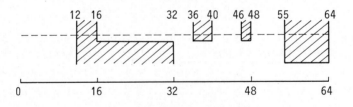

(a) Graphic representation

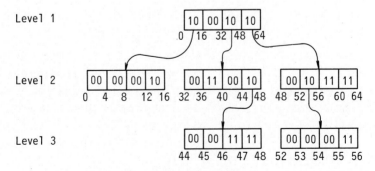

(b) Region bit-maps

Figure 15 Pattern resulting from applying the CHG HSL (Figure 14) to the NOW HSL (Figure 13)

The <u>circuit extraction</u> process usually consists of;
    (1) <u>Circuit element recognition</u>
    (2) <u>Connectivity analysis</u>
Most of these steps can be performed by a combination of basic pattern operations, as shown in Figure 3, and using algorithms similar to those described in Section 4.

5.1 Circuit element recognition

The structure of, and thus the recognition process of, circuit elements differs with element type and with technology used. However, a technology independent element recognition is possible by absorbing the technology dependency with preprocessing.

The following are typical recognition processes for MOS and bipolar transistors.

(1) MOS FET

    A Silicon-gate MOS FET can be illustrated as in Figure 16, as an example. A typical procedure to extract this from a set of mask pattern layers is as follows[2,11,16].

Step 1 (Preprocessing)
    (i)   <Channel region> = (<Diffusion> AND <Polysilicon>) SUB <Buried contact>
    (ii) <Actual diffusion> = <Diffusion> SUB <Channel region>

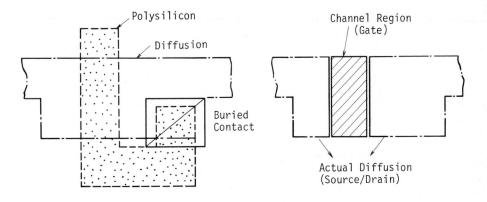

Figure 17 MOS FET extraction

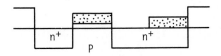

Figure 16 Si-gate MOS FET

Step 2 Find all sets of touching three polygons, as in Figure 17, one on
        <Channel region> and two on <Actual Diffusion>. Store the data into the
        database as the Gate and Source/Drains ports of MOS FETs.

Here, <name> notation means all polygons in the layer designated by the name.
AND, SUB etc. mean the pattern operations with its operand layers on the left and
right sides. Step 1 is technology dependent preprocessing and Step 2 is
technology independent element recognition process. To find touching polygons in
Step 2, scan line algorithms can be applied.

(2) Bipolar transistor

     An example of a bipolar transistor is shown in Figure 18. It can be recognized
     by following procedure, as an example[7,27,28,31].

Step 1 (Preprocessing)
     (i)   <Emitter contact> = <Contact> CONTAINED (<N+ Diff.> AND <Base Diff.>
                               AND <Isolation>)
     (ii) <Base contact> = <Contact> CONTAINED (<Base Diff.> AND <Isolation>)
     (iii) <Collector contact> = <Contact> CONTAINED (<N+ Diff.> AND <Isolation>
                               AND (NOT <Base Diff.>))

Step 2 Find all sets of three polygons, each on <Emitter contact>, <Base contact>

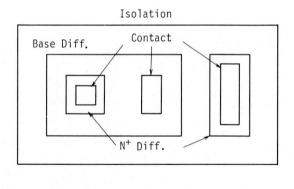

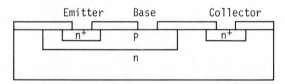

Figure 18 Bipolar transistor structure

and <Collector contact>, and store the data into the database as Emitter, Base and Collector ports of transistors.

Other circuit elements, such as lateral transistors, diodes, capacitors and resistors, are recognized in a similar way[26,27]. The recognition process for all elements in the design has the same time complexity as the basic pattern operation described in Section 4, namely $O(n^{1.5})$ or $O(n \log n)$.

5.2 Connectivity analysis

The purpose of this process is to determine the connectivity between nets, or electrically equivalent regions, and circuit element ports[12,15,16,31]. Touching and intersecting polygons in the same conducting layer are electrically equivalent. Electrical equivalence between polygons on different layers depends on the technology. Such relation between layers can be described by a graph, as in Figure 19[16] for the case of Si-gate MOS technology, as an example. An edge between layers in this graph means that polygons on the two layers are electrically equivalent, if those polygons overlap each other. For example, in Figure 16, a polygon on the polysilicon layer and a polygon on the Diffusion layer are electrically connected via a through hole in the Buried contact layer.

The connectivity analysis is performed by the following steps.

Step 1   For all edges in Figure 19, search for intersecting pairs of polygons, each in the two layers. These pair relations are stored in a connection graph.

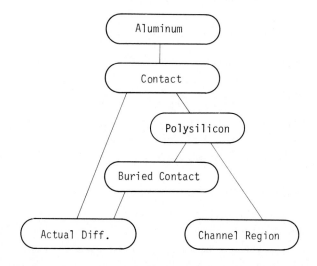

Figure 19   Layer relation for Si-gate MOS technology

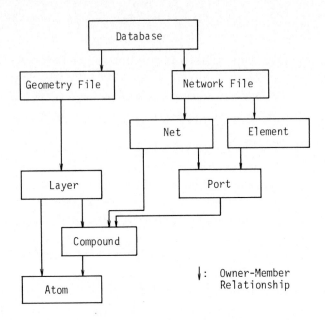

Figure 20   Database example for extracted circuit data

Step 2   Find all sets of connected components of this connection graph.   Store
         each set as a net.
Step 3   Find polygons which connect to element ports.   By retrieving the net and
         the element which those polygons belong to, interconnections between the
         net and element are obtained.

The overall time complexity for the above procedure is determined by Step 1, that
is similar tc the basic pattern operations.   The space complexity for the above
process is O(n), although improvement can be expected[21].

Extracted circuit data is stored in the database for use by various verification
tools, as described later.   An example of the data structure is as shown in
Figure 20[16], where the geometrical data and network data are connected by
pointers for ease of achieving various access ways.

5.3 Drawing circuit diagram

The extracted circuit diagram is plotted out for manual inspection or for
reference in analyzing the result of other verification tools.   There are two
types of drawings.   The first one is a layout oriented diagram, in which symbols
of circuit elements and wiring segments are placed around the same location as in

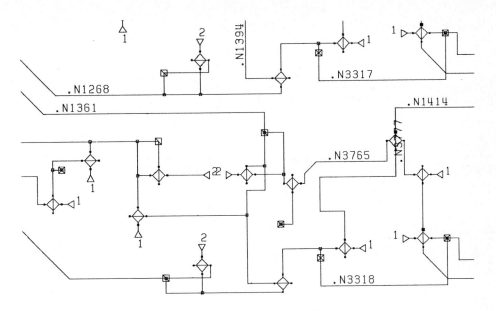

Figure 21   Layout oriented diagram

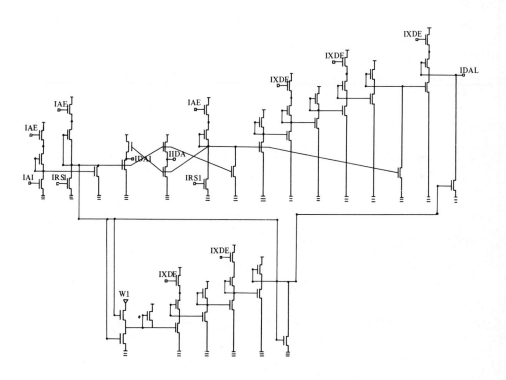

Figure 22   Function oriented diagram

the original layout pattern[16]. An example is shown in Figure 21. Another type
of drawing is a function oriented diagram, which is laid out considering signal
flow, circuit symmetry and so on[15]. An example is shown in Figure 22.

6. Topological or logical verification

There are three approaches proposed to verify extracted circuit connectivity or
logical function[36];

    (1) Electrical rule check (ERC)
    (2) Logic simulation
    (3) Circuit comparison

Approach (1) is most economical and easy to use, although it is effective only for
checking limited types of errors. On the other hand, more thorough verification
can be expected with the other two approaches, at expensive computation cost.

6.1 Electrical rule checking (ERC)

Most connectivity errors found in manually designed patterns are trival errors,
such as unreasonably connected or isolated nets. The purpose of the ERC is to
detect such errors[14,30]. Examples of such errors are;

    - Circuit elements with no current path to power supply and ground,
    - Illegally configured subcircuits, such as with no driver or load parts, with
      illegal combination of transistor type (P- or N-channel) and so on,
    - Short or open circuits between subcircuits.

In case of MOS circuits, the entire circuit can be partitioned easily into basic
subcircuits by separating at the nets which connect to MOS FET gates. Then, the
above mentioned checks can be performed for each subcircuit independently. Hence
the expected time complexity for this checking is linear.

6.2 Logic simulation

Simulating the extracted circuit is a practical approach to verify the resulting
function of mask patterns. The advantage of this approach over circuit comparison
is the flexibility allowed to the circuit and layout designers. There are
logically equivalent but topologically different circuit realizations for a given
logic function. One of them may be selected or changed during layout design in
order to optimize the design. Circuit comparison approach could make pathologies
in such cases. However, the logic simulation approach is free from such problems.

6.2.1 Logic gate extraction

Gate level simulation is advantageous in computing time over transistor level

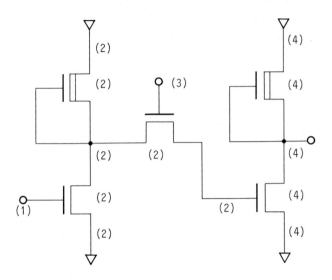

( ): Subcircuit Number

Figure 23 Circuit partitioning

simulation. Thus, <u>logic gate extraction</u> from transistor level circuit prior to the simulation is desirable[15,30].

The typical process for logic gate extraction for MOS circuits is as follows.

Step 1 Circuit partitioning
> The entire circuit is partitioned into subcircuits. The boundary nets are V_{DD}, Gnd, the nets connecting to FET gates, and the output candidate nets. The output candidate net is a net connecting to the source of a load transistor in the nMOS circuit, or a net connecting j both n channel and p channel transistors in CMOS circuit. Figure 23 is an example of circuit partitioning.

Step 2 Series-parallel reduction
> Each subcircuit is divided into three parts; Driver part, load part and transmission gates. Then, the series-parallel reduction is performed for the load and driver parts, respectively. Namely, transistors connected in series or in parallel are reduced to one transistor. This process is repeated and the relations between these transistors are recorded by a tree, such as shown in Figure 24 (b).

Step 3 Logic gate recognition
> By analyzing tree structure for the load and driver parts, the gate function can be recognized. For example, in nMOS circuit, a series

(parallel) vertex means the AND (OR) function and the load part inverts
the final stage. In normal CMOS gates, the load and driver parts have an
isomorphic tree structure with opposite vertex types and the corresponding
leaf vertices has the same nets. Figure 24 (c) shows the logic gates
recognized from Figure 24 (a). Remaining transistors, which are not
included in the recognized logic gates, are treated as transmission gates,
or switches.

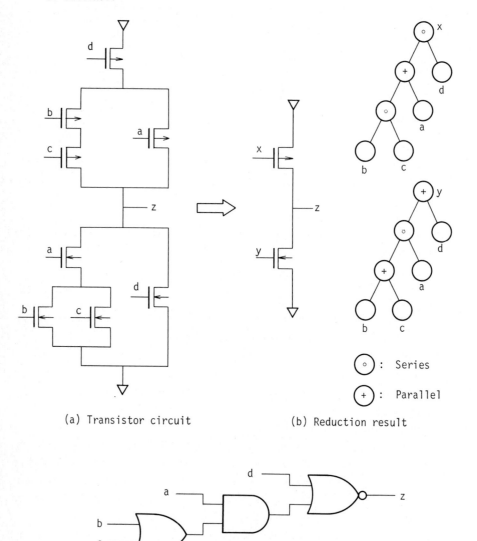

(a) Transistor circuit   (b) Reduction result

(c) Extracted Gates

Figure 24 Series-parallel reduction

Time complexity for MOS logic gate extraction is O(N), since the maximum sub-circuit size is usually limited.

6.2.2 Logic simulation

In a standardized design, a complete logic gate extraction is possible, but actual hand craft design often results in unrecognized subcircuits or remaining transistors. Such mixed level circuits are difficult to simulate for conventional logic simulators. Switch level or mixed level simulators are successfully applied to such circuits[35,36].

A problem in practical application is a long simulation time needed to assure circuit correctness. Another problem is localizing an error, which usually needs careful analysis of massive simulation results.

6.3 Circuit comparison

If an extracted circuit is completely isomorphic with the correct circuit from which the layout design originated, the connectivity of the layout is 100% verified. However, this approach also has problems. One problem is to extract logic gates to reduce the effective circuit size. As described before, practical circuits often result in unrecognized transistors. Another problem is caused from logically or electrically equivalent but topologically different circuits, which will cause false errors. Thus, the standardized design method is essential for circuit comparisons to be sufficiently practical.

There are two approaches for circuit comparison. One approach is to rely on the labels attached everywhere on the pattern for identifying the net or component names in the circuits[1,15]. Comparison process and error location are simple and easy in this approach. However, such labeling work is not only tedious but also itself error-prone.

Another approach is to apply graph isomorphism algorithms to compare the circuits globally[25,30,31]. General graph isomorphism is an intractable problem, whose complexity is not known. However, by utilizing various attribute information, which vertices carry in the actual circuit, algorithms with practically linear expected time complexity have been proposed.

An example of graph representation of a circuit is as shown in Figure 25. In this directed graph, edge direction is assigned depending on the connecting port types (gate, source or drain). A typical comparison algorithm consists of subsequent vertex partitioning of both graphs. Vertex grouping for the partition is determined by the number of in- or out- degree and by vertex attributes, such as element type (node or transistor type), label and so on. At each partition stage,

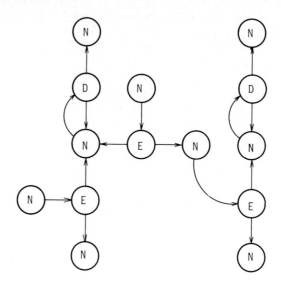

N: Net
D: D Type FET
E: E Type FET

Figure 25   Directed graph for comparison

if it is found that the number of vertices for the corresponding vertex groups are
not equal, or that corresponding vertices for two graphs are not consistent, the
two graphs are proved to be non-isomorphic.

7.   Electrical performance verification

Electrical characteristics, which will result from a set of mask patterns, can be
verified accurately by the circuit simulation based on circuit parameters
calculated from the pattern geometries[12].

For accurate simulation of relatively small circuits, circuit analysis programs,
such as SPICE-2[34], are used. Simulation model parameters for active devices as
well as passive device parameters are calculated. In case of MOS circuits, as an
example, L and W of MOS FET, and a set of parasitic capacitances formed by gate
oxide and p-n junctions, as shown in Figure 26, as well as wiring capacitances
caused by metal or polysilicon layers on the field oxide, are calculated.
Considering the boundary effects, the capacitance value is approximated by
($\alpha s + \beta l$), where s and l are area and perimeter length for a pertinent region,
respectively. More accurate treatment is also proposed[33]. Furthermore, the

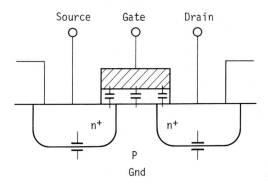

Figure 26 Parasitic capacitances for MOS FET

geometry deviation caused by a fabrication process can be compensated for by applying UNDER- or OVERSIZE operation as preprocessing[14].

In bipolar circuits, resistance calculation is important as well as active device parameters. In addition to simple $\rho\frac{L}{W}$ approximation for long straight resistors, more accurate calculation approaches for arbitrarily shaped resistors have been proposed and applied. Parasitic transistors, as well as parasitic resistances and capacitances, should sometimes be taken into account.

For larger logic circuits, macromodel timing simulator[37] and logic simulator with delay may be used after the logic gate extraction described previously. In case of the logic simulation of MOS circuits, the signal propagation delay is calculated from gate drivability and output capacitance, including wiring capacitance and input capacitance for the next stages.

8. Conclusion

Layout verification programs have become not only useful but also essential tools for VLSI development, especially for mass-production devices. Among them, design rule checking (DRC) and electrical rule checking programs are heavily used in industry today. However, increasing VLSI complexity requires further improvement in layout verification technology. For example, the followings are areas which will require more research or development.
(1) Incremental verification with quick response for use in an interactive design environment[38].
(2) Hierarchical verification method without pathology, realizing effectively sub-linear time complexity[39,40].
(3) Special hardware dedicated to layout verification, providing by far faster and more economical verification tools[41,42].

REFERENCES:

[1] Baird, H.D. and Cho, Y.E., "An Artwork Design Verification System", Proc.
     12th DA Conf., pp. 414-420 (June 1975).

[2] Lindsay, B.W. and Preas, B.T., "Design Rule Checking and Analysis of IC Mask
     Designs", Proc. 13th DA Conf., pp. 301-308 (June 1976).

[3] Yoshida, K., Mitsuhashi, T., Nakada, Y. et al, "A Layout Checking System for
     Large Scale Integrated Circuits", Proc. 14th DA Conf., pp. 322-330 (June
     1977).

[4] Wilcox, P., Rombeck, H. and Caughey, D.M., "Design Rule Verification Based on
     One Dimensional Scans", Proc. 15th DA Conf., pp. 285-289 (June 1978).

[5] Alexander, D., "A Technology Independent Design Rule Checker", Proc. 3rd
     USA-Japan Computer Conf., pp. 412(1978).

[6] McCaw, C.R., "Unified Shapes Checker - A Checking Tool for LSI", Proc. 16th
     DA Conf., pp. 81-87 (June 1979).

[7] Chang, C.S., "LSI Layout Checking Using Bipolar Device Recognition Technique",
     Proc. 16th DA Conf., pp. 95-101 (June 1979).

[8] McGrath, E.J. and Whitney, T., "Design Integrity and Immunity Checking",
     Proc. 17th DA Conf., pp. 263-268 (June 1980).

[9] Kozawa, T., Tsukizoe, A., Sakemi, J. et al, "A Concurrent Pattern Operation
     Algorithm for VLSI Mask Data", Proc. 18th DA Conf., pp. 563-570 (June 1981).

[10] Tsukizoe, A., Sakemi, J., Kozawa, T. et al, "MACH: A High Hitting Pattern
     Checker for VLSI Mask Data", Proc. 20th DA Conf., pp. 726-731 (June 1983).

[11] Dobes, I. and Byrd, R., "The Automatic Recognition of Silicon Gate Transistor
     Geometries: An LSI Design Aid Program", Proc. 13th DA Conf. pp. 327-335
     (June 1976).

[12] Preas, B.T., Lindsay, B.W. and Gwyn, C.W., "Automatic Circuit Analysis Based
     on Mask Information", Proc. 13th DA Conf., pp. 309-317 (June 1976).

[13] Baird, H.S., "Fast Algorithms for LSI Artwork Analysis", Proc. 14th DA Conf.,
     pp. 303-311 (June 1977).

[14] Losleben, P. and Thompson, K. "Topological Analysis for VLSI Circuits",
     Proc. 16th DA Conf., pp. 461-473 (June 1979).

[15] Nishiguchi, N., Kawanishi, H. et al, "PARADISE: A Circuit Diagram Generating
     System from LSI Mask Patterns", Proc. IEEE ICCC'82, pp. 312-315 (Sept. 1982).

[16] Mitsuhashi, T., Chiba T. and Takashima, M., "An Integrated Mask Artwork
     Analysis System", Proc. 17th DA Conf., pp. 277-284 (June 1980).

[17] Chapman, P.T. and Clerk, K. Jr., "The Scan Line Approach to Design Rule
     Checking: Computational Experiences", Proc. 21st DA Conf., pp. 235-241
     (June 1984).

[18] Bentley, J.L. and Ottman, T.A., "Algorithms for Reporting and Counting
     Geometric Intersections", IEEE Trans. Comp., Vol. 6-28, No. 9, pp. 643-647
     (Sept. 1979).

[19] Lauther, U., "An O(N log N) Algorithm for Boolian Mask Operations", Proc.
     18th DA Conf., pp. 555-560 (June 1981).

[20] Donath, W.E. and Wong, C.K., "An Efficient Algorithm for Boolian Mask
     Operations", Proc. IEEE ICCD'83, pp. 358-360 (Oct. 1983).

[21] Czymanski, T.G. and Van Wyk, C.J., "Space Efficient Algorithm for Boolian
     Mask Operations", Proc. 20th DA Conf., pp. 734-739 (June 1983).

[22] Baker, C.M. and Terman, C., "Tools for Verifying Integrated Circuit Designs", Lambda, Vol. 1, No. 3, pp. 22-30 (June 1980).

[23] Wilmore, J.A., "A Hierarchical Bit-Map Format for the Representation of IC Mask Data", Proc. 18th DA Conf., pp. 585-590 (June 1980).

[24] Wilmore, J.A., "Efficient Boolian Operations on IC Masks", Proc. 18th DA Conf., pp. 571-579 (June 1981).

[25] Ablasser, I. and Jäger, U., "Circuit Recognition and Verification Based on Layout Information", Proc. 18th DA Conf., pp. 684-689 (June 1981).

[26] Hofman, M. and Lauther, U., "HEX: An Instruction-Driven Approach to Feature Extraction", Proc. 20th DA Conf., pp. 334-336 (June 1983).

[27] Chiba, T., Takashima, M. and Mitsuhashi, T., "A Mask Artwork Analysis System for Bipolar Integrated Circuits", 1981 COMPCON FALL, pp. 175-185 (June 1981).

[28] Yoshida, J., Ozaki, T. and Goto, Y., "PANAMAP-B: A Mask Verification System for Bipolar IC", Proc. 18th DA Conf., pp. 690-695 (June 1981).

[29] Kubo, N., Shirakawa, I. and Ozaki, H., "A Fast Algorithm for Testing Graph Isomorphism", Proc. ISCAS, pp. 641-644 (1979).

[30] Takashima, M., Mitsuhashi, T., Chiba, T. et al, "Programs for Verifying Circuit Connectivity of MOS/LSI Mask Artwork", Proc. 19th DA Conf., pp. 544-550 (June 1982).

[31] Barke E., "A Layout Verification System for Analog Bipolar Integrated Circuits", Proc. 20th DA Conf., pp. 353-359 (June 1983).

[32] Mitsuhashi, T., Yamada, H. and Yoshida K., "An LSI Mask Artwork Verification System", Proc. IEEE ICCD'83, pp. 604-607 (Oct. 1983).

[33] McCormick, S.P. "EXCL: A Circuit Extractor for IC Designs", Proc. 21st DA Conf., pp. 616-623 (June 1984).

[34] Nagel, L.W., "SPICE 2: A Computer Program to Simulate Semiconductor Circuits", Univ. of Calif. Berkeley, ERL Memo, No. ERL-520 (May 1975).

[35] Byrant, R.E., "MOSSIM: A Switch-Level Simulator for MOS LSI", Proc. 18th DA Conf., pp. 786-790 (1981).

[36] Kawamura, M. and Hirabayashi, K., "Logical Verification of LSI Mask Artwork by Mixed Level Simulation", Proc. IEEE ISCAS, pp. 1021-1024 (May 1982).

[37] Kawamura, M., Takagi, H. and Hirabayashi, K., "Functional Verification of Memory Circuits from Mask Artwork Data", Proc. 21st DA Conf., pp. 228-234 (June 1984).

[38] Taylor, G.S. and Outerhout, J.K., "Magics' Incremental Design Rule Checker", Proc. 21st DA Conf., pp. 160-165 (June 1984).

[39] Whitney, T., "A Hierarchical Design Rule Checking Algorithm", Lamda, Vol. 2, No. 3, pp. 40-43 (June 1981).

[40] Tarolli, G.M. and Herman, W.J., "Hierarchical Circuit Extraction with Detailed Parasitic Capacitance", Proc. 20th DA Conf., pp. 337-345 (June 1983).

[41] Seiler, L., "A Hardware Assisted Design Rule Check Architecture", Proc. 19th Conf., pp. 232-238 (June 1982).

[42] Kane, R. and Sahni, S., "A Systoric Design Rule Checker", Proc. 21st DA Conf., pp. 243-250 (June 1984).

LAYOUT DESIGN AND VERIFICATION
T. Ohtsuki (Editor)
© Elsevier Science Publishers B.V. (North-Holland), 1986

Chapter 8

COMPUTATIONAL COMPLEXITY OF LAYOUT PROBLEMS

M.T. SHING*, T.C. HU**

*Department of Computer Science
University of California at Santa Barbara
Santa Barbara, CA 93106, U.S.A.*

**Department of Electrical Engineering and Computer Sciences*
*University of California at San Diego
La Jolla, CA 92093, U.S.A.*

1. Introduction

In this chapter, we shall describe several extremely simplified mathematical models of the layout problems. Then we will discuss the complexity of these problems and the algorithms to solve them. These mathematical models may be over-simplified. However, if we do not understand even these simplified models, we would never know whether we are doing our best or if we can do better in solving the real-world layout problems.

The first formal model for the VLSI layout was developed by Thompson [96,97]. The model is consistent with the VLSI design rules established by Mead and Conway [77] and is also similar to the widely used Manhattan wiring model. Intuitively, we can model a VLSI circuit by means of a graph. There is a natural one-to-one correspondence between the wires in the circuit and the edges in the graph. To establish the one-to-one correspondence between the devices in the circuit and the vertices in the graph, we can assume that the graphs are of bounded vertex-degrees and that vertices require only a constant area of silicon. We can also model a VLSI chip by means of a two-dimensional grid-graph. Assume that the chip has two layers for routing, one for the horizontal wires and one for the vertical wires. Since two parallel wires must be placed at a certain distance apart, and each wire has a minimum width, only limited number of horizontal and vertical wires are allowed in the chip. We call the spaces reserved for these wires as *horizontal* and *vertical tracks*. Hence, a chip can be thought of as a two-dimensional grid-graph where the nodes are the potential positions of the terminals and vias and the horizontal and vertical arcs are the places for the wires. A *layout* of a graph (or circuit) G is specified by an embedding which assigns vertices of G to nodes in the grid-graph and assigns edges of G to paths in the grid-graph. The paths of the layout are restricted to follow along the horizontal and vertical arcs in the grid-graph and are not allowed to overlap for any distance, except that a vertical path segment may cross a horizontal path segment. Using this model, we would like to solve problems like "*given a graph G, produce an area-efficient layout of G with minimax edge length*", or "*given a graph G, produce a layout with few wire crossings*" [1]. Unfortunately, many of these problems belong to the class of *NP*-

---

[1] See Appendix III for a list of the graph layout problems.

complete problems [2]. Since it is unlikely to find the optimum solutions of these problems efficiently, except for a few special cases, researchers have turned their attentions to the development of efficient heuristic algorithms to find near-optimum solutions for these problems.

In the simplest terms, the layout problem is to place many modules on a board in a non-overlapping manner, and then connect the terminals on various modules by mutually noninterfering wires according to a given wiring list [3]. Thus, we can divide the layout problem into two subproblems:

1. *Placement problem: How to place the modules on the board?*

2. *Routing problem: After the modules are placed, how to route the wires to connect all the nets on the wiring list?*

2. Placement Problems

The modules are usually of rectangular shapes and the terminals are on the perimeters of the modules. The board (a chip or a PC board) is a rectangle with Input-Output pads on its boundary. In order to emphasize the intrinsic mathematical problem, let us make some over-simplified assumptions:

Assumption 1. Every module can be placed anywhere on the chip. (In other words, we ignore the problems caused by long wires.)

Suppose we want to put a power line vertically across the middle of the chip and thus divide the chip into two regions of equal area. Then we have to solve a simple placement problem of deciding which modules to be put on the left region and which to be put on the right region. Obviously, the maximum number of modules that can be placed in a region is bounded above by the size of the region. However, since the modules and the regions may have different sizes and shapes, we may not be able to place all the modules in a region even if the total area occupied by the modules does not exceed the total area available in the region. (For example, we cannot put two 3×4 rectangles into a 5×5 square.) In fact, the problem of packing a set of rectangles into a larger rectangle optimally so that no rectangles overlap is called the *two-dimensional packing* problem and has been proved to be *NP-complete* [84]. Hence, we make another assumption.

Assumption 2. The shape of a module is not fixed, only the area of the module is known.

Assume that we have placed all the modules in the two regions. We want to connect all the nets. If a net has terminals in both regions, we say that the net is separated by the vertical line. For each net separated by the vertical line, we need at least one horizontal wire across the vertical line to connect the net. If the number of separated nets is

---
[2] See Appendix II for a brief description of the *NP-complete* problems.
[3] A *wiring list* consists of set of *nets*, where each *net* consists of a set of terminals from various modules to be connected together.

greater than the number of horizontal tracks available, we definitely cannot make the connections in the routing phase. Ideally, we want the number of separated nets to be as small as possible. Thus, we can model a VLSI circuit by a network as follows:

A network consists of set of vertices where each vertex represents a module in the circuit. Two vertices are connected by an arc if there is a net connecting terminals on the two corresponding modules. If there are k nets connecting the two modules, we associate a positive number k with the arc. We also associate a weight w_j with every vertex v_j to indicate the area of the corresponding module.

Problem 1:   *Partition the vertices of network into two parts of equal weights and minimize the total number of arcs connecting the two parts.*

2.1 Placement as a Set Partitioning Problem

For the moment, let us neglect the number of arcs connecting the two parts and concentrate on a simplified version of Problem 1.

Problem 1a: *Partition n positive numbers into two sets such that the difference between the two sums is minimized.*

Unfortunately, even this simplified problem, called the *partition* problem, is *NP-complete* [53]. In other words, for a network with n vertices, we may have to spend $O(2^n)$ time to consider all possible partitions if we insist on always getting the optimum solution. Since there exists no efficient algorithm to solve the partition problem, we should concentrate on developing efficient heuristic algorithms with good error-bounds. Let us consider the following heuristic algorithm.

Algorithm A:

Step 0.   Initialize the two sets, Set A and Set B, to empty.

Step 1.   Sort the n numbers from the largest to the smallest and put them in an ordered list L.

Step 2.   Remove the largest number from L and assign it into the set which currently has the smaller sum. Repeat the procedure until all n numbers are assigned.

Here we can immediately raise two questions:

(i)   *How efficient is the heuristic algorithm, how long does it take to find the partition?*

(ii)   *Does the algorithm always give the optimum solution? If not, how much is the error.*

To answer question (i), we can analyze the time complexity of the algorithm as follows [4]:

It takes $O(n \ log \ n)$ time to sort n number from the largest to the smallest in Step (1). Then we have to execute Step (2) n times. It only takes constant amount of time to remove the largest number from an ordered list and assign it to a set. Hence we need $O(n)$ time to execute Step (2). Thus, the Algorithm A has a worst-case running time of $O(n \ log \ n)$.

If we apply the heuristic procedure on the following six numbers 12,10,9,8,7,4, we would put 12,8,4 in one set and 10,9,7 in the other set. The difference between the two sums is 2. However, the optimum solution is to put 12,9,4 in one set and 10,8,7 in the other set, and the difference between the two sums will be 0. Since Algorithm A does not always finds the optimum solution, what is the maximum error produced by Algorithm A? To answer such question, we need a more rigorous definition for the error.

Let W_L = the total weight of the larger set obtained by the heuristic algorithm;

W_S = the total weight of the smaller set obtained by the heuristic algorithm;

R_A = the ratio $W_L \ / \ W_S$;

W_l = the total weight of the larger set in the optimum solution;

W_s = the total weight of the smaller set in the optimum solution;

R_o = the ratio $W_l \ / \ W_s$.

We can define the error-bound E either as (i) $(W_L - W_S) \ / \ (W_l - W_s)$ or as (ii) $R_A \ / \ R_o$. Measure (i) is perhaps the most direct measure of near-equality. However, it suffers from two potential disadvantages. First, it is not convenient in deriving meaningful worst-case bounds on the performance of approximation algorithms relative to the optimization algorithms. Second, performance according to this measure is not discounted by the "size" of the problem (i.e. the average subset sum), which is a property commonly desired in many applications. For these reasons, measure (ii) is used here and Algorithm A has a worst-case error bound of 7/5 [15]; and this bound is tight as seen from the example 3,3,2,2,2.

Recently a new heuristic algorithm is introduced by [54].

Algorithm KK:

Step 0.   Initialize the two sets, Set A and Set B, to empty.

Step 1.   Sort the n numbers from the largest to the smallest and put them in an ordered list L.

Step 2.   Replace the largest two numbers in L by their difference. Repeat this step until there are only two numbers left in the list.

Step 3.   Put one number into Set A and put the other into Set B. (The difference between the these two numbers is the difference of the final solution.)

---

[1] See Appendix I for a brief description of the time complexities and Big-O notations.

Step 4.   Replace those numbers in the sets which are not in the original list as follows:

Suppose x is in Set A and x is the difference of y and z (y ≥ z). Remove x from Set A, add y to Set A and add z to Set B.

Repeat this step until only numbers in the original list are present in both sets.

This algorithm also has a worst-case time complexity of O(n *log* n). If we apply this algorithm to the six numbers 12,10,9,8,7,4, it does give the optimum solution shown in Figure 1 while Algorithm A would give a difference of 2.

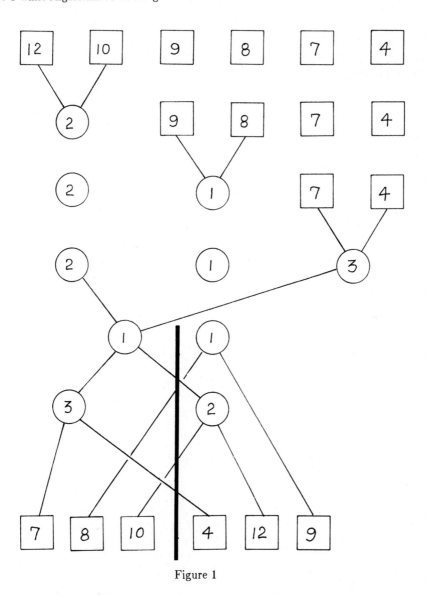

Figure 1

However, the KK algorithm still has a worst-case error bound of 7/5. (Take, for example, the numbers 3,3,2,2,2). Even though the KK algorithm performs no better than Algorithm A in the worst-case, we still prefer it over Algorithm A because it has been shown in [54] that the KK algorithm has a better error bound on the average.

2.2 Placement as a Min-Cut Problem

Now, let us assume that all the vertices have equal weights and consider another *extreme* version of Problem 1, namely, *find a partition which minimizes the total number of arcs connecting the two parts of the network.*

To describe the problem properly, let us go back to the network model. Here, the network consists of a set of vertices v_i and a set of arcs e_{ij} connecting v_i and v_j. Furthermore, there is a positive number c_{ij}, called the *arc capacity*, associated with every arc e_{ij}. (Here, c_{ij} corresponds to the number of nets connecting the two modules represented by v_i and v_j.) Suppose the vertices of the network is partitioned into a set X and its complement $\bar{X}$. The set of all the arcs connecting X and $\bar{X}$ is called a *cut* and is denoted by $(X \mid \bar{X})$. The sum of all arc capacities in the cut is called the *cut capacity* and is denoted by $C(X \mid \bar{X})$. For example, the network shown in Figure 2 has a cut $(\{v_1,v_2,v_6\} \mid \{v_3,v_4,v_5\})$ and $C(\{v_1,v_2,v_6\} \mid \{v_3,v_4,v_5\}) = 13$.

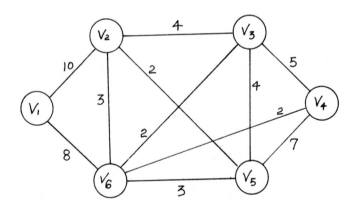

Figure 2

It so happens that the cut $(\{v_1,v_2,v_6\} \mid \{v_3,v_4,v_5\})$ has the minimum cut capacity among all cuts that disconnect the network into two parts. Such a cut is called a *minimum cut* and it corresponds to a partition which minimizes the number of separated nets.

Problem 1b: *Find the min cut of a network.*

If we want to disconnect a given pair of vertices by a cut of minimum capacity, we can use the standard maximum flow algorithms in network flow theory [25, 43]. Such algorithms have a worst-case running time of $O(n^3)$. If we want to disconnect the network in an arbitrary manner as long as the capacity of the cut is minimum, we can use the multi-terminal flow algorithm [33] which takes $O(n^1)$ time. For sparse networks and networks with special structures, special multi-terminal network flow algorithms which take $O(n)$ to $O(n^2)$ time are available. (See for example [45, 47].)

Note that the minimum cut separating a pair of vertices may not partition the network into two equal parts. In fact, it may divide the network so unevenly that one part consists of only 1 vertex and the other part consists of n-1 vertices. Hence, solving Problem 1b won't lead to optimum solutions for Problem 1; instead, it can be used as heuristics for the following problem.

Problem 1c: *Among those cuts which separates the network into two components of equal size (i.e. with equal number of vertices in each component), find the one with minimum cut capacity.*

Problem 1c is called the *graph bisection* problem (also known as the *min-cut* problem) and it is known to be *NP-complete* [29]. Since Problem 1c is a special case of Problem 1, it is not likely that Problem 1 can be solved optimally by efficient algorithms.

Because Problem 1 is essential to many other problems such as the design of Programmable Logic Arrays [44], a large number of heuristics have been developed for finding near-optimal partitions. (See for example [9, 10, 23, 56, 86, 89].)

Among all these heuristics, the most publicized algorithm is the one due to Keringhan and Lin [56]. They first partition the vertices into two sets of approximately equal weights and then exchange the vertices between the two sets to reduce the cut capacity. The algorithm is easy to implement but the exact error bound is not known.

2.3 Placement as an Assignment Problem

Although Problem 1 seems to be a basic problem which occurs in many other design automation problems, the network model does have the following drawbacks.

1.   Since each module has a definite shape, the weight w_i alone is not enough to describe the module.

2.   Not all modules are free to be placed anywhere in a region. Some modules have to be placed closed to each other and some modules have to be placed adjacent to certain I/O pads.

3.   Even if all n modules have equal weight (i.e. equal size and shape), there are usually more than n potential positions for the modules. These positions are called *slots*. The vertical power line may divide the slots evenly into two regions but this does not mean that we have to partition the modules evenly in the optimum placement.

4.   The criteria based on *min-cut* may not be the best criteria. An alternate criteria
      might be the *length* of all the wires needed to connect the modules. Since we do not
      know the actual wire-length at the time of placement, several distance functions are
      used to estimate the actual wire-length.

Let (x_i, y_i) and (x_j, y_j) be the coordinates of two pins to be connected together by a
wire and d_{ij} to be the estimated wire-length. We can define d_{ij} by one of the following distance functions.

$$\text{(i) } \sqrt{(x_i-x_j)^2 + (y_i-y_j)^2}, \text{ or (ii) } (x_i-x_j)^2 + (y_i-y_j)^2, \text{ or (iii) } |x_i-x_j| + |y_i-y_j|.$$

Distance function (i) gives the Euclidean distance but it is hard to compute. Furthermore
it does not reflect the real wire-length since the wires always run horizontally and vertically. Distance function (ii) is a nice convex function and is easy to compute, and distance function (iii) gives the Manhattan distance which does reflect the real wire-length.
Both distance functions (ii) and (iii) are often used in following placement problem.

Problem 2   *Place n modules on a two-dimensional plane such that the total wire-length,*
                indicated by the cost $\sum_{1\leq i\neq j\leq n} a_{ij}d_{ij}$, *is minimized.*

          (Here, a_{ij} is the number of wires needed to connect the two modules P_i and P_j.)

Problem 2 is called the *quadratic assignment* problem. It is again known to be *NP-
complete* [29]. Heuristic algorithms for the quadratic assignment problem can be found
in [31, 34].

3. The Routing Problem

After the modules are placed in chip, all terminals also have fixed positions. The
routing problem is to connect all the nets on a wiring list. In a gate-array chip, the size
of the chip is fixed and the main objective in circuit layout is the feasibility of routing. In
a standard-cell chip and a custom-designed chip, we can always increase the number of
tracks to ensure the feasibility of routing. Hence, the main objective is to to minimize the
area of the chip in addition to being able to connect all the nets. Here, we shall describe
a model for routing in the gate-array chips. Such model can be generalized easily to handle routing problems in the other chips.

The routing problem is to connect all nets in a grid-graph like the one shown in Figure 3. The nodes denote the potential positions of the terminals or vias and the dotted
lines are the horizontal or vertical tracks.

Two nets are to be connected in Figure 3. Net 1 has three terminals at positions
(1,4) (i.e., first row, fourth column), (2,1) and (4,2) and Net 2 has two terminals at positions (1,2) and (4,3). A possible way of connecting the two nets are shown as solid lines in
the figure. Note that the connections of the two nets cross at the position (2,2). In reality, the horizontal and vertical wires are at two different layers and hence there is no
crossing in the actual wiring. On the other hand, if we connect the two nets as shown in
Figure 4, we need a via for Net 1 at position (3,2) and a via for the Net 2, also at position

(3,2), and this is not allowed.

The appropriate graph model for routing in two-layer chips should really be a double-layer grid-graph as shown in Figure 5. However, since the double-layer representation is very tedious, we shall use the normal grid-graph shown in Figure 3 instead and keep in mind that the situation in Figure 4 should be avoided.

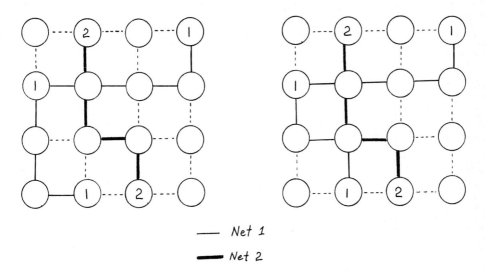

Figure 3   Figure 4

Since there are too many potential positions for terminals and vias, we generally partition the grid-graph into many sub-grid-graphs, called *global cells* (see [72, 80, 93, 98]) and the original wiring list is condensed into a list of connections between the global cells. Again, we can represent the global cells by nodes of a grid-graph G. Two nodes are connected by an arc in G if the corresponding global cells are adjacent. Associated with every arc is an non-negative number, called the arc capacity, which indicates the number of tracks available between the two global cells. For example, the sixteen-node grid-graph in Figure 3 can be condensed into a grid-graph of four global cells as shown in Figure 6.

In each global cell, the positions of the terminals within the cell are ignored and we are only interested in connecting the terminals between global cells. If we should decide to connect the terminals as shown in Figure 7a, we shall connect them in the original grid-graph as shown in Figure 7b.

There are basically three kinds of routing problems: (1) *global routing*, (2) *detailed routing*, and (3) *channel routing*. Global routing refers to the connecting of all the nets between global cells; while detailed routing refers to the connecting of all the nets between the nodes in the original grid-graph. Channel routing can be regarded as a special case of detailed routing in which all the terminals are situated at the upper and lower boundaries of the grid-graph. Not surprisingly, all these three problems are *NP-complete*. [5].

---

[5] See Appendix IV for the list of *NP-complete* routing problems.

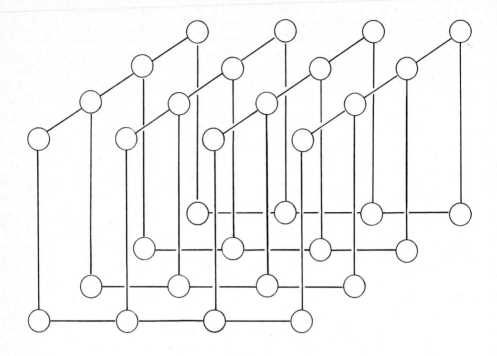

Figure 5

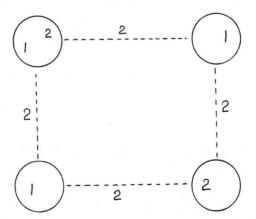

Figure 6

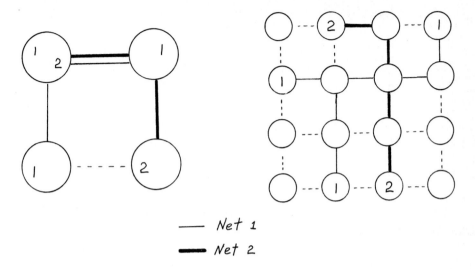

Figure 7a   Figure 7b

3.1 Global and Detailed Routing

Mathematically, both global and detailed routing problems are the basically the same, except that in global routing, we do not have to worry about how wires change directions inside a global cell.

If a net has two terminals, the problem of connecting the two terminals is a *shortest-path* problem. If the net has three or more terminals, the problem of connecting three or more terminals is a *minimum rectilinear Steiner tree* problem. Finally, the problem of embedding all the nets (the paths and the trees) in the grid-graph is a *compatibility* problem. So we have to solve:

(1) *the shortest-path problem;*

(2) *the minimum rectilinear Steiner tree;* and

(3) *the compatibility problem of embedding all the nets.*

3.1.1 The shortest-path problem

Consider a network $N = (V,E)$ consisting of a set of vertices and a set of arcs. Each arc e_{ij} has a length d_{ij} which denotes the cost of routing a wire through the arc. A sequence of vertices and arcs, say

$$(v_s, e_{si}, v_i, e_{ij}, v_j, \cdots, v_t)$$

is said to form a *path* from v_s to v_t. The *length* of the path is the sum of the lengths of all arcs in the path. The *shortest-path* problem is "*to find a path of minimum length joining two given vertices*". For convenience, we assume that $d_{ij} = \infty$ if there is no arc leading from v_i to v_j and $d_{ii} = 0$ for all i's. Assume that (i) all lengths d_{ij} are positive, (ii) some of the arcs are directed, and (iii) a path with many arcs may be shorter than a path with less number of arcs, we have the following three conditions:

(i) $d_{ij} > 0$ for all i,j

(ii) $d_{ij} \neq d_{ji}$ for some i,j

(iii) $d_{ij} + d_{jk} < d_{ik}$ for some i,j,k

Note that the problem become very trivial if condition (iii) does not hold. Under these three assumptions, Dijkstra has developed the first shortest-path algorithm in 1959 [19]. This algorithm not just finds the shortest paths from a given vertex v_s to another given node v_t, it actually gives the shortest path from v_s to all other nodes in the network. Hence, we call such an algorithm a *single-source shortest-path* algorithm.

To find the shortest paths between all $\binom{n}{2}$ pairs of vertices in a network, we can use Warshall-Floyd's algorithm [101] which takes $O(n^3)$ time. The Warshall-Floyd algorithm does not need all $d_{ij} \geq 0$ as long as there does not exist a cycle with negative length in the network. For a large space network, we can use decomposition algorithm for shortest paths [8] which treats a part of the network at a time.

In most single-source shortest-path algorithms, we basically construct a rooted tree from v_s to all other vertices in the network. There is a unique path from the root to every node in the tree, and the unique path is also a shortest path from the root to the node.

Different algorithms have different ways of expanding and modifying the tree. In Breadth-First-Search (BFS) type algorithms [43] (like the Lee-Moore method or its modifications [35, 39, 64, 79, 92]), we always expand the tree by selecting a node which is closest to v_s and repeat the process until the node v_t is included into the tree. The BFS algorithms has a worst-case running time of $O(n^2)$ and they always find a shortest path if one exists. The major drawback of these methods is that it will always pick a shorter path which may have many vias instead of a longer path with fewer vias, even though the latter path is more preferable in many practical situations.

In the Depth-First-Search (DFS) [43] type algorithms (like the line-expansion methods [41, 78]), we keep expanding the rooted tree along a horizontal (or a vertical) path, and then among the vertical (horizontal) paths which intersect the horizontal (vertical) path, pick the one which has the best *chance* of reaching v_t in the cheapest way and expand the tree along that path. The exact order was not specified and when v_t is reached, the path may *not* be of minimum cost. The line-expansion method has the advantage of finding a path quickly. However, it does not guarantee to find a path from v_s to v_t even if such a path exists. To cure such problem, backtracking is needed in case we fail to reach v_t along the current path [40].

More recently, Hu and Shing has proposed a new routing method, called the "α-β" routing [48]. This router allows the user to define a cost α for each unit of the wire used and a cost β for each 90° degree turn (i.e. a via) along the path, then it finds a minimum-cost path between a pair of nodes where the cost of a path is given by the sum of the costs of the wire plus the additional costs of turning around corners. Note that when $\beta \neq 0$, the other single-source shortest-path algorithms no longer work, because we have to construct a *directed acyclic graph* instead of a rooted tree.

3.1.2 The Minimum Rectilinear Steiner Tree Problem

A tree connecting all the nodes of a network is called a *spanning* tree. The length of the tree is the sum of the lengths of all the edges in the tree. The *minimum spanning tree* problem is *to find a spanning tree of minimum length*. This problem is quite easy and efficient algorithms for finding minimum spanning tree can be found in [60, 82]. For a network with n vertices and m arcs, these algorithms have a worst case running time of O(m *log* m) and O(n²) respectively.

Now, let us modify the problem slightly. Let N = (V,E) be a graph representing the chip and P be a non-empty set of nodes representing the terminals in a net. (P is a subset of V.) The shortest way of connecting all the terminals in P must be a tree which spans (connects) all the vertices in P. Such tree, which may contain nodes not in P, is called a *Steiner* tree. The *Steiner* problem is to find a Steiner tree with minimum total edge-length. It turns out that this modified problem is *NP-complete* [53]. Detailed survey of the Steiner problem can be found in [26].

When the underlying graph is a two-dimensional rectilinear grid-graph and all the arcs are of the same length, we have a special type of the Steiner problem, called the *rectilinear Steiner* problem (RSP). Besides applications in wire layout problems in VLSI circuit design [14, 36, 37, 65, 102], RSP also finds applications in the design of communication networks [27] and electrical systems in buildings [75]. Unfortunately, RSP is also *NP-complete* [28]. Because of its wide variety of practical applications, researchers have come up with a lot of efficient heuristic algorithms [49, 50, 51, 58, 65, 76, 102]. Interested readers should refer to [26, 50, 90] for detailed survey and empirical study of these algorithms.

Note that most nets in a VLSI circuit have only five terminals or less. Like all other *NP-complete* problems, RSP can be solved easily for small input sizes [1, 22, 35].

3.1.3 The Compatibility Problem

We have seen that to connect a single net of two terminals, we have to solve a shortest path problem. To connect a single net of three or more terminals, we have to solve a minimum rectilinear Steiner problem. Since most nets have five terminals or less, we can find the minimum rectilinear Steiner tree easily. (A bus or a power line may connect many terminals, but those large nets are usually treated separately and are placed in positions before all other nets are routed). The problem of global routing is then to embed the paths and trees in the grid-graph G. Two of the most commonly asked questions are:

(i)   In what order should the nets be routed?

(ii)   What is a systematic way of routing and re-routing of the nets?

Recently, Ting and Tien [98] suggested that all nets should be routed as if they were the first net to be routed, i.e. each net has equal chance to find its best pattern. This would make some of the cell boundaries overcrowded (or overflowed). These boundaries as well as the nets using these boundaries are then identified and represented as vertices of a bipartite graph. The nets which use many overflowed boundaries are re-routed. If the re-routing causes new overflowed boundaries, the same procedure will be repeated until the level of overflow can be tolerated.

The circuit routing problem is very much like a traffic problem. The terminals are the origins and destinations of the traffic. The wires connecting the terminals are the traffic, and the channels are the streets. If there are more wires than the number of tracks in a given channel, some of the wires have to be re-routed just like the re-routing of traffic.

For the real traffic problem, every driver wants to go to his destination in the quickest way, and he may try a different route every day. Eventually, every driver selects his best possible route and the traffic pattern become stabilized. Intuitively, we can do the same for the circuit routing problem. For example, we can associate a non-negative cost to each arc. This cost should reflect how congested the arc is. (The cost is set to infinity if the number of tracks available equals zero.) The cost of routing a path is then the sum of the costs of all the arcs along the path; and the cheapest cost of routing all the nets will correspond to a feasible way of laying out the circuits on the chip. The main question is: *What kind of cost function should we assign to each arc?*.

Neglecting the size of the problem, we can formulate the global routing problem as a very large linear integer program [46] [6], very much like a multi-commodity network flow problem [24, 42]. The cost of each arc is then given by the *shadow price*. The shadow price is a *better* cost function than most of the existing heuristic cost functions because most of the existing heuristic cost functions depend on parameters associated with the arc itself; whereas the *correct* cost function, like the *shadow price*, depends not only on the arc itself, but also on its adjacent arcs, in fact, on the arcs adjacent to its adjacent arcs.

3.1.4 Global Routing as an Integer Linear Program

There are many ways to connect a net, each way is a Steiner tree connecting the given terminals in the net. We associate a variable y_j to each tree which connects a net. The variable y_j is equal to 1 if that particular tree is used and y_j is equal to 0 if that particular tree is not used. For example, if there are three ways to connect the first net and five ways to connect the second net, we will set

$$y_1 + y_2 + y_3 \qquad\qquad = 1 \qquad\qquad\qquad (1)$$
$$y_4 + y_5 + y_6 + y_7 + y_8 = 1$$

Note that there is one equation for each net since only one tree is needed to connect a given net.

Now, there are many different ways to connect a given net.

In principle, we can represent all possible ways by a (0,1) matrix $[a_{ij}]$. For a grid graph with m arcs, there will be m rows in $[a_{ij}]$, with the ith row corresponding to the ith arc and each column corresponding to different ways of connecting a net. If there are a thousand nets and there are ten different ways to connect each net, then there will be ten thousand columns in the matrix and we cannot enumerate all possible ways of connecting all the nets in practice. Fortunately, we do not need to know all elements of the matrix to solve the linear integer program.

---

[6] Burstein *et al* [11, 12] have also proposed a different linear programming formulation of the wire routing problem and gave a divide-and-conquer solution for the problem.

For any column j, the ith entry, a_{ij}, equals 1 if that particular connection uses the ith arc; otherwise a_{ij} equals 0. The fact that the number of wires in any arc must be bounded above by its arc capacity can be expressed as

$$\sum a_{ij}\, y_j \le c_i \qquad (i = 1,...,m; \; j = 1,...) \qquad (2)$$

where c_i is the arc capacity of the ith arc. (Note that some c_i may be zero if the corresponding regions are forbidden. Note also that we leave the range of the index j open because we do not know how many ways to connect the nets.)

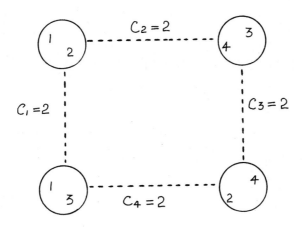

Figure 8

In Figure 8, we show a grid-graph with four nodes and four arcs, where $c_i \equiv 2$ for i=1,...,4. There are four nets to be connected and all nets are two-terminal nets. Net 1 connects the node in the upper left corner to the node in the lower left corner, Net 3 connects the node in the upper right corner to the node in the lower left corner, etc. Each net can be connected in two different ways and the two ways of connecting net j will be denoted by y_j and y_j'. The eight different ways of connecting the four nets are shown in the following (0,1) matrix:

	y_1	y_1'	y_2	y_2'	y_3	y_3'	y_4	y_4'
c_1	1	0	0	1	0	1	0	1
c_2	0	1	1	0	0	1	0	1
c_3	0	1	1	0	1	0	1	0
c_4	0	1	0	1	1	0	0	1

For the moment, let us assume that there are p nets and a total of n possible ways of connecting all the nets. We shall denote the set of y_j's which correspond to various ways of connecting the kth net by N_k.

If we write down all the constraints (1) and (2), we have the typical structure of a linear integer program (3) suitable for decomposition algorithms.

$$
\begin{aligned}
\max \quad & \sum b_j\, y_j & (j = 1,2,...,n) \\
\text{subject to} \quad & \sum_{y_j \in N_k} y_j = 1 & (k = 1,2,...,p) \\
& \sum a_{ij}\, y_j \le c_i & (i = 1,2,...,m) \\
& 0 \le y_j \le 1, & \text{integers.}
\end{aligned}
\tag{3}
$$

Note that the number of equality constraints in (3) is equal to the number of nets, and the number of inequality constraints of the type $\sum a_{ij}\, y_j \le c_i$ is equal to the number of arcs in the grid-graph G.

The constant b_j is the benefit of connecting the kth net by the jth tree. The constants b_j are the same for all $y_j \in N_k$, but the constants are different for different nets. For a net k with many terminals, we may want to assign a large positive constant, and for a two-terminal net with pins near to each other, we may want to assign a small constant. On the other hand, for a given net k we do not care which tree (y_j) is used to connect the net k. We can also assign all b_k to be the same, then the objective is simply to connect as many nets as possible.

Let π_i be the shadow price of the ith row under the current basis of the linear program (3), and let $\overline{b}_j$ be

$$
\overline{b}_j = b_j - \pi_i\, a_{ij}.
\tag{4}
$$

To solve the linear program (3), we will select a column j to enter the basis if $\overline{b}_j$ is positive.

Since the b_j's are the same for all spanning trees of any given net, the best way to connect the net corresponds to the column which maximizes $\overline{b}_j$; and among all possible ways of connecting the net, maximum $\overline{b}_j$ corresponds to minimum $\pi_i\, a_{ij}$. In other words, we want to find a minimum Steiner tree in a grid-graph where the arc lengths are defined by the shadow prices.

Once a new column enters the basis, and a pivot operation is performed, a new set of shadow prices is used for selecting the best way to connect the next net. This process is iterated until no more nets can be connected, or until the objective function is maximized. (Note that y_j are bounded, see [16]).

Note that the grid-graph model presented here is a simplified model which roughly corresponds to double metal layer gate-array chips. A real VLSI chip has multi-layers and the wires can be polysilicon and metal, which have different physical characteristics. Thus, all computer programs in industry are based on heuristic algorithms [61]. Intuitively, we will avoid the use of a region if that region is overcrowded, and diverse wires to a region which is currently empty or sparse. This amounts to assigning high prices to arcs in crowded regions and low prices to arcs in sparse regions. *This is exactly the role of shadow price.* If some regions are forbidden, we can simply let $c_i = 0$ in those places. The L. P. formulation has several defects even on this simple grid-graph model.

(1)   Too many columns in the matrix a_{ij}, too many to be written down explicitly.

(2)   Too many rows in the matrix a$_{ij}$.

(3)   Integer constraints on the y$_j$.

(4)   How to find a minimum Steiner tree with arcs of different lengths.

We can cure (1) by column-generating techniques [17, 24, 32] and cure (2) by *cut and paste* methods [46] similar to the partitioning methods suggested in [58]. (Here, we trade computer time and memory space with the possibility of obtaining suboptimal solutions.) (4) arises as a subproblem of generating the best column. This problem can be solved easily since most nets have five terminals or less.

For (3), we can relax the integer constraints on the variable y$_j$'s and solve the relaxed linear program. (Note that the y$_j$'s are bounded, see [16].) The relaxed L. P. solution may not be integers. However, since the number of nets is much larger than the number of arcs in global routing, most y$_j$'s will be at their integral upper bounds, which is exactly what we want. The few exceptions can be handled separately after we solve the relaxed L. P.

3.2 Channel Routing

Channel routing is a special case of the detailed routing. It arises in a common approach to break up the general components into more *manageable* subproblems [38]. A channel can be thought of as a rectangular grid, with all the terminals constrained to lie on the two horizontal sides of the grid and all the connection wires constrained to lie between the two horizontal sides. The *width* of a channel equals the number of horizontal tracks available for routing. In gate-array chips, the width of a channel is fixed and the the main objective is to connect all the nets. In standard-cell chips and custom-designed chips, we can always increase the width of a channel to ensure the feasibility of routing; hence, the main objective is to minimize the channel width while connecting all nets.

Given a two-layer chip with a channel width W and a wiring list, the problem *whether there exists a feasible routing for all the nets in the channel* is *NP-Complete* [94]. It remains *NP-complete* even if all the nets consist just of two terminals, one on each side of the channel [95]. Hence, a lot of research has been done in the development of good heuristic algorithms. (See, for example [4, 11, 18, 38, 87, 88, 103].) Among these heuristics, the most publicized algorithm is Deutsch's "Dogleg" channel router. His "difficult" channel has been used as a standard benchmark for all other heuristic algorithms. Based on the "divide and conquer" approach, Burstein and Pelavin have proposed a very successful hierarchical channel router recently. They have tested their router on Deutsch's difficult example and connected all the nets with 19 horizontal wiring tracks (the absolute minimum for that example), whereas all other known routers required 20 or more tracks. For a channel with length n and width w, this new router can route N nets in at most O(N n *log* m) time.

When there is only one wiring layer in the channel, we have a special case of channel routing, called *river routing*. River routing turns out to be a much easier problem than the two-layer channel routing. A necessary condition for connecting all the nets in a single-layer channel is that the nets can be connected in the plane without crossovers. Such condition is called *planar routability* and it can be checked in linear time [2]. Polynomial algorithms for solving various river routing problems can be found in [20, 71, 91, 99].

When there are three wiring layers in the channel, the problem of *finding an optimal dogleg routing in a 3-layer channel* is again *NP-complete* [74]. Readers can refer to [13] for heuristic algorithm.

4. Conclusions

The design of VLSI circuits has posed a lot of interesting and practical combinatorial problems. In this chapter, we have looked at one of these problems, namely the *layout* problem. This problem is very complicated and hard to analyze. Hence, simplified models are introduced to formalize the various layout problems.

Based on these models, we can show that most of the layout problems are *NP-complete*. Since it is unlikely to solve these problems efficiently, we should focus our attention in the development of efficient heuristic algorithms. The complexity and performance of most of the existing "real world" layout algorithms are hard to analyze. In order for these algorithms to be useful, more research is needed in analyzing the average-case complexity (efficiency) and performance (error bound) of these heuristic algorithms.

Currently, there are two groups of people working on the layout problems, the *engineers* working from the "real world" point of view and the *computer scientists* working from the "complexity and algorithms" point of view. May be we should also invite the operations researchers to work on these problems. It is quite natural to formulate the layout problems as large linear programs and we can definitely use the techniques in operations research to develop more efficient and effective layout algorithms.

5. Appendix

I. Analysis of Algorithms

The most widely accepted performance measure for an algorithm is its running time, i.e., the time it spends before producing the final answer. There are two problems in defining the running time of an algorithm: (1) it varies vastly from one computer to another and (2) it varies significantly from one input to another. We can circumvent the first problem by expressing the time requirement of an algorithm in terms of the number of elementary steps such as arithmetic operations, comparisons, branching instructions, and so on. To avoid the second problem, we can consider all inputs of a given size n, and define the *worst-case time complexity* of an algorithm for that input size as the worst-case behavior of the algorithm on any of these inputs. Likewise, we can define the *average-case time complexity* of an algorithm for that input size as the average-case behavior of the algorithm on any of these inputs. Hence, we can express the time complexity $T(n)$ of an algorithm as a function of the input size n.

In studying the complexity of an algorithm, we are often interested only in the behavior of the algorithm for very large inputs, because it is these inputs that enable us to determine the limits of the applicability of the algorithm. As the input size increases, we often find that it is the growth rate of the complexity of an algorithm that ultimately determines how big a problem we can solve on a computer. We can talk about growth rates of functions using the following definitions introduced by Knuth [57]:

Let $f(n)$, $g(n)$ be functions from the non-negative integers to the positive reals.

(i) We write $f(n) = O(g(n))$ if there exists constants c and n_0 such that, for all $n \geq n_0$, $f(n) \leq cg(n)$.

(ii) We write $f(n) = \Omega(g(n))$ if there exists constants c and n_0 such that, for all $n \geq n_0$, $f(n) \geq cg(n)$.

(iii) We write $f(n) = \Theta(g(n))$ if there exists constants c, c' and n_0 such that, for all $n \geq n_0$, $cg(n) \leq f(n) \leq c'g(n)$.

Using the above definitions, an algorithm is said to be a *polynomial algorithm* if its worst-case time complexity $T(n)$ is bounded by a polynomial of the input size, i.e. $T(n) = O(n^p)$ for some constant $p \geq 0$.

II. The complexity of a problem

Quite often, we would like to know whether there exists an efficient algorithm for a given problem. We can roughly classify the problems into four different classes according to their degrees of difficulty.

1. *Undecidable* problems (unsolvable problems): These are the problems that cannot be solved by any algorithm. For example, it was proved that there exists no algorithms which can automatically decide whether a program will halt on a computer.

2. *Intractable* problems (provably difficult problems): These are the problems which no polynomial algorithm can possibly be developed to solve them. In other words, they can at best be solved by exponential algorithms, i.e. algorithms with a worst-case time complexity of $O(c^{n^p})$ for some constants $c > 1$ and $p > 0$.

3. *NP*-problems (Nondeterministic Polynomial problems): For a problem to be in this class, we simply require that, if there exists a correct solution to the problem, such solution can be constructed and verified in polynomial time. Roughly speaking, this class includes all problems that have exponential algorithms but we have not proved that they cannot be solved by polynomial algorithms.

4. *P*-problems (Polynomial problems): This class includes all problems that have polynomial-time algorithms.

Today there is a general consensus among computer scientists that only polynomial algorithms are of practical use in solving computational problems. Unfortunately, a lot of the computational problems which we have to solve everyday belong to the class of *NP*-problems. Among the *NP*-problems, there exists a subset of problems, called the *NP-complete* problems, which are hardest in the following sense:

(i) No *NP-complete* problems can be solved by any known polynomial algorithms (in spite of the persistent efforts by many brilliant researchers for several decades).

(ii) We can transform any instance of an *NP*-problem to an instance of an *NP-complete* problem in polynomial time. Hence, if there is a polynomial algorithm for any *NP-complete* problem, there are polynomial algorithms for all *NP*-problems.

Since it is unlikely to solve the *NP-complete* problems in polynomial time, except for a few special cases, researchers have come up with polynomial-time heuristic algorithms to find the near-optimum solutions for these problems. (Interested readers should refer to [29] for more details.)

III. The Graph Layout Problems

A *layout* of a graph G is an embedding which assigns the vertices of G to the nodes of a two-dimensional grid-graph, and assigns the edges of G to the paths in the grid-graph. In graph layout problems, we want to find an efficient layout (or embedding) of a given graph G in a two-dimensional grid under various optimality criteria. Many of these problems are known to be *NP-complete*. Here, we shall list a few examples of the *NP-complete* graph layout problems. Interested readers should refer to [7, 66] for more details.

Problem III.1   Given a graph G, produce an area-efficient layout for G.

The *layout area* is perhaps one of the most desirable cost measure to be minimized. Here, we define the area of a layout as the smallest rectangle which encloses all the nodes and wire segments of the embedding. Dolev, Leighton and Trickey [21] have shown that determining the minimum layout area of a forest of trees is *NP-complete*. Other upper bounds and lower bounds on the area required to layout an arbitrary graph can be found in [7, 66, 67, 68, 69, 70, 96, 97, 100].

Problem III.2   Given a graph G, produce an area-efficient layout for G with minimax edge length.

Besides area, *speed* is another critical factor in chip performance. Here, we would like to find a layout such that the length of the longest edge in that layout is minimum over all layouts for G. Bhatt and Cosmadakis [5] have shown that computing the minimax edge length of a tree is *NP-complete*. Bounds on the minimax edge length of various trees can be found in [6, 81].

Problem III.3   Given a graph G, produce a layout for G with few wire crossings.

Since it is undesirable to have a large number of wire crossings in a VLSI layout, we would like to have a layout for G which minimize the number of wire crossings. The *crossing number* of a graph is defined to be the minimum number of wire crossings in any drawing of the graph on the plane. Garey and Johnson [30] have shown that determining the cross number of bipartite graphs is *NP-complete*. Upper and lower bounds on crossing numbers can be found in [67, 68].

To cope with these *NP-complete* problems, researchers have developed a lot of efficient heuristic algorithms. Many of these heuristic algorithms are based on the divide-and-conquer paradigm which uses either the *planar separator theorem* [73] or the *bifurcator* [7] to partition a given graph recursively into smaller subgraphs, solve the graph layout problem for each subgraph, and then patch the layouts together to obtain a near-optimal layout for the original graph, see, for example, [7, 69, 70, 100].

IV. The Routing Problems

Here is a brief list of *NP-Complete* routing problems. Interested readers should refer to [52] for more details.

Problem IV.1: Planar interconnection on a grid.

Given a set $P = \{(x_1,y_1),...,(x_n,y_n)\}$ of disjoint pairs of integer-coordinate grid points, is there a way of connecting each pair of the grid points in P by a path made up of grid segments, so that no two paths intersect?

This problem has been proved *NP-complete* independently by Raghavan *et al* [83], Richards [85] and Kramer *et al* [59]. It remains *NP-complete* even if orthogonal intersections are allowed [59], or if each interconnecting path must be "monotone" in the horizontal direction (i.e. must proceed from left to right without turning back) [55].

Problem IV.2: Two-layer Multiple Module Routing.

Given 6 (i) a collection $\{M_1, \ldots, M_n\}$ of non-intersecting modules (i.e., rectangles specified by their four integer-coordinate corners), (ii) a set P of terminals (i.e., grid points on the perimeters of the modules) partitioned into nets $N_1, \ldots, N_m$, and (iii) an integer bound B,

is there a (reserved layer) routing for the nets in a rectangle R of area B, i.e. to connect the pins in each net by a rectilinear Steiner tree such that (i) none of the grid segments of the Steiner trees is contained in any of the modules, (ii) all the modules and all the segments of the Steiner trees are included in the rectangle R, and (iii) no two trees have non-orthogonal intersections?

This problem was proved to be *NP-complete* by Szmanski [94]. It remains *NP-complete* even if there are only two modules and each net contains only two terminals [95]. Approximation algorithms and special-case polynomial algorithms can be found in [3, 62, 63].

Problem IV.3: Two-layer Channel Routing.

Given 6 (i) a channel which can be thought of as a rectangular region bounded by the lines $y=0$ and $y=W$, and (ii) a set of grid points $P = \{(a_i, W): 1 \leq i \leq p\}$ union $\{(b_j,0): 1 \leq j \leq q\}$, partitioned into nets $N_1, \ldots, N_m$,

is there a (reserved layer) two-layer channel routing for the nets, i.e., to connect the pins in each net by a rectilinear Steiner tree such that (i) all the grid segments of the Steiner trees lie in the region $y>0$ and $y<W$ and (ii) no two trees have non-orthogonal intersections?

This problem was proved to be *NP-complete* by Szymanski [94] It remains *NP-complete* even if all nets consist of just two points, one of the form (a_i, W) and one of the form $(b_j,0)$ [95] or even if all are two-terminal nets and each net T_i is only allowed to contain one horizontal line segment (i.e. sequence of horizontal grid segments) [62, 63].

Problem IV.4: Three-layer channel routing.

   Given an ensemble of edge-disjoint paths connecting the (two-terminal) nets in a channel routing instance, is there a way of assigning layers to path segments so that a valid 3-layer knock-knee routing is obtained.

This problem was proved to be *NP-complete* by Lipski [74]. See [13] for heuristic algorithms.

References

1. A.V. Aho, M.R. Garey, and F.K. Hwang, "Rectilinear Steiner Trees: Efficient Special Case Algorithms," *Networks*, vol. 7, pp. 37-58, 1977.

2. A. Amir, "A direct linear-time planarity test for unflippable modules," unpublished manuscript, MIT, 1984.

3. B.S. Baker, "A provably good algorithm for the two module routing problem," unpublished manuscript, 1982.

4. B.S. Baker, S.N. Bhatt, and F.T. Leighton, "An Approximation Algorithm for Manhattan Routing," *Proc. 15th Annual ACM Symposium on Theory of Computing*, pp. 477-486, 1983.

5. S.N. Bhatt and S. Cosmadakis, "The Complexity of Minimizing Wire Lengths in VLSI Layouts," unpublished manuscript, M.I.T., Cambridge, Mass., 1982.

6. S.N. Bhatt and C.E. Leiserson, "Minimizing the Longest Edge in a VLSI Layout," MIT VLSI Memo 82-86, 1982.

7. S.N. Bhatt and F.T. Leighton, "A Framework for Solving VLSI Graph Layout Problems," *J. of Computer and System Sciences*, vol. 28, pp. 300-343, 1984.

8. W.J. Blewett and T.C. Hu, "Tree Decomposition Algorithm for Large Networks," *Networks*, vol. 7, no. 4, pp. 289-296, 1977.

9. M.A. Breuer, "Min-cut placement," *J. Design Automation and Fault Tolerant Computing*, vol. 1, no. 4, pp. 343-362, 1977.

10. T. Bui, S. Chaudhuri, T. Leighton, and M. Sipser, "Graph Bisection Algorithms with Good Average Case Behavior," *Proc. 25th Annual Symposium on Foundations of Computer Science*, pp. 181-192, 1984.

11. M. Burstein and P. Pelavin, "Hierarchical Channel Routing," *Proc. 20th Design Automation Conference*, pp. 591-597, 1983.

12. M. Burstein and P. Pelavin, "Hierarchical Wire Routing," *IEEE Trans. on CAD*, vol. CAD-2, no. 4, pp. 223-234, 1983.

13. Y.K. Chen and M.L. Liu, "Three-Layer Channel Routing," *IEE Trans. on CAD*, vol. CAD-3, no. 2, pp. 156-163, 1984.

14. N. Christofides, *Graph Theory: An Algorithmic Approach,* Academic Press, New York, 1975.

15. E.G. Coffman, Jr. and M.A. Langston, "A Performance Guarantee for the Greedy Set-Partitioning Algorithm," *Acta Informatica*, vol. 21, pp. 409-415, 1984.

16. G.B. Dantzig and R.M. VanSlyke, "Generalized upper bounding techniques for linear programming," *J. Computer and System Sciences*, vol. 1, pp. 213-226, 1967.

17. G.B. Dantzig and P. Wolfe, "The decomposition algorithm for linear programming," *Econometrica*, vol. 29, no. 4, pp. 767-778, 1961.

18. D.N. Deutsch, "A dogleg channel router," *Proc. 13th Design Automation Conference*, pp. 425-433, 1976.

19. E.W. Dijkstra, "A Note on Two Problems in Connection with Graphs," *Numerische Mathematik*, vol. 1, pp. 269-271, 1959.

20. D. Dolev, K. Karplus, A. Siegel, A. Strong, and J.D. Ullman, "Optimal wiring between rectangles," *Proc. 13th Annual ACM Symposium on Theory of Computing*, pp. 312-317, 1981.

21. D. Dolev, F.T. Leighton, and H. Trickey, "Planar embeddings of planar graphs," MIT LCS Tech. Report 237, 1983.

22. S.E. Dreyfus and R.A. Wagner, "The Steiner problem in Graphs," *Networks*, vol. 1, pp. 195-207, 1972.

23. C.M. Fiduccia and R.M. Mattheyses, "An Almost Linear Algorithm for Partitioning Networks," unpublished manuscript, 1982.

24. L.R. Ford and D.R. Fulkerson, "Suggested computations for maximal multi-commodity network flows," *Management Science*, vol. 5, no. 1, pp. 97-101, 1958.

25. L.R. Ford and D.R. Fulkerson, *Flows in Networks*, Princeton University Press, New Jersey, 1962.

26. L.R. Foulds and V.J. Rayward-Smith, "Steiner Problem in Graphs: Algorithms and Applications," Tech. Report CSA2, School of Computing Studies and Accountancy, University of East Anglia, Norwich NR4 7TJ, England, 1983.

27. H. Frank and I.T. Frisch, "Network Analysis," in *Large Scale Networks: Theory and Design*, ed. F.T. Boesche, pp. 19-27, IEEE Press, New York, 1976.

28. M.R. Garey and D.S. Johnson, "The Rectilinear Steiner Tree Problem is NP-Complete," *SIAM J. Applied Mathematics*, vol. 32, pp. 826-834, 1977.

29. M.R. Garey and D.S. Johnson, *Computers and Intractability: A Guide to the Theory of NP-Completeness*, Freeman, San Francisco, CA, 1979. For an updated list of *NP-Complete* problems, see D. S. Johnson, "The NP-Completeness Column: An Ongoing Guide," *Journal of Algorithms*, 1981-present.

30. M.R. Garey and D.S. Johnson, unpublished manuscript, 1982.

31. P.C. Gilmore, "Optimum and Suboptimum Algorithms for the Quadratic Assignment Problems," *J. of SIAM*, vol. 10, no. 2, pp. 305-313, 1962.

32. R.E. Gomory, "Large and non-convex problems in linear programming," *Proc. Symposium in Applied Math.*, vol. 15, 1963.

33. R.E. Gomory and T.C. Hu, "Multi-terminal Network Flows," *J. Siam*, vol. 9, no. 4, pp. 551-570, 1961.

34. K.M. Hall, "An r-Dimensional Quadratic Placement Algorithm," *Management Science*, vol. 17, no. 3, pp. 219-229, 1970.

35. S. Hanan, "On Steiner's problem with rectilinear distance," *SIAM J. of Appl. Math.*, vol. 14, no. 2, pp. 255-265, March 1966.

36. M. Hanan and J.M. Kurtzburg,, "Placement Techniques," in *Design Automation of Digital Systems*, ed. M. Breuer, Prentice-Hall, Englewood Cliffs, New Jersey, 1972.

37. M. Hanan, "Layout, Interconnection and Placement," *Networks*, vol. 5, pp. 85-88, 1975.

38. A. Hashimoto and J. Stevens, "Wiring routing by optimizing channel assignment within large apertures," *Proc. 18th Design Automaton Workshop*, pp. 155-169, IEEE,

1971.

39. S. Heiss, "A path connection algorithm for multi-layer boards," *Proc. 5th Design Automation Conference*, pp. pp. 6-14, 1968.

40. W. Heyns, W. Sansen, and H. Beke, "A line-expansion algorithm for the general routing problem with a guaranteed solution," *Proc. 17th Design Automation Conference*, pp. 243-249, 1980.

41. D. W. Hightower, "The interconnection problem -- a tutorial," *Computer*, vol. 7, no. 4, pp. 18-32, April 1974.

42. T. C. Hu, *Integer Programming and Network Flows,* Addison-Wesley, Reading, Mass., 1969.

43. T. C. Hu, *Combinatorial Algorithms,* Addison-Wesley, Reading, Mass., 1982.

44. T.C. Hu and Y.S. Kuo, "Optimum Reduction of Programmable Logic Array," *20th Design Automation Conference*, pp. 553-558, 1983.

45. T.C. Hu and M.T. Shing, "Multi-terminal Flows in Outerplanar Networks," *J. of Algorithms*, vol. 4, no. 3, pp. 241-261, September 1983.

46. T. C. Hu and M. T. Shing, "A Decomposition Algorithm for Circuit Routing," to appear in *J. Math. Programming.* A revised version will also appear in the book *VLSI: Circuit Layout Theory,* ed. T.C. Hu and E.S. Kuh, IEEE Press, 1985.

47. T.C. Hu and M.T. Shing, "Decomposition Algorithms in Large Sparse Networks," Tech. Report TRCS84-08, Computer Science Dept., UCSB, Santa Barbara, CA, 1984.

48. T.C. Hu and M.T. Shing, "The α-β Routing," to appear in the book *VLSI: Circuit Layout Theory,* ed. T.C. Hu and E.S. Kuh , IEEE Press, 1985.

49. F.K. Hwang, "On Steiner minimal trees with rectilinear distance," *J. SIAM Appl. Math.*, vol. 30, pp. 104-114, 1976.

50. F.K. Hwang, "The Rectilinear Steiner Problem," *Design Automation and Fault-Tolerant Computing*, vol. 2, no. 4, pp. 303-310, 1978.

51. F.K. Hwang, "An O(n log n) Algorithm for Suboptimal Rectilinear Steiner Trees," *IEEE Trans. on Circuit and Systems*, vol. CAS-26, no. 1, pp. 75-77, 1979.

52. D.S. Johnson, "The NP-Completeness Column: An Ongoing Guide," *Journal of Algorithms*, vol. 3, pp. 381-395, 1982. See also Vol. 5, pp.147-160 of the same journal.

53. R.E. Karp, "Reducibility among combinatorial problems," in *Complexity of Computer Computations*, ed. R.E. Miller and J.W. Thatcher, pp. 85-103, Plenum Press, New York, 1972.

54. N. Karmarker and R.M. Karp, "The Differencing Method of Set Partitioning," Tech. Report UCB/CSD 82/113, University of California, Berkeley, California, 1982.

55. K. Karplus, "CHISEL: An extension to the programming language C for VLSI layout," Tech. Report STAN-CS-82-959, Dept. of Computer Science, Stanford University, Stanford, CA, 1982.

56. B.W. Kernighan and S. Lin, "An Efficient Heuristic Procedure for Partitioning Graphs," *Bell System Technical J.*, vol. 49, no. 2, pp. 291-307, 1970.

57. D.E. Knuth, "Big Omicron and Big Omega and Big Theta," *SIGACT News*, pp. 18-24, ACM, April 1976.

58. J. Komloś and M.T. Shing, "Probabilistic Partitioning Algorithms for the Rectilinear Steiner Problem," TRCS84-11, Computer Science Dept., University of California, Santa Barbara, CA., 1984.

59. M.R. Kramer and J. van Leeuwen, "Wiring-routing is *NP-complete*," Tech. Report RUU-CS-82-4, Dept. of Computer Science, University of Utrecht, Utrecht, the Netherlands, 1982.

60. J.B. Kruskal, Jr., "On the shortest spanning tree of a graph and the traveling salesman problem," *Proc. AMS*, vol. 7, pp. 48-50, 1956.

61. E.S. Kuh (ed.), "Special issue on routing in microelectronics," *IEEE Trans. on CAD*, vol. CAD-2, no. 4, 1983.

62. A.S. LaPaugh, "A polynomial time algorithm for optimal routing around a rectangle," *Proc. 21st Annual Symposium on Foundations of Computer Science*, pp. 282-293, 1980.

63. A.S. LaPaugh, "Algorithms for Integrated Circuit Layout: An Analytic Approach," Tech. Report MIT-LCS-TR-248, Doctoral Dissertation, Dept. of Electrical Engineering and Computer Science, MIT, Cambridge, MA, 1980.

64. C. Y. Lee, "An algorithm for path connections and its applications," *IRE Trans. Electronic Computer*, pp. 346-365, Sept. 1961.

65. J.H. Lee, N.K. Bose, and F.K. Hwang, "Use of Steiner's Problem in Suboptimal Routing in Rectilinear Metric," *IEEE Trans. on Circuit and Systems*, vol. CAS-23, no. 7, pp. 470-476, 1976.

66. F.T. Leighton, "Layouts for the Shuffle-Exchange Graph and Lower Bound Techniques for VLSI," Ph.D. Thesis, M.I.T., Cambridge, Mass., 1981.

67. F.T. Leighton, "New lower bound techniques for VLSI," *Proc. 22nd Annual Symposium on Foundations of Computer Science*, pp. 1-12, 1981.

68. F.T. Leighton, "A layout strategy for VLSI which is provably good," *Proc. 14th Annual ACM Symposium on Theory of Computing*, pp. 85-98, 1982.

69. C.E. Leiserson, "Area-efficient layouts (for VLSI)," *Proc 21st Annual Symposium on Foundations of Computer Science*, pp. 270-281, 1980.

70. C.E. Leiserson, "Area-efficient VLSI Computation," Ph.D. Thesis, Carnegie-Mellon University, Pittsburg, PA., 1981.

71. C.E. Leiserson and R.Y. Pinter, "Optimal placement for river routing," in *VLSI Systems and Computations*, ed. H.T. Kung, R. Sproull and G. Steele, pp. 126-142, Computer Science Press, Rockville, Maryland, 1981.

72. J.T. Li and M. Marek-Sadowska, "Global Routing for Gate Array," *IEEE Trans. on CAD*, vol. CAD-3, no. 4, pp. 298-307, 1984.

73. R.J. Lipton and R.E. Tarjan, "A Separator Theorem for Planar Graphs," *SIAM J. Appl. Math.*, vol. 36, pp. 177-189, 1979.

74. W. Lipski, Jr., "An *NP-complete* geometric problem related to three-layer channel routing," unpublished manuscript, 1982.

75. J. MacGregor-Smith and J.S. Liebman, "Steiner Trees, Steiner Circuits, and the Interference Problem in Building Design," *Engineering Optimization*, vol. 4, pp. 15-36, 1979.

76. J. MacGregor-Smith, D.T. Lee, and J.S. Liebman, "An O(n log n) Heuristic Algorithm for the Rectilinear Steiner Minimal Tree Problem," *Engineering Optimization*, vol. 4, pp. 179-192, 1980.

77. C. Mead and L. Conway, *Introduction to VLSI Systems,* Addison-Wesley, Reading, Mass., 1980.

78. K. Mikami and K. Tabushi, "A computer program for optimal routing of printed circuit connections," *IFIPS Proc.*, pp. 1475-1478, 1968.

79. E. F. Moore, "The Shortest Path through a Maze," *Proc. Intl. Symposium on Theory of Switching, part II*, pp. 285-292, 1959.

80. R. Nair, S.J. Hong, S. Liles, and R. Villani, "Global wiring on a wire routing machine," *Proc. 19th Design Automation Conference*, 1982.

81. M. Paterson, W. Ruzzo, and L. Snyder, "Bounds on minimax edge for complete binary trees," *Proc. 13th Annual ACM Symposium on Theory of Computing*, pp. 293-299, 1981.

82. R.C. Prim, "Shortest Connection Networks and Some Generalizations," *Bell System Technical Journal*, vol. 36, pp. 1389-1401, 1957.

83. R. Raghavan, J. Cohoon, and S. Sahni, "Manhattan and Rectilinear wiring," unpublished manuscript, 1982.

84. V.J. Rayward-Smith and M.T. Shing, "Bin Packing," *Bulletin of the Institute of Mathematics and its Applications*, vol. 19, pp. 142-148, July/August 1983.

85. D. Richards, "Complexity of single-layer routing," *IEEE Trans. on Computers*, vol. C-33, no. 3, pp. 286-288, 1984.

86. R.L. Rivest, "The PI (placement and interconnect) system," *Proc. 19th Annual Design Automation Conference*, 1982.

87. R.L. Rivest, A.E. Baratz, and G. Miller, "Provably good channel routing algorithms," in *VLSI Systems and Computations*, ed. H.T. Kung, R. Sproull and G. Steele, pp. 153-159, Computer Science Press, Rockville, Maryland, 1981.

88. R.L. Rivest and C.M. Fiduccia, "A greedy channel router," *Proc. 19th Design Automation Conference*, 1982.

89. A. Sangiovanni-Vincentelli, L. Chen, and L. Chua, "An efficient heuristic cluster algorithm for tearing large-scale networks," *IEEE Trans. Circuits and Systems*, vol. CAS-24, no. 12, pp. 709-717, 1977.

90. M. Servit, "Heuristic Algorithms for Rectilinear Steiner Trees," *Digital Processes*, vol. 7, pp. 21-32, 1981.

91. A. Siegel and D. Dolev, "The separation for general single-layer wiring barriers," in *VLSI Systems and Computations*, ed. H.T. Kung, R. Sproull and G. Steele, pp. 143-

152, Computer Science Press, Rockville, Maryland, 1981.

92. J. Soukup, "Circuit Layout," *Proc. IEEE*, vol. 69, no. 1, pp. 1281-1304, Oct. 1981.

93. J. Soukup and J.C. Royle, "On Hierarchical routing," *J. Digital Systems*, vol. 5, no. 3, 1981.

94. T.G. Szymanski, "Dogleg channel routing is *NP-complete*," unpublished manuscript, Bell Laboratories, Murray Hill, 1982.

95. T.G. Szymanski and M. Yannakakis, private communication, 1982.

96. C.D. Thompson, "Area-time complexity for VLSI," *Proc. 11th Annual ACM Symposium on Theory of Computing*, pp. 81-88, 1979.

97. C.D. Thompson, "A Complexity Theory for VLSI," Ph.D. thesis, Carnegie-Mellon University, Pittsburgh, PA., 1980.

98. B.S. Ting and B.N. Tien, "Routing Techniques for Gate Array," *IEEE Trans. on CAD*, vol. CAD-2, no. 4, pp. 301-312, 1983.

99. M. Tompa, "An optimal solution to the wire-length problem," *Proc. 12th Annual ACM Symposium on Theory of Computing*, pp. 161-176, 1980.

100. L.G. Valiant, "Universality considerations in VLSI circuits," *IEEE Trans. Computer*, vol. C-30, pp. 135-140, 1981.

101. S. Warshall, "A Theorem on Boolean Matrices," *J. ACM*, vol. ol. 9, pp. 11-12, 1962.

102. Y.Y. Yang and O. Wing, "Optimal and Suboptimal Solution Algorithms for the Wiring Problem," *Proc. IEEE International Symposium on Circuit Theory*, pp. 154-158, 1972.

103. T. Yoshimura and E. Kuh, "Efficient Algorithms for Channel Routing," *IEEE Trans. on CAD*, vol. CAD-1, no. 1, 1982.

LAYOUT DESIGN AND VERIFICATION
T. Ohtsuki (Editor)
© Elsevier Science Publishers B.V. (North-Holland), 1986

Chapter 9

COMPUTATIONAL GEOMETRY ALGORITHMS

Takao ASANO*, Masao SATO**, Tatsuo OHTSUKI**

* *Department of Mechanical Engineering, Sophia University
Tokyo, Japan*
** *Department of Electronics and Communication Engineering
Waseda University, Tokyo, Japan*

1. Introduction

In the last couple of decades we have seen an explosive growth in the field of geometric algorithms [1]. Efficient algorithms for a convex hull of points, point-location problems, intersection problems, Voronoi diagrams, etc., have been proposed, and they founded the theory of computational geometry. Computational geometry, as it stands nowadays, is concerned with the computational complexity of geometric problems within the framework of analysis of algorithms. The theoretical computational geometry has been advanced by contributions of computer scientists and aimed at applications such as geographic information processing, computer graphics, pattern recognition, structural data base, etc.

In parallel with the above advancement, electrical engineers, in the last decade, have studied large-scale 2-dimensional geometrical problems, which involve polygonal patterns with 10^6 to 10^8 edges, aiming at VLSI layout verification. Although there are a lot of common problems extending over both of the fields, algorithms for VLSI layout verification and those for the other applications have been studied independently, until very recently.

Now it is interesting to see where the difference of motivations lies between the computer scientists and the VLSI engineers. In the applications listed in the first paragraph, the algorithms are based on an underlying assumption that whole data lies within primary memory, or the data currently processed in the primary memory is fetched from a random access file in some peripheral storage. This is particularly typical in information retrieval kind of applications. Also algorithms in theoretical computational geometry have been analyzed aiming at best worst-case computational complexity, except the bucketing method (see Section 2.5).

On the other hand, algorithms for layout verification are usually based on the underlying assumption that the whole data is accommodated in a sequential file and that only those patterns within a few consecutive thin slits of the plane can be processed in the main memory. This limits algorithmic techniques to those based on slit-by-slit sequential processing. Furthermore it is assumed, as it really is, that the patterns are uniformly distributed throughout the plane. This leads to an underlying assumption that, for each horizontal or vertical slice line, there are $O(\sqrt{n})$ patterns intersecting it, where n is the number of all the polygonal patterns. Because of the background, the algorithms developed in this field tend to aim at fast expected running time. It should be noted here that the time complexity discussed in layout verification indicates the total time spent in the c.p.u. and that the space complexity indicates the capacity of the primary memory needed to perform algorithms. Thus an $O(n)$ space algorithm in theoretical computational geometry should be translated as an $O(\sqrt{n})$ algorithm for layout verification, if it is based on a single plane sweep (see Section 2.4), which is similar to the work-list method (see Chapter 7).

What will be described in this chapter is not a comprehensive compilation of computational geometry algorithms, but the authors' selection of algorithms among those developed in the field of theoretical computational geometry, which are also useful (or might be useful with further investigations) to VLSI layout verification and layout design. For example, the problems concerned with geometrical patterns defined by rectilinear line segments are emphasized here. Also we only describe algorithms acceptable to deal with the large-scale problems encountered in VLSI design. Namely those running in linear or almost linear time ($O(n)$, $O(n \log n)$, $O(n \log^2 n)$, etc.) with linear space are primarily discussed. Furthermore, algorithms based on the plane sweep method are stressed because they are convertible to sequential-file-based algorithm for VLSI layout verification.

In Section 2, basic data structures and algorithmic techniques in computational geometry are reviewed. In Section 3, efficient algorithms for intersection problems amongst points, line segments, rectangles, etc. are described. They are directly applied to the basic Boolean operations (AND, OR, etc.) and topological operations (CONTAINED, NOT CONTAINED and MEET) for checking geometrical or logical design errors (Chapter 7). Section 4 is devoted to geometrical search or transformation problems defined in a term of a number, representing length, width, spacing, etc. The results there are used for the minimum width/spacing design rule check, and as a preprocessing for gridless routing described in Section 6. In Section 5, decomposition and covering of polygonal regions by rectangles or trapezoids are discussed. These problems arise in various stages of layout design, including verification and artwork generation. In Section 6, recent results for "gridless" routing are discussed, which are aimed at accommodating complicated design rules as in the near future multi-metal-layer technologies, in which the existing grid-based routing algorithms will be useless.

2. Basic Data Structures and Algorithmic Techniques

Several techniques have been developed in order to solve geometric problems in the plane (i.e., 2-dimensional geometric problems) efficiently. Projection is one of the most popular techniques. It reduces a 2-dimensional geometric problem to a sequence of 1-dimensional geometric problems so that current computers can solve the problem efficiently. There, the following three types of 1-dimensional geometric problems are the most fundamental:

(A) for a set D of numbers and a query interval I, enumerate all numbers of D contained in the interval I;

(B) for a set D of intervals and a query number a, enumerate all intervals of D containing the number a; and

(C) for a set D of intervals and a query interval I, enumerate all intervals of D intersecting the interval I.

Set D may be updated by insertions and deletions. Based on efficient data structures manipulating these 1-dimensional geometric problems, many efficient 2-dimensional algorithms have been proposed.

In this section we first describe some of such data structures. Then we give two basic algorithmic techniques for 2-dimensional problems: one is the plane sweep method and the other is the bucketing method. The plane sweep method reduces a 2-dimensional geometric problem to a sequence of 1-dimensional ones. The bucketing method reduces a 2-dimensional geometric problem to one of quite small size by localizing the problem. Algorithms with fast expected running time can be obtained by this method.

2.1 Balanced tree

We consider the problem (A): for a set D of numbers and a query interval I, enumerate all numbers of D contained in the interval I. Set D is updated by insertions and deletions. Each item of D has a key (value). Then the following operations are fundamental for the problem.

ACCESS(a,D): return the item of D having key a;

INSERT(a,D): replace D with D $\cup$ {a}

DELETE(a,D): replace D with D - {a}

MIN(D)   : return the smallest key of D.

Set D can be represented by a list. But, then the above operations require $O(|D|)$ time in the worst case. If D is represented by a binary search tree, on the other hand, then each of the above operations can be done in time proportional to the

height of the tree.   Here, a <u>binary search tree</u> is a binary tree with key values
of its nodes arranged in a <u>symmetric order</u>,   i.e., for each node with key value x,
every node in its left (right) subtree has a key less (greater) than x.

The symmetric ordering of keys leads to an efficient way of searching the
tree.   What is needed in ACCESS(a,D) is to find the item with key a,   and when
a $\notin$ D,   the same procedure can be used to locate the position for the item being
inserted for INSERT(a,D) operation.   In DELETE(a,D) operation, the deleted node is
replaced by the rightmost node of its left subtree,   which is the node to be
searched.   In MIN(D),   all we need is to search the leftmost node.   In all the
operations,   the desired node can be found by means of such a top-down search that
only single node is traversed at each level of the tree.   Thus all the operations
can be done in time proportional to the height of the tree.

The motivation of using a balanced tree is to bound the worst case search
time within O(log|D|) by balancing the heights of left and right subtrees at each
node.   Several balanced trees (AVL-trees [2], 2-3 trees [3], B-trees [4], etc.)
have been proposed, and all of them require only O(n) memory space.   Among them,
AVL-tree, the simplest one is described here.

<u>AVL-tree</u> is a binary search tree such that for every node the heights of its two
subtrees differ by at most 1,   where a missing subtree is regarded to have height
-1. Examples are given in Fig. 2.1.

Now it is clear that the top-down search for every operation above requires
only O(log n) time for an n node AVL-tree.   However, a <u>rebalancing</u> scheme is
involved in INSERT and DELETE operations in order to maintain the AVL-tree
structure.   An overhead for this purpose is to attach to each node a <u>balance</u>
<u>factor</u> as the height of its right subtree minus the height of its left subtree.

Let us now consider what may happen when a new node is inserted in an AVL-
tree.   By retreating along the search path for the INSERT operation and by
checking the balance factor at each ancestor, we find the root x of the subtree
with the balance criterion violated.   Assume that the new node is inserted in the

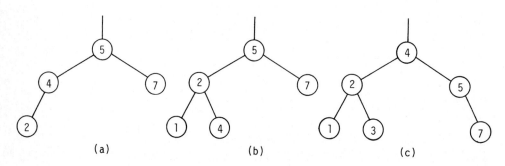

(a)   (b)   (c)

Fig. 2.1 AVL-trees.

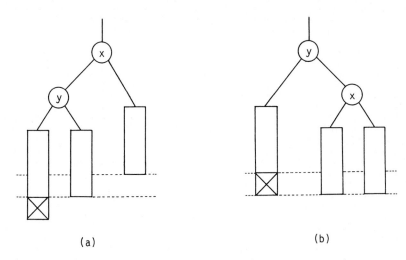

(a)   (b)

Fig. 2.2   Rebalancing for Case 1 (single rotation).

left subtree of x,   and let y be its root.   There are only two essentially
different cases needing individual treatment.

   In Case 1, the left subtree of y includes the inserted node as shown in Fig.
2.2(a), in which rectangular boxes denote subtrees and the height added by the
insertion is indicated by crosses.   A simple transformation,   called <u>single
rotation</u>, restores the desired balance as shown in Fig.   2.2(b).   Note that the
only movements allowed are those occurring in the vertical direction.   For
example, Case 1 is characterized by inserting a key 1 in the tree of Fig. 2.1(a),
and the clockwise rotation leads to the tree of Fig. 2.1(b).

   In Case 2, the right subtree of y includes the inserted node as shown in Fig.
2.3(a).   Let z be its root, then the subtrees of z must have different heights.
In this case a more complicated transformation,   called <u>double rotation</u> restores
the desired balance as shown in Fig.   2.3(b).   To be more precise,   the counter-
clockwise single rotation with respect to the subtree rooted by y leads to Case 1,
and then the clockwise single rotation leads to the tree of Fig.   2.3(b).   For
example, if a key 3 is inserted in the tree of Fig. 2.1(b), the double rotation is
needed.   The outcome is the tree (c).

   The rebalancing operations necessary are entirely expressed as sequences of
pointer re-assignments, resulting in either a single or a double rotation of the
two or three nodes involved.   In addition to pointer rotation, the respective node
balance factors also have to be adjusted.   Evidently,   rebalancing caused by the
insertion of an item can be performed with --- in the worst case --- O(log n)
operations.

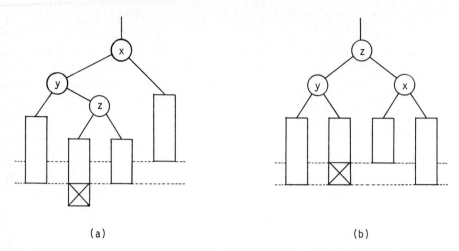

(a)   (b)

Fig. 2.3 Rebalancing for Case 2 (double rotation).

Rebalancing caused by the deletion of an item can also be performed with $O(\log n)$ operations.   An essential difference is that, whereas insertion of a single key results in at most one (single or double) rotation, deletion may require a rotation at every node along the search path.   For more details, see, e.g., [2].

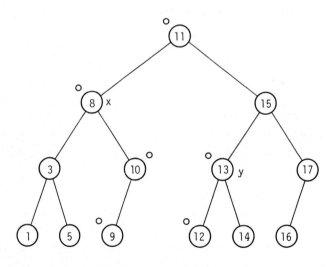

Fig. 2.4 Search for the problem (A). Nodes marked with "o" are enumerated for a query interval $I=[6,13]$.

By an AVL-tree representing D, the problem (A) can be solved as follows: for a query interval I=[a,b], searching from the root to leaves, find nodes x and y such that all nodes in the left subtree of x (y, resp.) have keys less than a (b, resp.) and all nodes in the right subtree of x (y, resp.) have keys greater than a (b, resp.), and then enumerate keys of all nodes between x and y (see Fig. 2.4). Thus, all numbers of D contained in the interval I can be enumerated in O(log n + k) time, where k is the number of enumerated numbers.

2.2 Segment tree

A segment tree devised by Bentley [5] is a data structure suited for manipulating the problem (B): for a set D of intervals and a query number a, enumerate all intervals of D containing the number a. Set D is updated by insertions and deletions. Segment tree has been widely used in efficient algorithms for the orthogonal object problems such as reporting intersecting pairs among a set of rectilinear rectangles.

The segment tree is based on the idea of representing a set of intervals on the line by a balanced binary tree, called an <u>interval tree</u>. The underlying assumption here is that the endpoints of all the segments to be processed are known beforehand. Without loss of generality, we can assume that these endpoints occupy the integer coordinates 0,1,2,...,m, where m ≤ 2n and n denotes the number of intervals. Because the essential property of the endpoints is their relative ordering and not their absolute values.

For an integer interval [a,b], an <u>interval tree</u> I(a,b) is defined recursively as follows (Fig. 2.5).

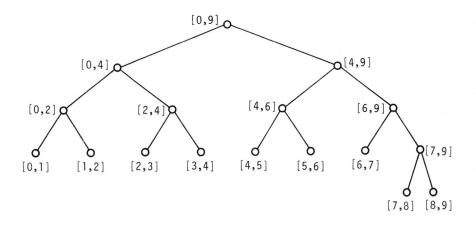

Fig. 2.5 Interval tree I(0,9).

(1) The root is associated with interval [a,b],

(2) the left subtree is an interval tree $I(a, \lfloor\frac{a+b}{2}\rfloor)$ and the right subtree is an interval tree $I(\lfloor\frac{a+b}{2}\rfloor,b)$ if $2 \le b-a$, and

(3) the left and right subtrees are empty if $b-a \le 1$.

With a set of intervals $I_1,I_2,\ldots,I_n$, we can associate the interval tree $I(0,m)$; $m \le 2n$ such that the leaves represents the m consecutive unit intervals $[0,1],[1,2],\ldots,[m-1,m]$. For each interval $A \in \{I_1,I_2,\ldots,I_n\}$, we mark with A every node in the tree whose interval is totally covered by A, on the condition that its father is not also covered by A. This marking is called the _canonical covering_ of A. Fig. 2.6 illustrates an example of the canonical coverings of intervals, based on the interval tree of Fig. 2.5.

A _segment tree_ $S(0,m)$ for a set of intervals $I_1,I_2,\ldots,I_n$, is the interval tree $I(0,m)$ with each node a list of intervals representing the canonical coverings attached. To establish the $O(\log n)$ running time for inserting an interval (finding its canonical covering), it is sufficient to observe that the

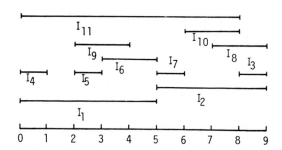

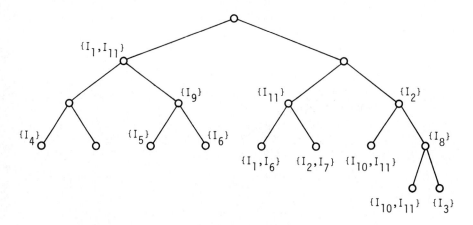

Fig. 2.6 Segment tree S(0,9) with canonical coverings.

top-down search visits at most four nodes and marks at most two nodes on each level of the tree.   Note that the depth of the tree is of $O(\log n)$, since $m \leq 2n$. The problem of deleting an interval from a segment tree is more subtle than insertion or query.   But it still can be done in $O(\log n)$ time with slight modification of the data structure [5].

We will now see how the problem (B) is solved by means of a segment tree. Let x be an internal node which is associated with an interval containing the query integer a.   Then integer a is also contained in one of the intervals with which two sons of x are associated.   Thus we can find a path from the root to a leaf such that every node on the path is associated with an interval containing integer a.   Then we only have to enumerate intervals of D stored with these nodes on the path. Since depth of the path is $O(\log n)$ and each interval of D is stored with at most one node on the path, we can enumerate the intervals of D containing integer a in $O(\log n + k)$ time, where k is the number of intervals containing integer a.

Segment trees, with some sophisticated auxiliary data structures [6], [7], achieve better asymptotic time complexity than any other existing data structures for solving some dynamic intersection search problems (see Sec. 4).   The disadvantage, on the other hand, lies in its space complexity of $O(n \log n)$ and in nontrivial overhead.

2.3 Priority search tree

A priority search tree proposed by McCreight [8] is a data structure suited for manipulating the problem (C):   for a set D of intervals and a query interval I, enumerate all intervals of D intersecting the interval I. Set D is updated by insertions and deletions.   It is one of the most powerful data structures in computational geometry and originally a data structure for representing a dynamic set D of points (x,y) in the plane such that the following operations can be done efficiently:

(1) delete a point from D;

(2) insert a point into D; and

(3) for arbitrarily given x_1, x_2, and y, enumerate points of D contained in the rectangle $[x_1,x_2] \times [0,y]$.

A priority search tree for a set D of n points is recursively constructed as follows (see Fig. 2.7).

(1) The root is associated with the point (x_0,y_0) with minimum y value.   Divide the half plane above the horizontal line $y=y_0$ into two subregions by an upward half line $x=x'$, where x' is determined in such a way that the numbers of points in the right and left subregions differ at most by one.   The value of x',   called

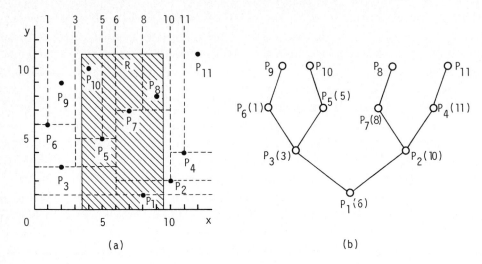

(a)   (b)

Fig. 2.7 Set D of points (a) and the priority search tree (b).

discriminator is attached at the root together with the coordinates (x_0, y_0).   The discriminators are indicated in the parenthesis at each node of tree of Fig. 2.7(b).   Note that, if there is no point above the horizontal line $y=y_0$, the discriminator is void.

(2) The right (left) subtree represents the right (left) subregion above the horizontal line $y=y_0$.

The key properties of a priority search tree are:

(1) it is a Heap [3] with respect to the y values of the points of D, and

(2) it is a binary search tree (AVL-tree) with respect to the discriminators.

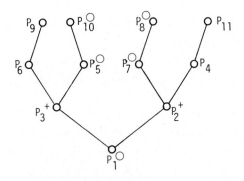

Fig. 2.8 Search for query rectangle $R = \{(x.y) \mid 3.5 < x < 9.5 \text{ and } y < 11\}$.
o : enumerated, + : checked but not enumerated.

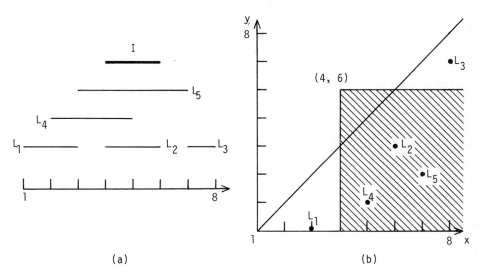

Fig. 2.9 Set of intervals with a query interval (a) and set of points with a
query rectangle (b).

Evidently, an n node priority search tree is of depth O(log n) and requires O(n)
memory, and hence update operations of insertion and deletion can be done in O(log
n) time.

By a priority search tree, we can efficiently enumerate all points of D
inside a query rectangle R with the lower horizontal segment on the x-axis.   We
enumerate points inside R by depth-first-search from the root and backtrack when
the region corresponding to a node of the tree has no point of D in R.   Fig. 2.8
shows how the points in the rectangle (indicated by shaded region) of Fig. 2.7(a)
is searched. By this search, we can enumerate all points inside R in O(log n + k)
time, where k is the number of points inside R.

As an application of a priority search tree, we consider the problem (C):
for a dynamic set D of intervals and a query interval I, report all the intervals
of D intersecting I.   This problem can be transformed into the above-mentioned
query problem by means of the following abstraction.   With each interval [a,b] in
D, we associate a point (b,a) in the plane.   Then the interval [a,b] intersects
I=[x,y] if and only if point (b,a) is inside the rectangle [x,∞)X[0,y] (Fig. 2.9).
Thus we can apply a priority search tree and solve the problem (C) in O(log n + k)
search time and in O(n) space.

2.4 Plane sweep

The plane sweep method reduces a 2-dimensional geometric problem to a sequence of 1-dimensional ones.   Based on this method many efficient algorithms for 2-dimensional geometric problems have been proposed.   This method is similar to the work-list method in layout verification and known as scan-line method in computer graphics.   In the work-list method, the plane sweep is exploited in such a way that only geometric objects intersecting the current scan line are fetched from a sequential file which accommodates whole data.   In theoretical computational geometry, the plane sweep is used to speed up algorithms even if whole data is accommodated in main memory.   It can be outlined as follows. Given a set of objects in the plane, consider a vertical scan line which moves from left to right.   Only objects intersecting the scan line are examined and something to

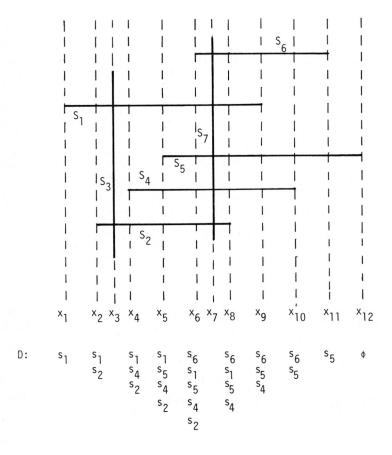

Fig. 2.10 Plane sweep method: reporting all the intersecting pairs of line segments.

be reported will be found. Objects lying to the left of the scan line have finished their roles and have been swept up. Objects lying to the right of it will play their roles in the future.

Now we shall illustrate the method by giving a specific example. Consider the problem of reporting all intersecting pairs among a given set of rectilinear (horizontal and vertical) line segments. Let $S=\{s_1,s_2,\ldots,s_n\}$ be the given set of line segments. For simplicity, we assume that no two horizontal (vertical) segments intersect. Let $x_1,x_2,\ldots,x_m$ ($m \leq 2n$) be the x-coordinates of endpoints of segments in increasing order (Fig. 2.10). In the method described below, vertical line $x=x_i$ is a current scan line and D denotes the set of segments of S intersecting the scan line.

0. D:= "empty"; and i:=1;
1. if i=m then stop;
2. let s_j be the segment such that x_i is the x-coordinate of its one endpoint (x_i,y_j);
3. if s_j is vertical then go to 5;
4. if (x_i,y_j) is the left endpoint of s_j then insert s_j into D, else delete s_j from D; i:=i+1; and go to 1;
5. report all segments of D intersecting s_j; i:=i+1; and go to 1.

If the horizontal line segments of D are ordered by their y-coordinates, then Step 5 of the above plane sweep method is equivalent to the problem (A): for set D of numbers and a query interval s_j enumerate all numbers contained in s_j. Thus, if D is represented by a balanced search tree described in Section 2.1, then the above algorithm based on the plane sweep method reports all intersecting pairs in O(n log n + k) time, where k is the number of all intersecting pairs [9], [1].

2.5 Bucketing

A technique for using "bucket" has recently been recognized to be one of the most powerful techniques to improve the expected efficiency of geometric algorithms [10]. Bucketing method is outlined as follows. For a given set of objects (points and/or line segments) in the plane,

(1) divide an entire region containing the objects into rectangles of equal size called <u>buckets</u> (Fig. 2.11),

(2) for each bucket, find a local solution within the bucket, and

(3) obtain a global solution of the given set of objects by merging the solutions within buckets.

Unless the distribution of objects is freakish, the number of objects belonging to a bucket is usually bounded by a constant if the entire region is partitioned into a number of buckets proportional to the number of objects.

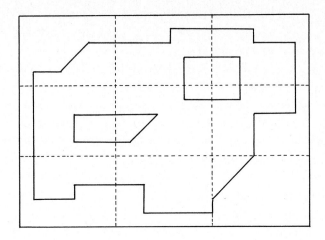

Fig. 2.11   A bucket partition.

Furthermore, in most geometric problems, the mutual influence between two objects
is likely to decrease with their distance, so that bucketing technique leads to
efficient algorithms for practical applications.

The bucketing method is similar to the slit method in layout verification if
the whole plane is divided by only vertical (or horizontal) lines; each band
region corresponds to a slit.   In typical problems in VLSI layout verification,
only the data in a few consecutive slits can be accommodated in main memory.
Therefore the efficiency of the bucketing method, when applied to VLSI layout
verification, depends on how the local solutions within buckets (slits) can be
merged based on a sequential slit-by-slit processing.

3.   Intersection Problems

The geometric intersection problems are most important in VLSI layout
verification and design.   The layout objects under consideration are polygons,
line segments, or points lying on the digitized plane. The intersection problems
are classified into two categories.   One of them, called <u>batched mode</u>, is to
report all the intersecting pairs among a given set D of objects.   A problem of
the other category, called <u>repetitive mode</u> or <u>on-line mode</u>, is to report all the
items or a particular item of D that intersect query objects. Generally speaking,
the batched mode problems are common in layout verification, but usually they are
treated as a sequence of on-line mode subproblems.   On the other hand, we often
encounter originally on-line mode problems in layout design.

The on-line mode problem is further divided into two classes: one is the static problem where the underlying set D is not changed; and the other is the dynamic problem where the underlying set D is updated by insertions and deletions. In the static problem, a given set D has to be preprocessed so that those segments intersecting the query segment can be enumerated efficiently. For this reason, algorithms are generally characterized in terms of the three attributes: (1) preprocessing time, (2) space and (3) search time. Similarly, in the dynamic problem, algorithms are generally characterized in terms of the three attributes: (1) update time, (2) space and (3) search time.

3.1 Line segment intersection

For the problem of reporting all intersecting pairs among a given set of n rectilinear line segments, we have already seen an $O(n \log n + k)$ time algorithm based on the plane sweep method. In this section we consider the general line segment intersection problem. First we present a method which determines with 5 multiplications whether given a pair of segments intersect or not. Next we consider the problem of reporting all intersecting pairs.

For a line segment L in the plane, let $z_1(L)=(x_1(L), y_1(L))$ and $z_2(L)=(x_2(L), y_2(L))$ denote two endpoints of L. For two given line segments L and M, let

$$\Delta = \begin{vmatrix} x_2(L)-x_1(L) & y_2(L)-y_1(L) \\ x_1(M)-x_2(M) & y_1(M)-y_2(M) \end{vmatrix} \qquad \Delta' = \begin{vmatrix} x_1(M)-x_1(L) & y_1(M)-y_1(L) \\ x_1(M)-x_2(M) & y_1(M)-y_2(M) \end{vmatrix}$$

$$\Delta'' = \begin{vmatrix} x_2(L)-x_1(L) & y_2(L)-y_1(L) \\ x_1(M)-x_1(L) & y_1(M)-y_1(L) \end{vmatrix}$$

Then L and M intersect if and only if $0 < \Delta'/\Delta < 1$ and $0 < \Delta''/\Delta < 1$, because, for L' and M' which are the (infinite) lines containing L and M respectively, the intersection point of L' and M' lies both inside L and inside M if and only if $0 < \Delta'/\Delta < 1$ and $0 < \Delta''/\Delta < 1$ (see Fig. 3.1). Thus, if we have computed Δ, Δ' and Δ'', we can determine whether L and M intersect or not. Actually, Δ, Δ' and Δ'' can be computed with only 5 multiplications as follows [11].

$a_1=x_2(L)-x_1(L),$ $a_2=x_1(M)-x_2(M),$ $a_3=x_1(M)-x_1(L),$

$b_1=y_2(L)-y_1(L),$ $b_2=y_1(M)-y_2(M),$ $b_3=y_1(M)-y_1(L),$

$c_1=a_1+a_3,$ $c_2=a_2-a_3,$

$d_1=b_1+b_3,$ $d_2=b_2-b_3,$ $d_3=b_1-d_2,$

$A_1=c_1 d_2,$ $A_2=c_2 d_1,$ $A_3=a_3 d_3,$

$A_4=a_1 b_2,$ $A_5=a_2 b_1,$

$\Delta=A_4-A_5,$ $\Delta'=-A_2-A_3+A_5,$ $\Delta''=-A_1-A_3+A_4$

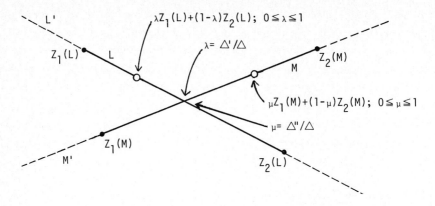

Fig. 3.1　Intersection of two line segments.

For a given set of n line segments, a naive algorithm can report all the intersecting pairs in $O(n^2)$ time, since all we have to do is to check the intersection for every pair of line segments. However, a more subtle algorithm based on the plane sweep with a pertinent data structure leads to the solution in almost linear time [9], [12]. The key consideration here is that two line segments intersect only when they are adjacent at some position of the scan line.

The algorithm here is similar to the one described in Section 2.4 for reporting all intersecting pairs of rectilinear line segments. But the following additional considerations are needed.

(1) Initially, the endpoints of the given line segments are sorted in increasing order of their x coordinates as in the algorithm in Section 2.4. In addition a priority queue is needed as a temporary storage for intersection points.

(2) While determining the next position of the scan line being moved, we must choose the leftmost point among both the endpoints and the intersection points lying to the right of the current scan line.

(3) Whenever intersection of a pair of line segments is reported, the intersection point is inserted in the priority queue.

(4) When the scan line is at the left endpoint of a line segment S, intersections between S and U and between S and L are checked, where U and L are the adjacent line segments above and below S, respectively, on the scan line.

(5) When the scan line is at the right endpoint of a line segment, intersection between its two adjacent line segments are checked.

(6) When the scan line is at an intersection point of two line segments S and T (assume that S is located above T to the left of the scan line), intersections between S and L and between T and U are checked, where U and L are as in (4).

Furthermore, the above-below relation between S and T must be interchanged before the next scan line is processed. The necessary operations for different cases are summarized in Fig. 3.2.

The above plane sweep method generates all the k intersecting pairs of n line segments in O((n + k)log n) time with O(n + k) space [9], if the line segments currently intersecting the scan line are managed by means of a balanced search tree, e.g., an AVL-tree. As an extreme case of this problem, existence of an intersecting pair can be judged in O(n log n) time, and it has been proved that the time complexity is optimal to within a constant factor [12].

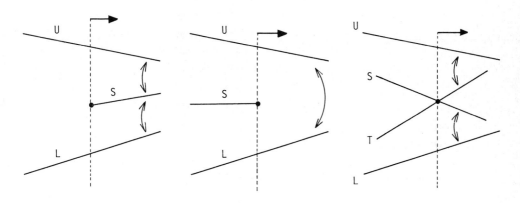

(a) at left endpoint,
* insert S
* check if S∩U
* check if S∩U
* generate intersections

(b) at right endpoint
* delete S
* check if U∩L
* generate an intersection

(c) at intersection
* check if U∩T
* check if L∩S
* generate intersections

Fig. 3.2 Plane sweep for the line segment intersection.

3.2 Rectangle intersection

In this section we will deal with another batched-mode intersection problem as follows.

Given n rectilinearly oriented rectangles in the plane, report all pairwise intersections.

Two rectangles are said to intersect either if
(1) their edges intersect, or if
(2) one entirely encloses the other.

The plane sweep based on an AVL-tree can report all pairwise intersections of rectangle edges (see the algorithm in Section 2.4). But the AVL-tree is not suitable to consider intersections of type (2). On the other hand, those intersections can be reported by means of a segment tree as follows. Initially we represent each rectangle by one of its interior point (e.g., its centroid, or one of its vertices), and sort those representative points by their x coordinates. If the segment tree is updated in such a way that it includes the projections onto the y-axis of those rectangles currently intersecting the scan line, then the solution of problem (B) in section 2, with respect to the scan line at each representative point of rectangles, reports the desired pairwise intersections [5].   But this method, on the contrary of that based on the AVL-tree, does not always report an intersection of type (1) (e.g., $R_2 \cap R_4$, $R_5 \cap R_6$, etc., of Fig. 3.3).

The plane sweep based on a priority search tree [8] favorably compares with those based on an AVL-tree or segment tree, since it can deal with both intersections of types (1) and (2) simultaneously.   All we have to do there is to

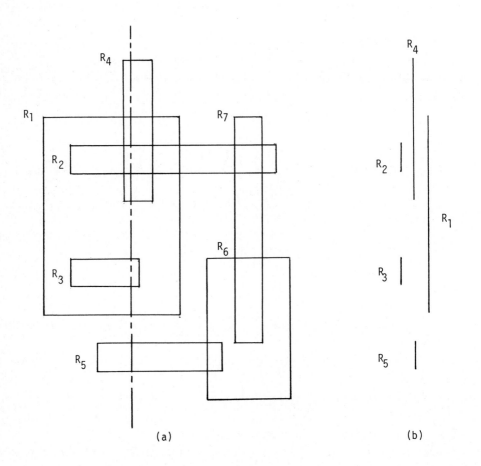

(a)   (b)

Fig. 3.3 Rectilinearly-oriented rectangles (a) and the projection onto y-axis of those intersecting the scan line (b).

sequentially solve problem (C) of Section 2 along with the scan line moved. Note that the set of (vertical) intervals are updated as in the algorithm based on segment tree, and that the query interval is given at the left (or right) edge of each rectangle (see Fig. 3.3). It is clear that the algorithm runs in O(n log n + k) time and in O(n + k) space, where k is the number of pairwise intersections of rectangles.

3.3 On-line line segment intersection

So far we have discussed batched mode intersection problems. In this section we will consider the following on-line intersection problems.

For a set D of vertical (horizontal) line segments and a horizontal (vertical, resp.) line segment query h,

(a) report all line segments of D intersecting h, or

(b) find the leftmost (lowermost, resp.) line segment intersecting h (see Fig. 3.4).

The above problems involve only rectilinear line segments, but are more subtle than the problem described in Section 2.4 as they are on-line problems. These problems plays an essential role in the "gridless" routing algorithms, which will be discussed in Section 3.6.

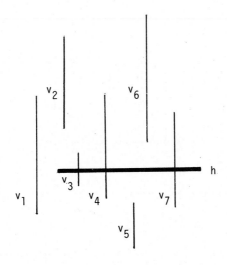

Fig. 3.4 On-line line segment intersection.
        h   : query line segment
$\{v_3, v_4, v_7\}$: solution of problem (a)
        v_3 : solution of problem (b)

Various algorithms for the above problem (a) have been proposed [7], [8], [13], [14], [15]. Note that algorithms for problem (a) usually lead to those for problem (b) by obvious modifications.   Among them, the best available algorithm for the static case is the one based on the technique called "filtering search" proposed by Chazelle [13].   This algorithm requires $O(\log n + k)$ search time, $O(n \log n)$ preprocessing time and $O(n)$ space, where k is the number of intersecting line segments.

For the dynamic problem in which distinct coordinate values of line segment endpoints are known in advance, McCreight [8] proposed an algorithm with $O(\log^2 n + k)$ search time, $O(\log n)$ update time and $O(n)$ space. This is considered to be the best available among those requiring only linear memory space.   We shall only describe here the McCreight's algorithm based on the priority search tree, since it seems to be most useful in VLSI layout.

The basic idea for exploiting priority search tree is that, if the set D of line segments are replaced by that of upward vertical half lines, then the problem can be handled by the algorithm described in Section 2.3.   Let two endpoints of the query segment h be $(a_1,$ b) and $(a_2,$ b); $a_1 < a_2$ and R be a rectangle defined by $a_1 \le x \le a_2$ and $y \le b$.   Then the half lines intersect h if and only if their endpoints lie inside R (see Fig. 3.5).   Therefore this problem can be solved in $O(\log n + k)$ time by means of a priority search tree (see Section 2.3).

To deal with the original line segments, we divide the problem into $O(\log n)$ subproblems concerning half lines by means of a number of horizontal slice lines

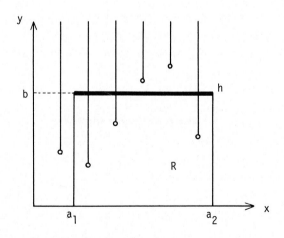

Fig. 3.5 Set of half lines and a line segment query.

as follows.   We bisects the plane by a horizontal slice line in such a way that the numbers of line segment endpoints above and below the bisector (slice line) are balanced.   Then we associate the root of a tree (external tree) with the line segments intersecting the bisector, and the right (left) subtree with those entirely lying above (below, resp.) the bisector.   The line segments cut by the bisector are represented by two priority search trees (internal trees), one representing the pieces of the cut segments below the bisector, and the other representing those above.   In each of these priority search trees, the line segments can be viewed as half lines, because they extend to the limit of the reduced y-coordinate.   We recursively continue the bisection until every line segment is cut by a bisector.   Note that the whole line segments are managed by a two-level tree;   the external one is a binary search tree with respect to y-coordinates of the bisectors and the internal ones are priority search trees representing the fictitious half lines (see Fig. 3.6).

For a horizontal segment query, there are $O(\log n)$ priority search trees which we have to look into to report line segments intersecting it.   Each such a tree corresponds to a recursive level of bisections.   Any line segment in the other priority search trees has no intersection with the query segments.   Thus, the solution of the problem can be obtained from the solutions of $O(\log n)$ subproblems of reporting half lines intersecting the horizontal segment query. Therefore, we can report all vertical line segments intersecting the horizontal segment query in $O(\log^2 n + k)$ time.   Furthermore, for a segment to be inserted into (or deleted from) D, the priority search tree to be inserted into (or deleted from) can be found in $O(\log n)$ time and it is updated in $O(\log n)$ time.

A modification of the above algorithm leads to an algorithm for problem (b), where $O(\log^2 n + k)$ search time is of course reduced to $O(\log^2 n)$.   In this case, the priority search tree is used to find the point with minimum x-coordinate among those inside the rectangle in Fig. 3.5.

For the general dynamic problem, where coordinate values of endpoints of segments are not known in advance, Edelsbrunner has presented an algorithm with $O((\log n)^2 + k)$ search time, $O((\log n)^2)$ update time and $O(n \log n)$ space [14].

For the restricted version of dynamic problems where line segments are updated by only insertions or only deletions, Imai and Asano [7] presented an algorithm which executes an intermixed sequence of $O(n)$ queries and updates in $O(n \log n + k)$ time, but it, on the other hand, requires $O(n \log n)$ space.   Also this algorithm involves a set of sophisticated data structures [15], [6].

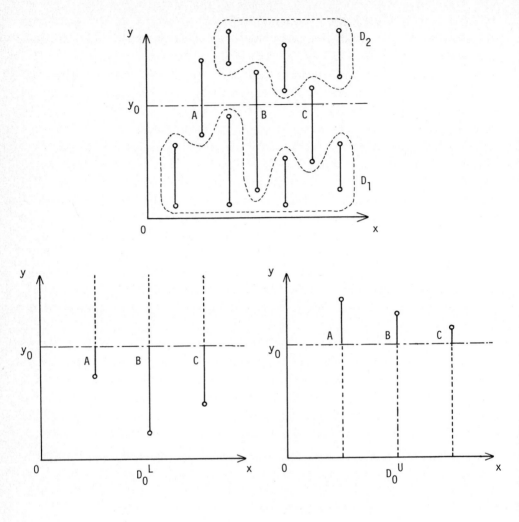

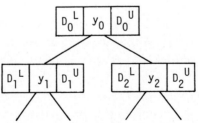

Fig. 3.6 From line segments to half lines.

3.4 Range search and point enclosure

In theoretical computational geometry, there are two basic on-line intersection problems between points and polygonal regions.

Range Search: for a given set of n points and a query polygon R, enumerate all the points lying inside R (Fig. 3.7).

Point Enclosure: for a given set of n polygons and a query point P, enumerate all the polygons including P (Fig. 3.8).

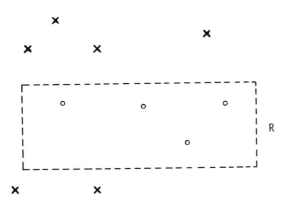

Fig. 3.7 Range search problem. For a query R, points marked o are enumerated.

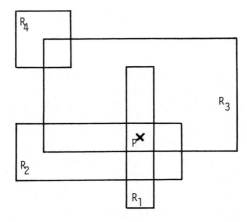

Fig. 3.8 Point-enclosure problem: For a query P, rectangles R_1, R_2 and R_3 are enumerated.

In VLSI mask pattern analysis, polygons are usually decomposed into rectilinearly-oriented rectangles. If polygons have (a small number of) non-rectilinear edges, the polygons include some trapezoids with two horizontal edges, and algorithms concerning rectangles can be used to deal with such trapezoids with an obvious modification. Therefore, we restrict our arguments here to intersection problems concerning only rectilinearly-oriented rectangles (we henceforth simply call them rectangles). If the polygons are restricted to such rectangles, the range search and the point enclosure problems can be regarded as the 2-dimensional versions of problems (A) and (B) described in Section 2, respectively. Note that the intervals in these problems are replaced by rectangles. Now the 2-dimensional version of problem (C) is the following.

On-line Rectangle Intersection: for a given set of n rectangles and a query rectangle R, enumerate all the rectangles intersecting R.

There are three cases for rectangles intersecting R:

(1) rectangles entirely included in R;
(2) rectangles entirely including R; and
(3) rectangles whose edges intersect those of R.

Thus on-line rectangle intersection problem can be divided into three subproblems: range search, point enclosure and the on-line line segment intersection.

We have already seen an algorithm with $O(\log^2 n + k)$ search time, $O(\log n)$ update time and $O(n)$ space for the dynamic line segment intersection problem. For the range search problem, an algorithm with $O(\log n + k)$ search time, $O(\log^2 n)$ update time and $O(n \log n)$ space is available. For the point enclosure problem, an algorithm with $O(\log^2 n + k)$ search time, $O(\log^2 n)$ update time and $O(n \log n)$ space is available. Both of them are based on priority search trees. Combining these three algorithms, we can obtain an algorithm with $O(\log^2 n + k)$ search time, $O(\log^2 n)$ update time and $O(n \log n)$ space for the dynamic rectangle intersection problem [8]. If the three subproblems are static, their search time can be reduced to $O(\log n + k)$ with $O(n \log n)$ preprocessing time and $O(n)$ memory, except that fast range search algorithms still require $O(n \log n)$ space. Therefore the static rectangle intersection problem can be solved with $O(\log n + k)$ search time and $O(n \log n)$ preprocessing time and space.

In view of the computational complexity of these algorithms, algorithms for on-line 2-dimensional problems are more involved than dynamic 1-dimensional problems or batched-mode 2-dimensional problems. We should recall here that the original design rule checking problems for VLSI's are batched-mode 2-dimensional problems, and a good strategy to deal with such a problem is to reduce it to a sequence of simple dynamic 1-dimensional problems.

For example, consider the following batched-mode range search problem.

For a given rectilinear polygonal region R with n vertices and a set D of m points, enumerate all the points inside R (see Fig. 3.9).
If this problem is solved as a sequence of O(n) on-line rectangle range search problems by decomposing R into O(n) rectangles, it would require O(m log m + n) space.   On the other hand, it can be treated as a sequence of the following dynamic 1-dimensional problems, saving memory space without increasing time complexity.

For a set D of n numbers and a query number a, find the least (greatest) number of D that are greater (less, resp.) than a.
This subproblem can be solved in O(log n) time by means of an AVL-tree with an obvious modification of the procedures for MEMBER operation.   Provided that a given m points are sorted in advance, which can be done in O(m log m) time, the plane sweep method leads to the solution of the original 2-dimensional problem in O((m + n)log(m + n)) time with O(m + n) space [16]. Note that, while the vertical scan line moving from left to right the y-coordinate is inserted (deleted) at the left (right resp.) endpoint of each horizontal edge of R and the above 1-dimensional subproblem is solved at each point of D.

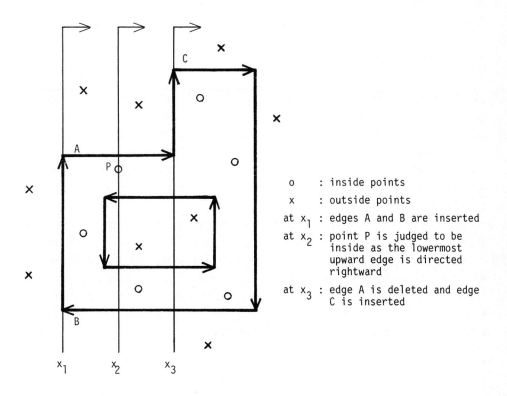

o   : inside points

x   : outside points

at x_1 : edges A and B are inserted

at x_2 : point P is judged to be inside as the lowermost upward edge is directed rightward

at x_3 : edge A is deleted and edge C is inserted

Fig. 3.9 Off-line range search problem.

So far we have been concerned with algorithms with excellent worst-case behavior.   If the positions of given points and the lengths of polygon edges are uniformly distributed, the bucketing method leads to algorithms with almost linear expected running time [17].

3.5 Applications

VLSI mask patterns consist of a large number of polygonal regions with primarily rectilinearly oriented edges and rarely a small number of non-rectilinear edges.   In the design rule check of such mask patterns, the polygonal regions are conventionally represented by a union of elemental figures, usually rectangles.   Thus algorithm, dealing with rectangles plays an essential role there.   Among basic pattern operations (see Chapter 7) for design rule check, Boolean operations and topological operations can be viewed as direct applications of the batched-mode rectangle intersection problem described in Section 3.2. Because the knowledge of all the pairwise rectangle intersections implies the complete information needed for any Boolean operations.   For example, the contour of n rectangles, equivalent to the result of OR operations, can be generated in O(n log n + k) time and O(n + k) space, where k is the number of vertices on the contour.   Furthermore, if the algorithm in Section 3.2 is slightly modified in such a way that the two types of pairwise intersections (edges intersecting and one containing another) are enumerated separately, then it leads to the results of the topological operation, such as CONTAINED and MEET (see Chapter 7).

In the initial stage of logical design rule check, we have to generate polygonal patterns realizing wiring nets in order to restore logic circuits.   In this case, the union (OR operation) of the rectangles constituting the wiring nets again, leads to the desired wiring patterns. Also the problem of checking whether the polygonal region for each net includes contacts correctly can be solved by algorithms for range search or point enclosure problems.   Note that a contact can be abbreviated by a point.

The algorithms described in this section can be applied not only to layout verification but also to gridless routing, which will be discussed in Section 6. As a subproduct of the rectangle intersection algorithms, the connected components of n rectangles can be obtained in O(n log n) time and O(n) space [18], [19]. The result may be used to analyze connectivity of pin-pairs of wiring nets. Moreover, the on-line line segment intersection problem plays a very important role in gridless routing algorithm as will be described in Section 6.

4. Dimension Checking Problems

This section deals with geometrical transformation or search problems specified by a positive number, representing width, spacing, length, etc. We call such geometrical problems <u>dimension checking problems</u>. A typical dimension checking problem is to calculate area and/or perimeter of a polygonal region, which is required to extract circuit parameters for electrical performance verification of layout patterns (see Chapter 7). This simple problem is not discussed here, since the area and perimeter of a polygonal region can be easily calculated by scanning the vertex sequences along its boundaries.

The first problem treated in this section, called <u>resizing problem</u>, is to expand or shrink given polygonal regions by a specified width. The other problems are searches for measuring minimum width and minimum space of a polygonal region, and for reporting portions of polygonal regions which violate prescribed minimum width/space design rules.

4.1 Resizing polygonal patterns

A mask pattern can be viewed as a set of polygonal regions on the digitized plane. The problem considered in this subsection, called <u>resizing</u>, is to either <u>expand</u> or <u>shrink</u> such polygonal regions by a specified resizing width. The polygonal regions are not limited to convex ones and may contain inner boundary loops, called "windows". Therefore the resizing operation does not always preserve the original pattern as shown in Fig. 4.1. The expanding operation may remove windows and notches, and unite multiple regions into one piece. Similarly, the shrinking operation may remove small pieces of regions and protrusions, and separate a region into several pieces.

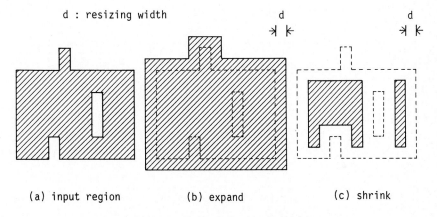

d : resizing width

(a) input region   (b) expand   (c) shrink

Fig. 4.1 Expansion and shrinking.

4.1.1 Basic properties of resizing

For a given set R of polygonal regions and a fixed width d, the strict meaning of <u>expansion</u> (<u>shrinking</u>) is to transform R into a new set R' in such a way that every external (internal) point of R within distance d from the boundary of R becomes inside (outside) of R'.   Then the boundary of R' in general includes circular arcs as shown in Fig. 4.2.   However, it is convenient in VLSI applications to modify the definition so that the resultant pattern is also polygonal as shown in Fig. 4.1. The resized pattern as in Fig. 4.1 can be derived from the result of strict resizing by removing each circular arc and by extending the corresponding pair of adjoining straight line segments until they intersect.

The modified definition of the resizing operation implies the following properties, which will be exploited in the algorithm to be described.

<u>Property 1</u>. Let R and R' be the original and the resized patterns, respectively.   Then R' can be obtained by (1) resizing R in the X direction only followed by (2) resizing the result in the Y direction only (see Fig. 4.3).

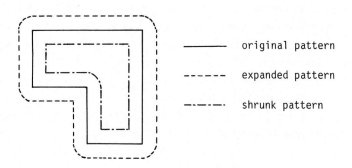

Fig. 4.2 Strict definition of expansion/shrinking.

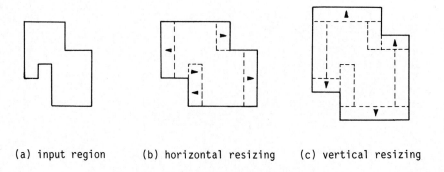

(a) input region   (b) horizontal resizing   (c) vertical resizing

Fig. 4.3 Horizontal and vertical resizing.

Property 2. For a set R of polygonal regions, let $\overline{R}$ be its complement (the set of regions obtained by interchanging the inside and outside of R). Then the region derived by expanding (shrinking) R by a fixed width d is equal to the complement of the region derived by shrinking (expanding) $\overline{R}$ by the same width d.

Property 3. Let n and n' be the numbers of vertices of the original and the resized patterns, respectively. Then n'≦n.

4.1.2 The XY-method: an O(n log n) resizing algorithm

For the resizing problem, one might think of a naive method of (1) moving the vertices according to a given resizing width, (2) scanning the vertex sequence along the boundaries, and (3) removing unnecessary portions of the boundaries caused by intersection of edges, as shown in Fig. 4.4. However, for a pattern with n vertices, this method may have to process $O(n^2)$ intersections of edges in the worst case and therefore the running time can not be better than $O(n^2)$.

An efficient algorithm for resizing a pattern with a set of polygonal regions is described here. For simplicity of descriptions, the given polygonal regions are restricted to be those rectilinear. Furthermore it is assumed that the input data have been pre-processed to ensure no crossing of edges so that none of the different connected regions overlaps to each other.

The resizing algorithm described below is called the XY-method, because it is a two-stage algorithm performing resizing operations in the X and Y directions separately. Property 1 guarantees that the desired pattern can be obtained in this way. We only describe here the procedure for resizing in the X direction, since the resizing in the Y direction can be done in the same way. Moreover, since by Property 2 shrinking is basically the same as expansion, expansion is only described here.

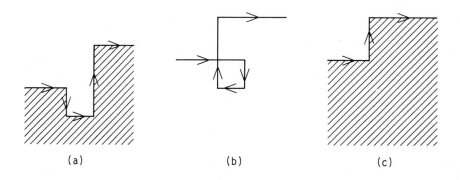

(a)   (b)   (c)

Fig. 4.4 A method to expand.

<u>An algorithm for expanding in the X direction.</u>

<u>Step 1</u>.   Sort the vertices according to increasing Y-coordinate order (when the Y coordinates are equal, sort according to increasing X-coordinate order). Here the plane, as shown in Fig. 4.5(a), is viewed as being divided into banded regions, or slits, by lines parallel to X-axis through these vertices.

<u>Step 2</u>.   For each banded region, determine whether the distance from a vertical side A in the outward direction to the neighboring boundary B is less than 2d or not.   If it is the case, the area between A and B is added to the inside of the region.

<u>Step 3</u>.   Expand by d in the X direction the regions resulted from Step 2.   In this step it is clear that there is no intersection of edges.

An example of the above procedure (Steps 1-3) is illustrated in Fig. 4.5.

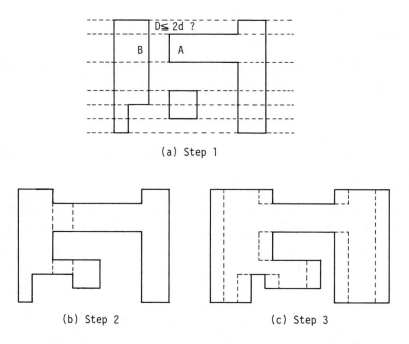

(a) Step 1

(b) Step 2   (c) Step 3

Fig. 4.5   Expansion of a rectilinear region in the X direction.

The above algorithm can solve the resizing problem in $O(n \log n)$ time with $O(n)$ space, where n is the number of vertices of the regions, by applying the plane sweep technique described in section 2.4.   Furthermore, this algorithm can be enhanced to resize polygonal regions that may have non-rectilinearly oriented edges, keeping the complexity unchanged [20].

A disadvantage of the XY-method lies in that it requires two plane sweeps (horizontally and vertically), which is not desirable when a large amount of data is accommodated in a sequential file. However, it is reasonable in practice to assume that the resizing width is small enough to limit the change of vertex coordinates within a few consecutive slits. Then the XY-method can be performed by means of a single plane sweep with a slight modification.

4.1.3 Applications of resizing

There are many applications of the resizing algorithm to VLSI pattern design, as follows.

(1) Mask pattern design verification

Mask patterns must be designed in such a way that minimum width and space rules are satisfied. However, due to partial changes in design or other human intervention, it often happens that in the final design result the mask artwork will violate such rules. It will be described in Section 4.2 how the resizing algorithm is exploited to check such design rule violations automatically.

(2) Notch/protrusion removal

Resizing algorithm can also be used to amend areas that violate design rules by removing notches automatically. For example, as in Fig. 4.6 notches are removed by expanding by some resizing width and then shrinking by the same width. Similarly, by first shrinking and then expanding, overly small protrusions can be removed. Note that notch removal must not be allowed to change the logical interconnection.

(3) Compensation for pattern generator characteristic errors

Pattern generators used for making mask patterns can be the cause of systematic errors in figure size, resulting in exposure with figures that are slightly larger or smaller than the input figure dimensions. In order to reduce these characteristic errors, resizing algorithm can be used to prepare slight compensating expansion or shrinking in the input data for the pattern generators.

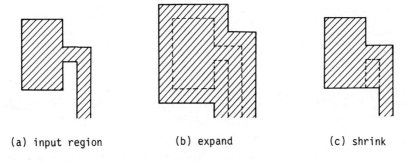

(a) input region   (b) expand   (c) shrink

Fig. 4.6 Notch removal.

(4) Proximity effect compensation

The use of a pattern generator for patterns with very high density can result in a proximity effect depending on the spacing between figures, relative figure sizes, etc. The result is that the figures actually laid on the mask are different from the designers' original intent, as in Fig. 4.7(a). One method of compensating for such proximity effect errors would be to locally modify the figure dimensions before exposure, as in Fig. 4.7(b). The previous algorithm, with pertinent modifications, can make this compensation possible.

(5) Extraction of a routable region

In Section 6, gridless routing algorithms will be proposed. These algorithms do not use a grid as in the maze method, but rather treat the problem as one of pattern processing in which the area to be used for interconnection wiring is taken as a figure itself. As a preliminary step for such methods, a way of extracting just the region that contains the pin pair for the wire being routed would be desirable. This should be done, as in Fig. 4.8, after considering the minimum interwire space and the minimum wire width, and then shrinking the regions.

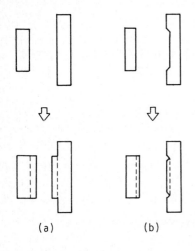

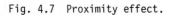

(a)   (b)

Fig. 4.7 Proximity effect.

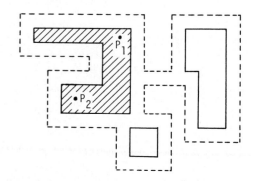

Fig. 4.8 Extracting a routable region.

4.2 Minimum width/space checking

One of the most important problems for geometrical verification of VLSI mask patterns is the minimum width/space checking. Since the minimum width of polygonal regions is equal to the minimum space of their complementary regions with respect to the plane, we only discuss here the minimum space problem. For simplicity of descriptions, we restrict our arguments to rectilinear regions.

4.2.1 Voronoi diagram

The geometrical structure, called Voronoi diagram [21], is a major concept in computational geometry.   Among mathematicians this structure is also referred to as Dirichlet tessellation or Thiessen tessellation.   Although this concept is not directly relevant to VLSI applications, it is introduced here as a preliminary process for the subsequent arguments.

For a set S of n points p_i (i=1,...,n) (called generators) on the plane, the Voronoi polygon of p_i is defined as

$$V(p_i) = \{p \mid d(p,p_i) \leq d(p,p_j) \text{ for all } j \neq i\},$$

where d denotes the Euclidean distance.   In other words, $V(p_i)$ is the locus of points which are closer to p_i than to any other point of S.   Then, the planar skeleton formed by the boundaries of $V(p_i)$ (i=1,...,n) is called the Voronoi diagram.   Fig. 4.9 illustrates a set of 16 points and its Voronoi diagram.

Many algorithms for constructing Voronoi diagram have been proposed in recent ten years.   There are two useful approaches.   One of them is to modify the diagram step by step as generators are added one by one.   Based on the idea, Rhynsburger [22] proposed an $O(n^2)$ algorithm.   Recently, Ohya et al. [23] improved this method to run in $O(n)$ expected time by means of the bucketing method described in Section 2.5.   The other approach is the divide-and-conquer technique, by means of which Shamos and Hoey [21] proposed an $O(n \log n)$ algorithm.   At the same time, Shamos [24] showed that $O(n \log n)$ is a worst-case lower bound, i.e., the above algorithm is optimal.

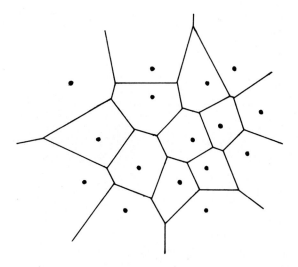

Fig. 4.9 Voronoi diagram.

In comparing the above two approaches, the second one based on the divide-
and-conquer should be more efficient when the given points are non-uniformly
distributed and when the whole data can be accommodated in primary memory. On the
other hand, when a large number of uniformly distributed points are given in
secondary memory, as in VLSI mask pattern verification, the first approach seems
to be more promising.

4.2.2 Measuring minimum space

When the polygonal regions under consideration are rectilinear, a pair of
either (1) convex vertices (Fig. 4.10) or (2) parallel edges (Fig. 4.11) possibly
constitutes the minimum space.   This property leads to an efficient algorithm for
evaluating the minimum space [25], which is outlined below.

Consider the Voronoi diagram with respect to the set of convex vertices of
the input pattern. Then, for each convex vertex p, its nearest convex vertex q is
adjacent to p [26].   Here two points p and q are said to be _adjacent_ if their
Voronoi polygons V(p) and V(q) have a common boundary edge. Therefore the point q
closest to p can be found by checking only those convex vertices which are facing
and adjacent to p.   For example, the desired convex vertex q nearest to p for the
input pattern of Fig. 4.12(a) is found as shown in Fig. 4.12(b).   Let d_1 be the
minimum value of such distances for all the convex vertices.

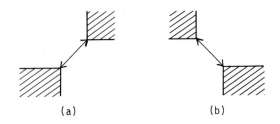

(a)   (b)

Fig. 4.10 Convex vertices facing to each other.

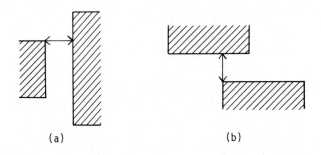

(a)   (b)

Fig. 4.11 Parallel edges facing to each other.

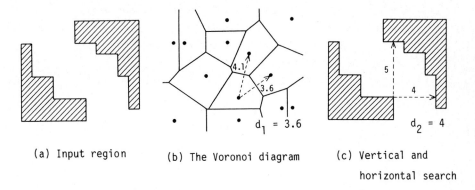

(a) Input region   (b) The Voronoi diagram   (c) Vertical and
                                                         horizontal search

Fig. 4.12 Minimum space algorithm. $d = \min\{d_1, d_2\} = 3.6$.

Next, the distance from every convex vertex to the horizontal and vertical edges facing to it (see Fig. 4.12(c)) can be obtained systematically by means of the plane sweep technique.   Note that this procedure can equivalently check all the facing parallel edge pairs which possibly provide the minimum space.   Let d_2 be the minimum value of such distances.   Now the smaller one of the values d_1 and d_2 is the desired minimum space.

The computational complexity of the algorithm is easily analyzed.   Let n be the number of vertices and n' ($\leq$n) be the number of convex vertices.   The Voronoi diagram is drawn in $O(n' \log n')$ time as mentioned in Section 4.2.1.   And, since there are only $O(n')$ adjacent vertex pairs to be checked in the Voronoi diagram, the value d_1 can be obtained in $O(n')$ time based on the diagram.   The other value d_2 can be obtained in $O(n \log n)$ time by means of the plane sweep.   Therefore the total running time of the algorithm is no more than $O(n \log n)$.

4.2.3 Reporting minimum space errors

We consider here the problem of reporting all the portions of a pattern which violate a given minimum space rule.   The XY-method described in Section 4.1.2 with a slight modification can be used to solve this problem as follows.

Step 1.   Expand the given regions by d in the X direction, where d is half the minimum spacing.

Step 2.   Expand the regions produced by Step 1 by d in the Y direction, where the strict meaning of expansion (recall 4.1.1) is applied.   In order to pertinently insert circular arcs, we have to locate the transition points from an arc or line segment to another.   Fig. 4.13 illustrates all the possible cases [25].

An example of performing the above algorithm is given in Fig. 4.14. If such expansion results in overlapped portions, they are reported as violating the given minimum space rule. It is obvious that the algorithm can also be performed in O(n log n) time as the one described in Section 4.1.2.

o : transition points

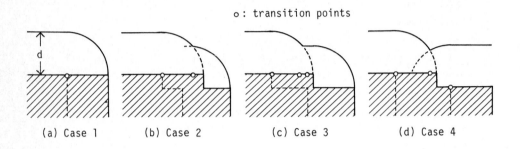

(a) Case 1   (b) Case 2   (c) Case 3   (d) Case 4

Fig. 4.13 Cases where transition points exist.

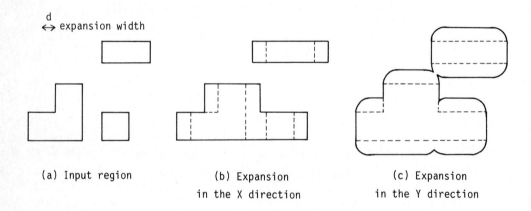

(a) Input region   (b) Expansion   (c) Expansion
                        in the X direction   in the Y direction

Fig. 4.14 Expansion for checking minimum space errors.

5. Partitioning and Covering Problems

The problem of partitioning geometric figure into or covering it by more fundamental ones arises in various applications. Triangles, convex polygons and rectangles have been chosen as the fundamental figures. For a simple polygon, i.e., a polygonal region without windows (holes), there have been several polynomial-time algorithms partitioning it into a minimum number of triangles, convex polygons, etc. [27], [28]. However, for a polygonal region with windows, in most case, the minimum partition problem has been shown to be NP-complete. As an exceptional case, a rectilinear region can be partitioned into a minimum number of rectangles in polynomial time [29]. On the other hand, even the simplest covering problem, i.e., covering a rectilinear region by rectangles, is known to be NP-complete.

Although various partitioning and covering problems have been investigated in the area of computational geometry, only those problems relevant to VLSI design are discussed here.

5.1 Partitioning a rectilinear region by rectangles

We consider here the problem of partitioning a rectilinear region into a minimum number of non-overlapping rectangles. This problem arises in automated VLSI mask generation. In case a pattern generator is used to expose masks, since its aperture is of rectangular shape, we need to partition a given pattern, which can be viewed as opaque polygonal regions, into a number of rectangles. As almost all edges of polygonal patterns in VLSI mask are horizontal or vertical, rectilinear polygonal regions are considered here as a basic case.

This partitioning problem has other applications to VLSI design. In the global routing (Chapter 5) for building block or gate array LSI's, the wire routing space is decomposed into regions of more manageable shape, which usually are rectangles. Similar decomposition is needed in analyzing complicated mask patterns for layout verification.

A rectilinear region R consists of an even number, say n, of vertices and edges. There are two kind of vertices; convex (90°) vertices and concave (270°) vertices. A chord of R connecting two cohorizontal or covertical, concave vertices is called a degenerate chord. R is said to be degenerate if a degenerate chord can be drawn, otherwise it is non-degenerate.

A simple partitioning is to draw an either horizontal or vertical cut-line at each concave vertex. Then, for k non-degenerate rectilinear regions with the total number of n vertices and w windows, the number P of resultant rectangles is given by

P = n/2 + w - k,

and no other partitioning leads to less number of rectangles [29], i.e., this partitioning is optimal. An optimal partitioning, with cut-lines restricted to those horizontal (or vertical) ones only as shown in Fig. 5.1, can be obtained by means of the plane sweep method, and the algorithm runs in O(n log n) time if the vertical (horizontal) edges currently intersecting the horizontal (vertical) scan line is managed by a balanced tree (see Section 2.1).

For partitioning degenerate rectilinear regions, some degenerate chords can be used as cut lines. Since the number of necessary rectangles is reduced by one if a degenerate chord is exploited as shown in Fig. 5.2, the strategy to obtain an optimal partitioning is to exploit as many independent degenerate chords as possible, where two chords (line segments) are said to be <u>independent</u> if they have no point in common.   Therefore the number of rectangles resulted from an optimal partitioning is given by

$$P = n/2 + w - k - d,$$

where d is the maximum number of independent degenerate chords [29].

An optimal choice of degenerate chords can be obtained by means of the <u>intersection graph</u>, which is derived by associating a node with each degenerate chord and by inserting a branch between each node pair if the corresponding pair of chords intersect.   For example, the set of degenerate chords indicated by dotted lines in Fig. 5.3(a) is represented by a graph shown in Fig. 5.3(b).   Then a maximal independent set of nodes leads to the desired choice of degenerate chords, which in this case is {A, C, D, E, H}.   Here a set of nodes is said to be <u>independent</u> if there is no branch connecting two nodes of the set.   The resultant optimal partitioning for this example is illustrated in Fig. 5.3(c).

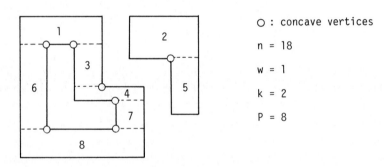

O : concave vertices

n = 18

w = 1

k = 2

P = 8

Fig. 5.1 Optimal partitioning of non-degenerate rectilinear regions with horizontal cut-lines only.

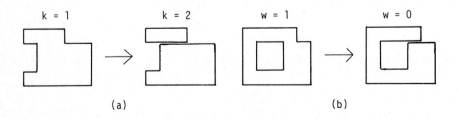

Fig. 5.2 Exploiting a degenerate chord.

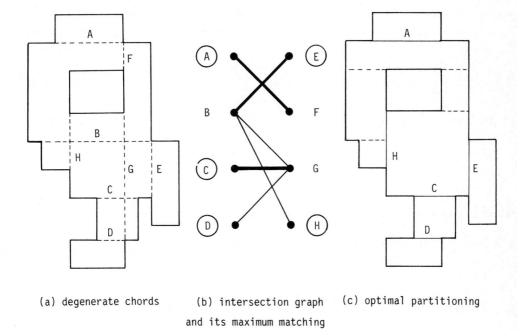

(a) degenerate chords   (b) intersection graph   (c) optimal partitioning

and its maximum matching

Fig. 5.3 Optimal partitioning of a degenerate rectilinear regions.

Since the intersection graph associated with the set of degenerate chords of rectilinear regions is bipartite and the maximum independent set problem is equivalent to the maximum matching problem, the optimal choice of degenerate chords can be found in $O(n^{2.5})$ time by means of the Hopcroft-Karp algorithm [30]. Recently, an $O(n^{1.5}\log n)$ time algorithm was proposed by exploiting the property that the bipartite graph is associated with a set of rectilinear line segments [31]. Once given rectilinear regions have been divided into several pieces by the degenerate chords as chosen in this way, then the remaining problem is to partition each piece (non-degenerate rectilinear region) into minimum number of rectangles, which can be solved by means of the simple algorithm as described earlier.

5.2 Covering a polygonal region by rectangles

When a pattern generator with a rectangular aperture is used, a non-rectilinear region must be represented by overlapping rectangles as shown in Fig. 5.4(a). Such partially overlapped exposure can be exploited even for a rectilinear region in order to represent it by less number of rectangles. For example, the patterns shown in Fig. 5.4(b) and (c) are represented by two overlapping rectangles, whereas three rectangles are needed if no overlapping is allowed.

The problem posed above can be mathematically stated as a covering problem: covering a polygonal region by a minimum number of rectangles. However, this problem, even if the given polygonal region is rectilinear, is known to be NP-hard [32], and polynomial-time algorithms are available for very restricted cases [33], [34]. Therefore some heuristic algorithms are used in practice so as to exploit cohorizontal and covertical edges as in Fig. 5.4(b) and bottle-neck parts as in Fig. 5.4(c).

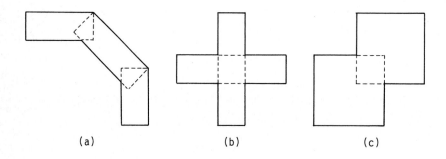

(a)   (b)   (c)

Fig. 5.4 Covering polygonal regions by rectangles.

5.3 Partitioning a polygonal region by trapezoids

Since last several years ago, electron beam (EB) exposure systems have been used to expose highly dense layout patterns, which may include diagonal edges, on a mask. In some of the EB exposure systems, a special-purpose hardware is implemented to expose at a time an arbitrary-sized trapezoid, which is represented by a bit-map. Therefore a trapezoidal partitioning is required to construct the input data for such exposure systems. Here, because of the hardware configuration, the parallel edges of all the trapezoids must be aligned horizontally. In the case where only horizontal cut-lines are allowed, the minimum trapezoid partitioning problem can be solved in $O(n \log n)$ time by means of the plane sweep method as in Section 5.1. An example of such a partitioning is given in Fig. 5.5. The number of resultant trapezoids for such a partitioning is given by

$P = n - h + w - k - d_H,$

where n, w and k are as in Section 5.1, h is the number of horizontal edges, and d_H is the number of horizontal degenerate chords.

In the general case where arbitrary-angled cut-lines are allowed, as is shown in Fig. 5.5(b) for example, the number of trapezoids may be decreased by exploiting non-horizontal cut-lines. However, this minimum trapezoid partitioning problem is known to be NP-hard [35].

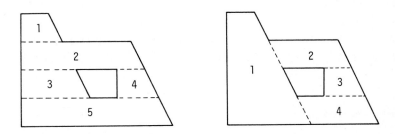

(a) partitioning by horizontal cut-lines   (b) optimal partitioning

Fig. 5.5 Trapezoidal partitioning of polygonal regions.

6. Gridless Routing

Since wire routing is one of the most important process in VLSI design, various routing algorithms have been proposed as described in Chapters 3, 4 and 5. In this section, a new family of routing algorithms based on computational geometry is reviewed. These algorithms are called <u>gridless routers</u> as they, unlike maze routers and channel routers, do not use grid graphs but directly consider polygonal regions as space for routing.

Among various routing algorithms, the <u>sequential routing</u>, as in maze-running and line-search algorithms, is the most basic approach.   In this approach, a wiring net or a pin pair is routed one after another.   A drawback of the existing sequential routing methods lies in its running time and memory requirement.   For example, the maze-running algorithm requires $O(N^2)$ time and memory space for an N X N grid to find a single connecting path.   On the other hand, the sequential routing based on computational geometry find a path in $O(n \log n)$ or $O(n \log^2 n)$ time with $O(n)$ memory, where n is the number of vertices of the polygonal region provided for wiring space. It should be noted here that $n \ll N^2$ usually.   Therefore the gridless routers run faster and require less memory than the grid-based sequential routing methods.   Furthermore they have flexibility in accommodating complicated design rules and maintaining a reasonable interface with mask pattern manipulation programs.

For simplicity of the following discussions, the routable regions and wiring paths to be found are assumed to be rectilinear. It should be noted here that the area used for the already-routed nets is considered to be unroutable and that the routable regions have been virtually shrunk (see Section 4) by taking wiring width and between-wire clearance into account. With this assumption, the routing of a two-terminal interconnection is considered as finding a wiring path connecting a pair of points in rectilinear regions. For a multi-terminal interconnection, the terminal points will be decomposed into a number of terminal pairs. For multi-layer interconnection problems, we measure the problem size n by the total number of polygon vertices for all the wiring layers. We assume that vias for interlayer connections can be placed at any intersection points of routable regions on two neighbouring layers.

The routing algorithms introduced in this section are separated into two classes in terms of computational complexity. Those of one of the classes exploit the plane sweep (Section 2.4) and balanced trees (Section 2.1), and run in $O(n \log n)$ time with $O(n)$ memory space. The algorithms of the other class are represented by repeated solutions of the on-line rectilinear line-segment intersection problem (Section 3.3) and exploit priority search trees (Section 2.3), which leads to $O(n \log^2 n)$ time, $O(n)$ space algorithms. An alternative way of implementing the second class algorithms is to use the data structure named layered segment tree [15] with some additional consideration [7]. Then the algorithm runs in $O(n \log n)$ time, but instead $O(n \log n)$ memory space is required.

6.1 Manhattan wiring

The problem treated here, called Manhattan wiring [36], [37], [7], is defined as follows. For given n pairs of terminal points on the plane, connect all the pairs by rectilinear paths with no more than one bend so that the paths do not intersect to each other. A solution to this problem is not directly applicable to VLSI design, but it may be used to estimate necessary wiring space or difficulty of routing.

For a pair of points, a rectilinear path connecting them with no more than one bend is called an M-wire. If the pair of points are co-horizontal or co-vertical, there is only one M-wire. Otherwise there are two one-bend M-wires. By associating a node with each M-wire, we obtain an intersection graph, where a branch is inserted between a pair of nodes if the corresponding pair of M-wires intersect. Then the Manhattan wiring problem resolves itself into the problem of discerning the existence of an independent node set of maximal possible size n in the intersection graph. An example of solving the problem by means of the intersection graph is illustrated in Fig. 6.1.

The Manhattan wiring problem can be reduced to a sequence of the on-line

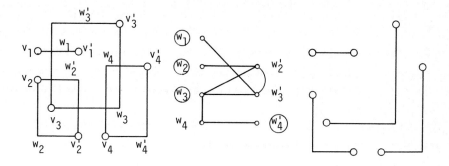

(a) four pairs of points   (b) intersection graph   (c) a Manhattan wiring
    and their M-wires   of M-wires

Fig. 6.1 Manhattan wiring.

intersection problems of rectilinear line segments described in Section 3.3. This observation leads to an $O(n \log^2 n)$ time, $O(n)$ space algorithm by virtue of the priority search tree.   The similar problem of determining the maximum number of point pairs which are Manhattan wirable on one layer is known to be NP-hard.   The problem of determining the minimum number of layers needed to Manhattan wire a set of point pairs is also NP-hard.

6.2 Line-search with a guaranteed solution

The classical line-search algorithm (see Chapter 3), aimed at finding a simple-shaped path faster than the maze algorithm, has been considered not to guarantee a solution even if one exists.   We present here an improved line-search algorithm, which always find a path, if one exists, in almost linear time with linear memory space [38].   The improved algorithm heavily depends on computational geometry.

We shall outline the algorithm along with a single-layer example in Fig. 6.2.

Suppose we have given an n vertex rectilinear region for routing area. Initially, we extend each edge of the routable region to both directions until it hits an obstruction or the external frame as shown in Fig. 6.2(a). Such a set of horizontal and vertical line segments are called escape lines in the line-search method.   In an actual implementation, such escape lines can be generated in advance by means of the plane sweep method.   If the edges of the routable regions are stored in an AVL tree, the initialization can be done in $O(n \log n)$ time with $O(n)$ space.   To find a path connecting a given pair of points S (source) and T (target), we store the set of horizontal and vertical escape lines H and V, respectively, into separate priority search trees. In addition, we generate two escape lines incident at the target T; one horizontal (denoted by h_T) and the other vertical (denoted by v_T). Similarly, we generate horizontal and vertical line segments h_S and v_S incident at the source S, and put them into a queue (see dotted lines in Fig. 6.2(a)). Note that these operations can be done in $O(\log^2 n)$ time by solving Problem (b) in Section 3.3.

A path connecting S and T, if exists, can be found by recursively continuing the following operations starting from h_S and v_S.

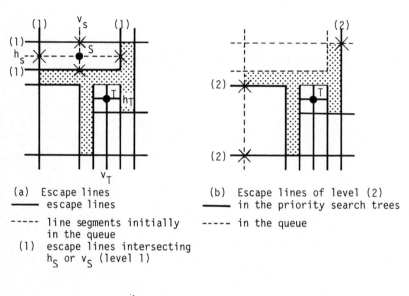

(a) Escape lines
——— escape lines
----- line segments initially in the queue
(1) escape lines intersecting h_S or v_S (level 1)

(b) Escape lines of level (2)
——— in the priority search trees
----- in the queue

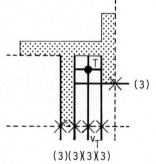
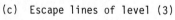
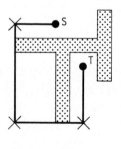

(c) Escape lines of level (3)

(d) The generated (minimum-bend) path

Fig. 6.2 The improved line search.

(1) Pick up a line segment from the queue and enumerate all the escape lines that intersect it.

(2) If one of them is either h_T or v_T, then backtrace to generate a path.

(3) Otherwise, delete the enumerated line segments from the priority search trees and put them into the queue.

In the example of Fig. 6.2, four escape lines (indicated by (1)'s) are found first as shown in Fig. 6.2(a). Next three escape lines (level(2)) are found as shown in Fig. 6.2(b). Further, five escape lines (level(3)) are found in the next stage as shown in Fig. 6.2(c). Now one of the generated escape line is v_T, thus the backtrace leads to the desired path as shown in Fig. 6.2(d).

By analyzing the above algorithm, we see that the intersection is checked at most once for each escape line. Since there are only O(n) escape lines, and such an intersection can be checked in $O(lon^2n)$ time per escape line by solving Problem (a) in Section 3.3, the total time for generating a path is of $O(n \log^2 n)$. Note that this algorithm always leads to a minimum bend path.

This algorithm runs in almost constant time ($O(\log^2 n)$) if only a path with a limited number of bends is to be searched. In practical applications, the algorithm can be modified in such a way that escape lines lying to the direction of a target point is emphasized. This modification leads to a better expected running time algorithm as in the fast maze algorithm [39].

The improved line-search algorithm can be generalized to multi-layer interconnection. The essential modification for this purpose is to store the set of escape lines in separate priority search trees classified by the available wiring layers, and the computational complexity described above remains unchanged.

6.3 Single-layer interconnection

The rectilinear region R for wiring space as shown in Fig. 6.3(a) may or may not be connected, and we do not know whether source S and target T belong to the same connected components of R. This can be easily checked and connecting path, if it exists, can be generated in the following way. First, partition R into O(n) rectangles by horizontally slicing it at each concave vertex and at the terminal points (marked " ● "). Here n is the number of vertices of R. Then any two neighboring rectangles belong to the same connected component of R. Hence, by means of a "search", starting from S, on the incidence relation between the rectangles and the slice line segments, which in this case is represented by the graph in Fig. 6.3(b), a path connecting S and T can be found whenever it exists. Since there are only O(n) rectangles and slice lines, the "search" terminates in O(n) time. The connecting path generated in this way is indicated by the bold line. In conclusion, since the rectangular partitioning requires O(n log n) time and O(n) space as described in Section 5, the time and space complexity of the above method is O(n log n) and O(n), respectively. This complexity is known to be optimal.

The horizontally-sliced rectangular partitioning as in Fig. 6.3 is suitable to find a "shortest" path in the sense that the total vertical length is minimized. Since any slice line segment has the same vertical distance from S, a "Dijkstra-type search" with temporary labels of slice line segments stored in Heap leads to such a "shortest" path in O(n log n) time.

As for another version of single-layer routers, the authors have presented a gridless wave propagation method for finding a rectilinear shortest path in O(n log^{2}n) time with O(n) memory [40]. Furthermore a minimum bend path can also be found within the same computational complexity by means of the improved line-search algorithm as have been described in Section 6.2.

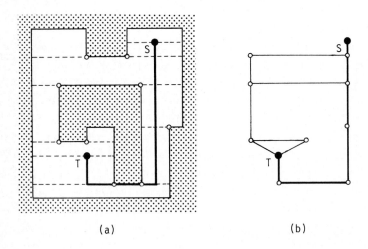

(a)   (b)

Fig. 6.3 Rectangular partitioning of a routable region (a) and its associated graph (b).

6.4 Two-layer interconnection

For two-layer interconnection problems, we denote by R_1 and R_2 the rectilinear regions which are routable in the first and second layers, respectively. And let n be the total number of vertices of R_1 and R_2. Furthermore it is assumed that vias for interlayer connections can be placed at any point in $R_1 \cap R_2$. We describe here two typical versions of two-layer interconnection algorithms.

6.4.1 Shortest path on one of the layers

The underlying constraint here is that the first layer allows only horizontal wires whereas the second layer only vertical ones. Then the following algorithm leads to a connecting path with the minimum total length of the second layer sub-paths [41]. Such a wiring path is desired for metal-polysilicon or double metal two-layer LSI chips.

Step 1. Partition the routable region R_1 of the first layer into $O(n)$ rectangles by horizontally slicing it at each concave vertex and at source point S and target point as shown in Fig. 6.4(a). Let H be the set of horizontal edges of all rectangles, where overlapping edges are identified as one edge by taking their union. Consider a graph G by associating a node with each item in H. The

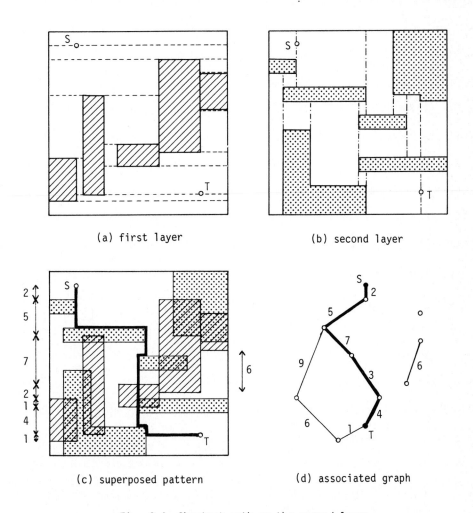

(a) first layer

(b) second layer

(c) superposed pattern

(d) associated graph

Fig. 6.4 Shortest path on the second layer.

branches to be inserted between node-pairs will be determined in Steps 4 and 5, and, for such an edge, a weight will be attached to represent the vertical distance between the corresponding pair of horizontal slice lines.

Step 2.   Let $\overline{R_1}$ be the complement of R_1, i.e. the unroutable region (obstruction) on the first layer. Partition $\overline{R_1}$ into $O(n)$ rectangles by vertically slicing it at each convex (concave with respect to R_1) vertex as shown in Fig. 6.4(a).

Step 3.   Partition the routable region R_2 of the second layer into $O(n)$ rectangles by vertically slicing it at each concave vertex and at source point S and target point T as shown in Fig. 6.4(b). Let V be the set of vertical edges of all rectangles, where overlapping edges are identified as one edge by taking their union.

Step 4.   For each rectangle A obtained in Step 1, insert a branch of G between the nodes corresponding to the horizontal edges of A if one of the following condition is satisfied.

(a)   The set of vertical line segments in V, which intersect with A, totally covers A in the y-coordinate as shown in Fig. 6.5(a).

(b)   No vertical line segments in V intersect with A but one of the rectangles obtained in Step 3 totally includes A as shown in Fig. 6.5(b).

Step 5.   For each rectangle B obtained in Step 2, insert a branch of G between the nodes corresponding to the horizontal edges of B if one of the following condition is satisfied.

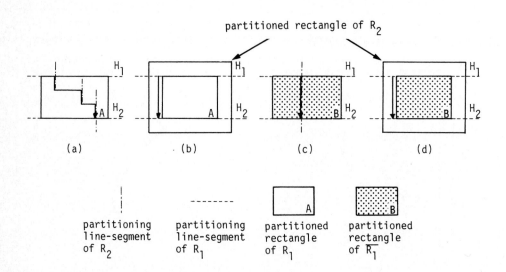

Fig. 6.5 Four conditions to insert a branch of G between the nodes corresponding to H_1 and H_2.

(c) There exists a vertical line segments in V, which intersects with both of the horizontal edges of B as shown in Fig. 6.5(c).

(d) Same as Step 4(b) except that A is replaced by B as shown in Fig. 6.5(d).

   Step 6.   Find a shortest path on graph G starting from S as shown in Fig. 6.4(d), which corresponds to the desired path with shortest vertical (second layer) length as indicated by a bold line in Fig. 6.4(c).

   The rectangular partitionings in Steps 1, 2 and 3 are done in $O(n \log n)$ time. Steps 4 and 5 can be executed in $O(n \log n)$ time by applying the plane sweep technique. Then, the Dijkstra's algorithm with a Heap finds a shortest path on graph G in $O(n \log n)$ time and the remainder of Step 6, i.e. to find the wiring path on each layer, is executed in $O(n)$ time. From the above consideration, this algorithm runs in $O(n \log n)$ time with $O(n)$ memory.

6.4.2 Minimum via interconnection
   It is assumed here that the connecting path can run either horizontal or vertical directions in both of the layers. In general, the routable region on each layer consists of a number of connected components as in Fig. 6.6(a). By associating a node with each connected component, the bipartite graph G_1 as in Fig. 6.6(b), on which the shortest path corresponds to the desired minimum via interconnection, is obtained. A key consideration here for performance improvement in the following algorithm is that, instead of G_1 with $O(n^2)$ branches, a tree-structured bipartite graph G_2 as in Fig. 6.6(c) which includes only $O(n)$ branches is considered [41].

   Step 1.   Partition R_1 and R_2 into connected components. In Steps 2, 5 and 6, a bipartite graph G by associating a node with each of them will be considered.

   Step 2.   For each pair of connected components of the routable regions of different layers, insert a branch between them if one totally includes the other.

   Step 3.   List up all the horizontal and vertical edges of R_1 and R_2, and store them in EDGE-LIST.

   Step 4.   Insert the connected components including the source point S into QUEUE. And delete all the edges of them from the EDGE-LIST.

   Step 5.   For each item u in the QUEUE, repeat the following operations. If there exists a connected component v of the routable region in the other layer which intersects with u, insert a branch between u and v. And delete all the edges of v from the EDGE-LIST.

   Step 6.   Find a path on graph G. Note that G is a tree graph and that the path is the shortest.

   The computational complexity of this algorithm depends on the data structure used for EDGE-LIST in Steps 3, 4 and 5. The essential part can be regarded as the

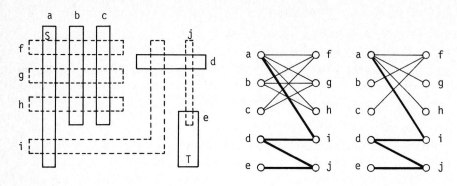

first layer, second layer

(a) routable regions on both layers (b) bipartite graph G_1 (c) bipartite graph G_2
with $O(n^2)$ branches.   with $O(n)$ branches.

Fig. 6.6 Minimum-via interconnection.

on-line line-segment intersection problem.   As described in Section 3.3,   the
EDGE-LIST managed by the priority search trees leads to an $O(n \log^2 n)$ time $O(n)$
space algorithm.

7. Conclusion

Various computational geometry algorithms relevant to VLSI layout design and
verification have been discussed in this chapter.   Although applications of
computational geometry to VLSI design have been limited to layout verification and
artwork generation until very recently,   gridless routing algorithms were
introduced in Section 6 as a new application area.   As design rules are becoming
of increased complexity,   the authors expect that computational geometry will be
used more extensively as a unified approach for both layout design and
verification.

References

[1] D.T. Lee and F.P. Preparata, "Computational Geometry --- a Servey," IEEE Trans. on Computers, vol. C-33, no. 12, pp. 1072-1101, 1984.

[2] N.Wirth, Algorithms + Data Structures = Programs, Prentice-Hall, Englewood Cliffs, 1976.

[3] A.V. Aho, J.E. Hopcroft, and J.D.Ullman, The Design and Analysis of Computer Algorithms, Addison-Wesley, Reading, Mass., 1974.

[4] D.E. Knuth, The Art of Computer Programming, Vol. 3, Sorting and Searching, Addison-Wesley, Reading, Mass., pp. 471-479, 1973.

[5] J.L. Bentley and D. Wood, "An Optimal Worst-Case Algorithm for Reporting Intersections of Rectangles," IEEE Trans. on Comput., vol. C-29, pp. 571-577, July 1980.

[6] H.N. Gabow and R.E. Tarjan, "A Linear-Time Algorithm for a Special Case of Disjoint Set Union," J. Computer and System Sciences, vol. 30, pp. 209-221, 1985.

[7] H. Imai and T. Asano, "Dynamic Segment Intersection Search with Applications," Proc. 25th Annu. IEEE Symp. on Foundations of Computer Science, Singer Island, Florida, pp. 393-402, 1984.

[8] E.M. McCreight, "Priority Search Trees," SIAM J. Comput., vol. 14, no. 2, pp. 257-276, May 1985.

[9] J.L. Bentley and T. Ottmann, "Algorithms for Reporting and Counting Geometric Intersections," IEEE Trans. on Comput., vol. C-28, pp. 643-647, Sept. 1979.

[10] T. Asano, M. Edahiro, H. Imai, M. Iri, and K. Murota, "Practical Use of Bucketing Techniques in Computational Geometry", in Computational Geometry, G.T.Toussaint, Ed., North-Holland, Amsterdam, The Netherlands, 1985, to appear.

[11] K. Yoshida, unpublished manuscript, Dept. Math. Eng. Instru. Phys., Univ. Tokyo, 1982.

[12] M.I. Shamos and D. Hoey, "Geometric Intersection Problems," Proc. 17th Annu. IEEE Symp. on Foundations of Computer Science, pp. 208-215, Oct. 1976.

[13] B.M. Chazelle, "Filtering Search: A New Approach to Query-Answering," Proc. 24th Annu. IEEE Symp. on Foundations of Computer Science, pp. 122-132, Nov. 1983.

[14] H. Edelsbrunner, "Dynamic Data Structures for Orthogonal Intersection Queries," Report 59, Institut f<r Informationsverarbeitung, Technische Universitεt Graz, 1980.

[15] V. Vaishnavi and D. Wood, "Rectilinear Line Segment Intersection, Layered Segment Trees, and Dynamization," J. Algorithms, vol. 3, pp. 160-176, 1982.

[16] M. Sato, M. Tachibana, S. Torii, and T. Ohtsuki, "An Algorithm for Partitioning a Set of Points on the Plane by a Rectilinear Region (in

[17] M. Edahiro, I. Kokubo, and T. Asano, "A New Point-Location Algorithm and its Practical Efficiency --- Comparison with Existing Algorithms," ACM Trans. on Graphics, vol. 3, no. 2, pp. 86-109, April 1984.

[18] H. Imai and T. Asano, "Finding the Connected Components and a Maximum Clique of an Intersection Graph of Rectangles in the Plane," J. Algorithms, vol. 4, pp. 310-323, 1983.

[19] H. Edelsbrunner, T. Ottmann, J.van Leeuwen, and D. Wood, "Connected Components of Orthogonal Geometric Objects," Unit for Comput. Sci., McMaster Univ., Hamilton, Ont., Canada, Rep. 81-CS-04, 1981.

[20] M. Sato, M. Tachibana, and T. Ohtsuki, "An Algorithm for Resizing Polygonal Regions and its Applications to LSI Mask Pattern Design," Electronics and Communications in Japan, vol. 67-C, no. 4, pp. 93-101, 1984. Translated from Trans. IECE, Japan, vol. 66-C, no. 12, pp. 1132-1139, 1983.

[21] M.I. Shamos and D. Hoey, "Closest-Point Problems," Proc. 16th Annu. IEEE Symp. on Foundations of Computer Science, pp. 151-162, Oct. 1975.

[22] D. Rhynsburger, "Analytic Delineation of Thiessen Polygons," Geographical Analysis, vol. V, pp. 133-144, 1973.

[23] T. Ohya, M. Iri, and K. Murota, "A Fast Voronoi-Diagram Algorithm with Quaternary Tree Bucketing," Information Processing Letters, vol. 18, no. 4, pp. 227-231, 1984.

[24] M.I. Shamos, "Geometric Complexity," Proc. 7th Annu. ACM Symp. on Theory of Computing, pp. 224-233, 1975.

[25] M. Sato, "A Minimum Width/Interval Algorithm for Polygonal Regions (in Japanese)," Institute of Electronics and Communication Engineering of Japan, Report CAS84-116, pp. 5-11, 1984.

[26] S.E. Howe, "Estimating Regions and Clustering Spatial Data: Analysis and Implementation of Methods Using the Voronoi Diagram," Ph.D. Thesis, Brown University, Providence, 1978.

[27] B.M. Chazelle and D.P. Dobkin, "Decomposing a Polygon into its Convex Parts," Proc. 11th Annu. ACM Symp. on Theory of Computing, pp. 38-48, 1979.

[28] T. Asano, T. Asano, and R.Y. Pinter, "Polygon Triangulation: Efficiency and Minimality," J. Algorithms, to appear.

[29] T. Ohtsuki, "Minimum Dissection of Rectilinear Regions," Proc. IEEE Internat. Symp. on Circuits and Systems, Rome, pp. 1210-1213, 1982.

[30] J.E. Hopcroft and R.M. Karp, "An $n^{5/2}$ Algorithm for Maximum Matchings in Bipartite Graphs," SIAM J. Computing, vol. 2, pp. 225-231, 1973.

[31] H. Imai and T. Asano, "Efficient Algorithms for Geometric Graph Search Problems," SIAM J. Computing, to appear.

[32] S. Chaiken, D. Kleitman, M. Saks, and J.Shearer, "Covering Regions by Rectangles," SIAM J. Alg. Disc. Meth., vol. 2, no. 4, pp. 394-410, Dec. 1981.

[33] M. Yamashita, T. Ibaraki, and N. Honda, "The Minimum Cover Problem of a Rectilinear Region by Rectangles-(2) (in Japanese)," Institute of Electronics and Communication Engineering of Japan, Report AL-84-16, pp. 13-24, 1984.

[34] D.S. Franzblau and D.J. Kleitman, "An Algorithm for Constructing Regions with Rectangles: Independence and Minimum Generating Sets for Collections of Intervals," Proc. 16th Annu. ACM Symp. on Theory of Computing, pp. 167-174, 1984.

[35] T. Asano, T. Asano, and H. Imai, "Partitioning a Polygonal Region into Trapezoids," J. ACM, to appear.

[36] R. Raghavan, J. Cohoon, and S. Sahni, "Manhattan and Rectilinear Wiring," University of Minnesota, Technical Report 81-5, 1981.

[37] S. Masuda, S. Kimura, T. Kashiwabara, and T. Fujisawa, "On Manhattan Wiring Problem (in Japanese)," Institute of Electronics and Communication Engineering of Japan, Report CAS83-20, pp. 25-32, 1983.

[38] T. Ohtsuki, "Gridless Routers --- New Wire Routing Algorithms Based on Computational Geometry," Internat. Conf. on Circuits and Systems, China, 1985.

[39] J. Soukup, "Fast Maze Router," Proc. 15th Design Automation Conf., pp. 100-102, 1978.

[40] I. Kojima, M. Sato, and T. Ohtsuki, "A Shortest Path Algorithm on a Planar Rectilinear Region (in Japanese)," Institute of Electronics and Communication Engineering of Japan, Report CAS83-205, pp. 45-50, 1984.

[41] T. Ohtsuki and M. Sato, "Gridless Routers for Two-Layer Interconnection," Proc. IEEE Internat. Conf. on Computer Aided Design, pp. 76-78, Nov. 1984.

AUTHOR INDEX

SUBJECT INDEX